Hydroscience and Engineering

This Springer book series Hydroscience and Engineering addresses multidisciplinary advancements in water-related science and engineering practice. Main scope of the book series includes fundamental research and cutting-edge engineering technology in water resources and river/reservoir management in relation with hydraulic structures, erosion control, sediment transport, river basin management and planning, flood control, geological/hydrological risk assessment, offshore and costal engineering. The book series focuses on developing innovative, sustainable and environment-friendly materials and structures to improve infrastructural functionality and sustainability, and to enhance adaptability to changing environmental conditions. Recent advancements and successful insights obtained from engineering practice in China and worldwide will also be thoroughly introduced.

The overarching goal of the book series is to provide combined knowledge on foremost research and solid engineering expertise on hydroscience and engineering from an interdisciplinary standpoint. The individual book volumes in the series are thematic, each specializing a focused topic, such as a new dam type and innovative dam construction materials. As a collection, the book series provides valuable resources targeting on a wide spectrum of audience, including researchers in academia, practicing engineers in engineering community and students who work on expanding their knowledge in the related areas.

Yan Xiang · Zhimin Fu · Chengdong Liu ·
Kai Zhang · Yakun Wang · Guangze Shen

On-site Inspection and Dam Safety Evaluation

Springer

Yan Xiang
Nanjing Hydraulic Research Institute
Nanjing, China

Zhimin Fu
Hohai University
Nanjing, China

Chengdong Liu
Nanjing Hydraulic Research Institute
Nanjing, China

Kai Zhang
Nanjing Hydraulic Research Institute
Nanjing, China

Yakun Wang
Nanjing Hydraulic Research Institute
Nanjing, China

Guangze Shen
Nanjing Hydraulic Research Institute
Nanjing, China

ISSN 2730-9002 ISSN 2730-9010 (electronic)
Hydroscience and Engineering
ISBN 978-981-97-3986-8 ISBN 978-981-97-3987-5 (eBook)
https://doi.org/10.1007/978-981-97-3987-5

This Springer imprint is published by the registered company Springer Nature Singapore Pte Ltd.
The registered company address is: 152 Beach Road, #21-01/04 Gateway East, Singapore 189721, Singapore

Preface

Reservoir dam is not only an important engineering measure to regulate the spatiotemporal distribution of water resources and optimize its allocation but also an essential part of the river flood control engineering system. Dams act as guarantee system for improving the ecological environment on the one hand and play an important role in national flood control and ensuring security in water supply, food, energy, and ecology on the other. Dam's role in public welfare is irreplaceable when it comes to ensuring national water security and supporting national strategy. Reservoir dams are major national infrastructure. Their safety concerns national public security. Dams' safe operation and full function matter directly to national interests.

There is an obvious contradiction between safety requirements of reservoir dams and management techniques and means in China. The influencing factors include dam's complex construction history, large quantity of reservoirs, intricacy of reservoir projects, random operating environment, extreme climate, earthquake, aging of projects, and weak management, etc. Mismanagement results in dam breaks and accidents occasionally. The operational risk of reservoir dams is becoming more and more asymmetric with rapid economic and social development and increasing public attention and requirements on dam safety. Practices show that for hydraulic structures with potential safety hazards, engineering physical examination can be carried out on each building of the reservoir. For the safe and normal operation of project, dam safety evaluation, correct maintenance, repair, and reinforcement measures can be taken in time according to the results of physical examination. Therefore, the safety appraisal of the reservoir dam has become a necessary guarantee measure for engineering safety and is of great significance to ensure the safe operation of reservoir dams. Regular dam safety evaluation is vital to ensure dam safety.

In this book, after extending the flood series, safety inspection, geological exploration, observation data analysis, function adjustment, calculation review, an objective reflection of the impact of changing environment, project aging, state evolution on dam safety will be formed, providing scientific basis to establish a safety evaluation system throughout the life cycle of dams. For different project scale, risk severity, management level, and basic data completeness, "differential" safety evaluation methods and disposal countermeasures are put forward, making the technical route of safety evaluation more reasonable and operable. By systematically sorting out and integrating existing technologies at different stages of the dam life cycle, including planning, design, construction, safe operation, decommissioning, and establishing internal relevant relationships, a complete system of existing dam safety management technologies has been formed.

This book is compiled on the basis of a combination of China's current water conservancy industry standards and project management practices, the author's years of experience in the water conservancy industry, as well as the engineering guidelines prepared by other countries. The book has ten chapters with detailed contents and methods of dam safety appraisal, including: Overview (Chap. 1), Dam Safety On-site Inspection and Test (Chap. 2), Analysis on Safety Monitoring Data (Chap. 3), Evaluation on Construction Quality (Chap. 4), Evaluation on Operation Management (Chap. 5), Evaluation on Flood Control Ability (Chap. 6), Evaluation on Seepage Safety (Chap. 7), Evaluation on Structure Safety (Chap. 8), Evaluation on Seismic Safety (Chap. 9), and Evaluation on Metal Structure Safety (Chap. 10). It provides reference on reservoir dam operation for management personnel. This book is for practitioners, engineers, and students related to water conservancy projects, as well as all those who are interested in dam safety evaluation. To understand this book, a certain knowledge of water conservancy and hydropower engineering is needed.

Yan Xiang is responsible for writing the main contents of Chaps. 1, 2 and 4, and assisting in writing the contents of Chaps. 6–8; Zhimin Fu is responsible for writing the main contents of Chaps. 5–7, and assisting in writing the contents of Chaps. 4 and 8; Chengdong Liu is responsible for writing the contents of Chaps. 3 and 5; Kai Zhang is responsible for writing the contents of Chaps. 2, 8 and 9, and assisting in completing the contents of Chaps. 7 and 10; Guangze Shen is responsible for writing the contents of Chaps. 9 and 10, and assisting in the completion of Chaps. 3 and 7; Yakun Wang is responsible for writing the contents of Chaps. 1, 2 and 10, as well as sorting out the final contents of the book. The compilation of this book has received help and support from Liushan Tang, Bo Dai, Xin Yang, Ying Meng, Fengjie Zhou, Yi Li, Xixi Ge, Guoxuan Yang, Zhaomin Zhu, Minzhi Yang, Jia Li, Yuhao Xu, and others from the Nanjing Hydraulic Research Institute. We would like to express our gratitude.

All authors express their gratitude to the Nanjing Hydraulic Research Institute for its strong support of this book. During the writing process, this book received funding from National Key R&D Program of China (2022YFC3005405), National Natural Science Foundation of China (U23B20150), Major Science and Technology Projects of the Ministry of Water Resources (SK-2022096), Key R&D Program of Yunnan Province (202203AA080009), Reservoir Dam Safety and Management Innovation Team Fund Project of Ministry of Water Resources (Y722003, Y723008), and NHRI Publishing Fund. We would like to express our gratitude.

Since there are many factors involved in the safe operation of the reservoir dam, and writers' information and knowledge are limited, there may be some inadequacies and omissions in the book. Please put forward your valuable suggestions for further improvement in the future. We welcome all readers of this book to make valuable comments.

Nanjing, China

Yan Xiang
Zhimin Fu
Chengdong Liu
Kai Zhang
Yakun Wang
Guangze Shen

Contents

Chapter 1
Overview

1.1 Background

Flood and drought disasters have always been serious problems affecting economic and social development [1, 2]. "Improving water conservancy and eliminating water disasters" has always been a major issue in governing a country [3]. Reservoir dams help regulate the spatial and temporal distribution of water resources and optimize the allocation of water resources, being an important part of flood control system, and an important infrastructure for economic and social development [4–8]. As an important part of national water security and public security, the guarantee of dam safe operation bears on flood control security, water supply security, energy security, ecological security and public security [9]. The composition and influencing factors of dam safety are shown in Fig. 1.1.

1.1.1 Current Situation of Reservoir Dams in China

In China, the water retaining structures of artificial water storage projects with a storage capacity of over 10×10^4 m^3 are called reservoir dams. According to the *2020 Statistic Bulletin on China Water Activities*, 98,566 reservoirs of various types have been built in China [10, 11], with a total storage capacity of 930.6 billion m^3. There are 774 large reservoirs (0.8%) with a total storage capacity of 741 billion m^3. There are 4098 medium-sized reservoirs (4.2%) with a total storage capacity of 117.9 billion m^3, and 93694 small reservoirs (95.1%). By the end of 2020, the installed capacity of China's reservoirs had been 367.71 million kW [12]. China's reservoir dams have the characteristics in five aspects as shown in Fig. 1.2.

(1) **High numbers**. There are more than 98,000 reservoirs in China, which makes China a country with the largest number of reservoir dams in the world [13].

Y. Xiang et al., *On-site Inspection and Dam Safety Evaluation*, Hydroscience and Engineering, https://doi.org/10.1007/978-981-97-3987-5_1

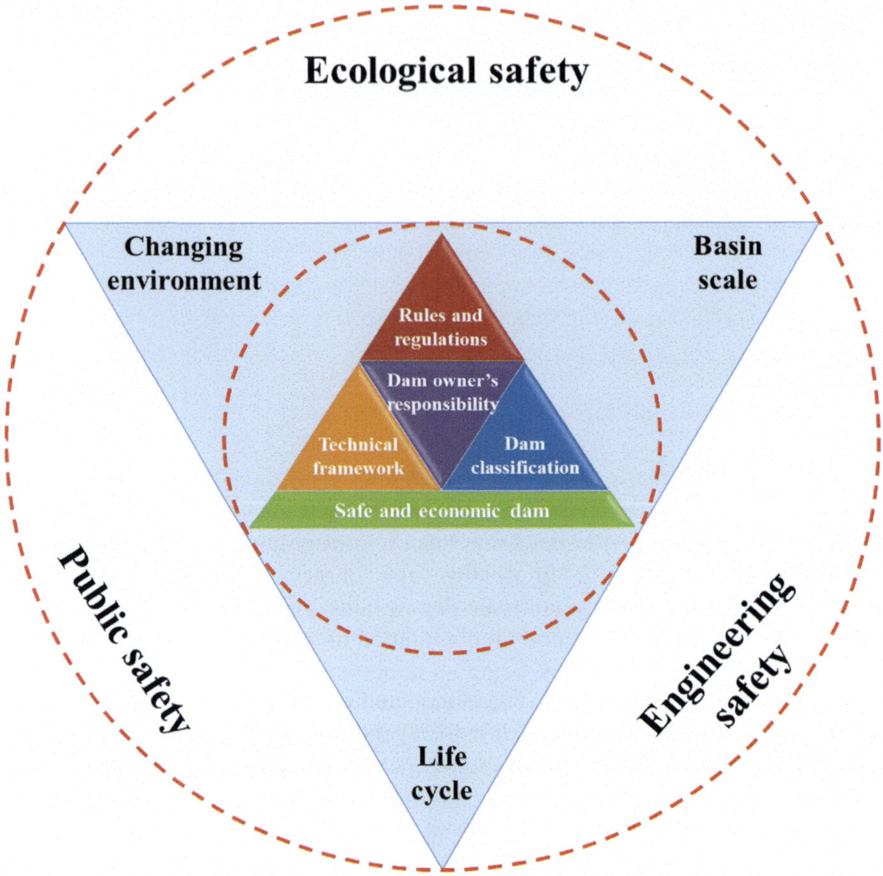

Fig. 1.1 Dam safety components

(2) **Abundant small reservoirs**. 95% of the existing reservoirs in China are small reservoirs [13].

(3) **Many earth-rock dams**. More than 92% of dams in China are earth-rock dams [14].

(4) **Many old dams**. There are many old dams (long dam age). 80% of existing reservoir dams in China were built in the 1950s–1970s [15].

(5) **Many high dams**. There are 77 dams with a dam height of over 200 m built in the world, among which 20 dams are in China, accounting for 26%. There are 22 high dams with a dam height of over 200 m under construction all over the world, among which 15 dams are in China, accounting for 68%, ranking the first [16].

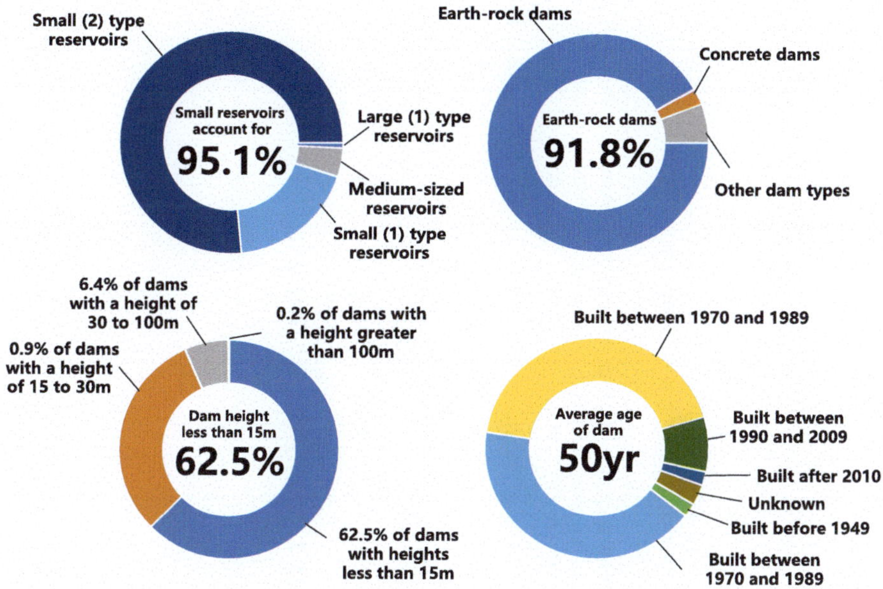

Fig. 1.2 Characteristics of reservoir dams in China

1.1.2 Classification of Reservoir Dams

A dam project includes permanent water retaining structures, discharge facilities, water conveyance and navigation structures, metal structures, as well as reservoir bank near the dam. Reservoir dams may be classified according to reservoir capacity, flood control importance, benefit indexes such as flood control, irrigation and power generation, and importance in national economy, and the project scale and engineering classification are determined.

The scale of a dam project is divided into **Large (1), Large (2), Medium, Small (1) and Small (2)** scales, and the corresponding projects are divided into 5 categories, i.e. **Class I, Class II, Class III, Class IV, and Class V** projects. See Table 1.1 for details, in which the criteria for classifying the importance of cities, industrial and mining enterprises and water supply objects are shown in Table 1.2 [17].

Then, according to the scale of a project, the level of dam structure is determined, corresponding to the project grade. Dam is divided into 5 levels, namely, 1-level dam, 2-level dam, 3-level dam, 4-level dam and 5-level dam. At the same time, it is stipulated that for Class II and III dams, if the dam height exceeds a certain height, the dam grade can be increased by one level, as shown in Table 1.3, while the reservoir engineering grade and flood standard of the dam are not increased. For dams with a height of more than 200 m, the design criteria shall be specially studied and demonstrated, and submitted to the competent authorities for review and approval.

Table 1.1 Classification standard for reservoir project scale and grade

Engineering grade	Reservoir		Control flood			Waterlogging control	Irrigation	Water supply		
	Project scale	Total storage capacity (10^8 m^3)	Importance of protected towns and industrial and mining enterprises	Area of protected farmland (10^4 Mu)	Equivalent economic scale of protected areas (10^4 persons)	Waterlogging control area (10^4 Mu)	Irrigation area (10^4 Mu)	Importance of water supply objects	Annual water inflow (10^8 m^3)	Installed capacity (MW)
I	Large (1)	≥ 10	Particularly important	≥ 500	≥ 300	≥ 200	≥ 150	Particularly important	≥ 10	≥ 1200
II	Large (2)	10–1.0	Important	500–100	300–100	200–60	150–50	Important	10–3	1200–300
III	Medium	1.0–0.10	More important	100–30	100–40	60–15	50–5	More important	3–1	300–50
IV	Small (1)	0.10–0.01	General	30–5	40–10	15–3	5–0.5	General	1–0.3	50–10
V	Small (2)	0.01–0.001		< 5	< 10	< 5	< 0.5		< 0.3	< 10

Table 1.2 Classification standard for importance of cities and towns, industrial and mining enterprises and water supply objects

Importance of towns, industrial and mining enterprises, and water supply objects	Cities and water supply objects	Industrial and mining enterprises	
	Permanent population (10^4 persons)	Scale	Monetary indicators (10^8 yuan)
Particularly important	≥ 150	Extra large	≥ 50
Important	150–50	Large	50–5
More important	50–20	Medium	5–0.5
General	< 20	Smell	< 0.5

Table 1.3 Criteria for dam classification and upgrading

Engineering grade	Dam grade	Promotion standard	
		Dam type	Dam height (m)
I	1	No standards, determined based on actual situation	
II	2	Earth rock dam	90
		Concrete dam and masonry dam	130
III	3	Earth rock dam	70
		Concrete dam and masonry dam	100
IV	4	No standards, determined based on actual situation	
V	5		

1.1.3 Safety Management of Reservoir Dams

China promulgated *Regulations on Safety Management of Reservoir Dams* in 1991, and then successively issued a series of relevant laws, regulations and technical standards, such as *Measures for Safety Appraisal of Reservoir Dams, Measures for Safety Management of Small Reservoir Dams, Guidelines for Safety Evaluation of Reservoir Dams*, etc., on the basis of *Water Law, Flood Control Law* and *Regulations for Safety Management of Reservoir Dams.* See Table 1.4 for the comparison of the starting year of dam safety legislation in different countries. Laws and regulations are relatively complete in China. Complementary departmental regulations, technical standards support and dam safety regulations and technical standards system conform to the national conditions.

China continues to strengthen the establishment of the concept of dam safety management, as shown in Fig 1.3. Before 1980s, the management of reservoir dams in China mainly relied on administrative means. Administrative documents of varied forms were issued for guidance and implementation of management of reservoirs. Then in 1980s, relevant technical standards v and since then the dam management has entered a period when the management work was standardized taking technical

Table 1.4 Comparison of starting year of dam safety legislation in different countries

Country	Laws and regulations	Starting year
China	*Regulations on Safety Management of Reservoir Dams*	1991
USA	*National Dam Inspection Act*	1972
	National Dam Safety Act	2012
Switzerland	*Dam Safety Act*	1957
	Regulations on Reservoir Safety	1998
UK	*Reservoirs Act*	1930

standards as domination. Various kinds of standards issued at this period laid foundation for the technical standards for management of reservoirs at present. In January of 1988, *Water Law of the People's Republic of China* was issued and implemented. This marked a new process for establishment of laws and regulations and technical standards for management of reservoir dams. In March of 1991, *Regulations for Safety Management of Reservoir Dam* was issued, which laid a legal foundation for dam management. It is stipulated that the competent authority of dams shall establish a system for regular safety inspection and evaluation [18]. After that, a series of rules and regulations, normative documents and technical standards from various departments were issued one after another and a system of laws and regulations and standards for reservoir dam management has been gradually established, taking *Water Law* and *Flood Control Law* as basis, *Regulations for Safety Management of Reservoir Dam* as the core and rules and regulations and normative documents from various departments and technical standards as support. For this purpose, the Ministry of Water Resources, China issued *Method for Evaluation of Reservoir Dam Safety* (SHUI GUAN No. 86-1995) in 1995, which was amended in 2003. Therefore, a system for regular evaluation of dam safety has been established and fully implemented [19]. At present, China has established a comprehensive dam safety management department, as shown in Fig. 1.4, which clarifies the responsibilities of each department.

Now, the regular safety evaluation of reservoir dams has become one of the important systems for reservoir dam safety management, and is the important basis for acknowledgement and identification of dam safety performance and for taking such safety measures as scientific regulation, controlled operation, reinforcement or degradation and abandonment of dams (Fig. 1.5).

Dam safety appraisal include three basic procedures, that is, **evaluation on dam safety**, **technical review on the safety evaluation of dams** and **examination and approval of comments on dam safety evaluation** [20]. It shows that the evaluation on dam safety is the main technical work for dam safety appraisal. Dam safety evaluation comprises nine main items as follows:

(1) On site safety inspection, testing and detection
(2) Supplementary engineering geological survey
(3) Observation data analysis
(4) Engineering quality evaluation
(5) Operation management evaluation

Fig. 1.3 Development process of dam safety laws and regulations

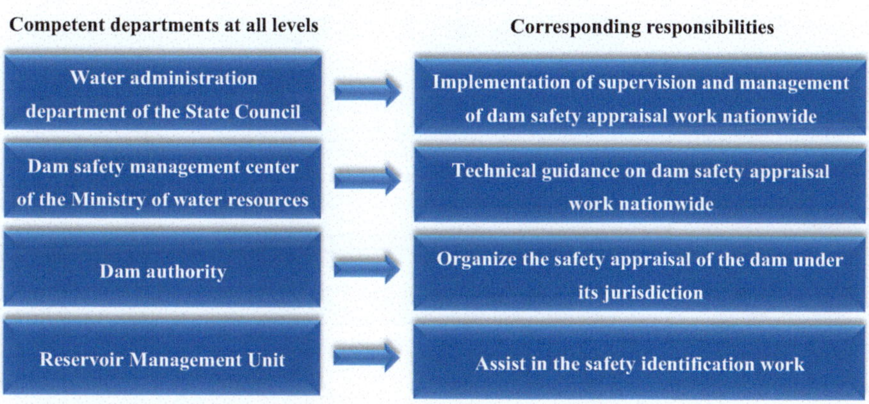

Fig. 1.4 China's reservoir and dam management department and its corresponding responsibilities

(6) Flood standard review
(7) Dam seepage safety, structural safety, and seismic safety review
(8) Metal structural safety review
(9) Dam safety comprehensive evaluation (Fig. 1.6).

Fig. 1.5 Whole life process management of dams in China

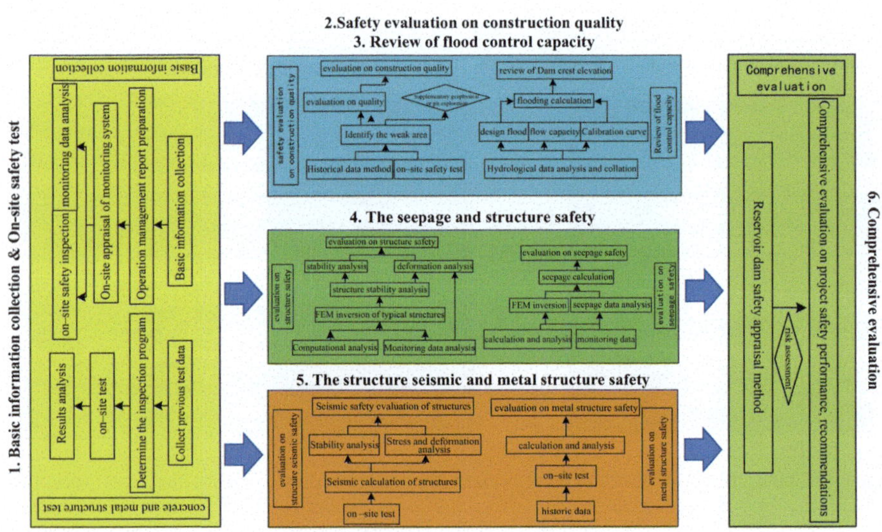

Fig. 1.6 Basic process of dam safety evaluation

1.2 Dam Safety Appraisal Procedure

1.2.1 Responsibilities

In the process of reservoir dam safety assessment, organizational institutions, implementation institutes and identification and validation departments must work together to ensure the smooth implementation of reservoir dam safety assessment.

China has a large number of dams with multiple ownership of different dams, and different departments manage the safety of these dams. China's reservoir dams are mainly divided into reservoir dams (water conservancy) and hydroelectric dams (electricity) according to the different departments to which they belong, with reservoir dams accounting for the vast majority. There are also a few dams in transport, construction and agriculture sectors. Reservoir authority or reservoir management agency should be identified as organizational unit for the development of dam safety assessment plan and organization and implementation, and raise funds for dam safety assessment, and be commissioned to identify an agency to carry out dam safety assessment.

Design institute or scientific research institute with relevant qualifications undertakes the identification of implementation agency, which will be responsible for the following specific tasks: organizing and conducting on-site safety inspection; collecting data, conducting quality exploration, safety inspection and other work; conducting dam safety assessment; and submitting dam safety assessment report.

State-level water conservancy administrative departments supervise and manage dam safety assessment work nationwide. The Dam Safety Management Centre of the Ministry of Water Resources provides technical guidance for dam safety assessment nationwide. The water conservancy administrative department of the local people's government at or above county level supervises and manages the dam safety assessment work under its jurisdiction. Provincial water conservancy administrative department is responsible for the reservoir dam safety appraisal of large reservoirs affecting the national safety or medium-sized reservoirs with a dam height of more than 50 m, municipal (local) level water conservancy administrative department for that of other medium-sized reservoirs affecting the county safety or small reservoirs with a dam height of more than 30 m, and county water conservancy administrative department for other small reservoirs dam safety appraisal. The basin organization and the Ministry of Water Resources shall review and approve the dam safety appraisal opinions of the reservoirs under their direct jurisdiction, with specific work of organizing dam safety appraisal meeting, and approving and issuing appraisal report.

1.2.2 Preliminary Preparation

Many of the reservoir dams built in the 1950s and 1970s are the projects built without survey, design and construction quality control, or the projects where design, construction and operation were carried out in parallel. In addition, weak management, incomplete basic information and data on project situation, construction and reconstruction, operation and maintenance, inspection and monitoring (the data/information/documents are probably kept in different departments) [21] have caused great difficulties in the scientific evaluation of dam safety.

For this reason, the relevant basic data/information should be collected before dam safety assessment is carried out. The missing or problematic data/information shall be verified and supplemented by such approaches as site visits, geological survey, safety inspection and testing to ensure that the working basis for dam safety assessment is appropriate and sufficient. In particular, selected calculation parameters must represent the actual state of a dam.

Relevant baseline data shall be collected and verified in order to perform dam safety assessment. If basic data do not meet the requirements of dam safety assessment, additional geological investigations and safety tests are required to supplement and complete the data.

1.2.3 The First Evaluation of Dam Safety and the Follow-Up Dam Safety Evaluation

The first dam safety assessment is carried out within 5 years of completion and acceptance of a project. Thereafter, the evaluation shall be carried out every 6–10 years [22]. In the event of a catastrophic flood, strong earthquake or serious accident or abnormality affecting the safety of a dam, a special safety assessment shall be organized.

The initial dam safety evaluation shall provide an overall assessment of dam safety in accordance with the requirements of this book. In subsequent dam safety evaluation, emphasis shall be placed on quality deficiencies and safety problems revealed during operation, and special demonstration required. For those dams with safety monitoring data, evaluation shall focus on analyzing the monitoring data to understand the safety performance of the dams.

For small reservoirs in general, economic conditions are poorer and there is little basic data/information available. Therefore, dam safety assessment may be simplified as appropriate. For small reservoirs with obvious deficiencies but insufficient baseline data, a team of experts may be organized by the reservoir authority to verify the safety class on the basis of an on-site safety inspection.

In general, the initial dam safety assessment for large and important medium-sized reservoirs shall include specific analysis and demonstration reports on on-site safety inspection, safety tests, engineering geological survey, monitoring data analysis, construction quality, operation management, flood control safety, seepage safety, structural safety, seismic safety, and metal structure safety. Then, a comprehensive dam safety assessment report is prepared by integrating the above specific analysis and demonstration reports. Subsequent dam safety evaluation for small reservoirs may be simplified as appropriate.

1.3 The Collection of Basic Data on Reservoir Dams

The collection of basic data on reservoir dams is important for carrying out dam safety assessments [23]. Data on the following area must be collected and sorted as required [22, 24].

- Watershed, hydrology, meteorology
- Dam engineering features
- Engineering geology
- Dam design and construction
- Dam safety monitoring
- Dam operation and maintenance
- Previous safety assessment and rehabilitation
- Other relevant data.

The basic data shall reflect dam's current situation [25]. In particular, constantly changing data shall be collected. These include extension of hydrological series, changes in reservoir functions, protection targets of flood prevention measures, anti-earthquake standard, sediment deposition, reservoir storage capacity, reservoir features, reservoir regulation, downstream scouring, etc. [26].

The accuracy and reliability of the collected basic data shall be analyzed [27]. Data with obvious errors or systematic deviations shall be corrected or rejected. In case of inadequate data, additional efforts shall be made for more information by means such as site visit, field inspection, geological survey, safety test, defect monitoring, etc. [28].

1. **Data on watershed, hydrology, meteorology**

Data in these areas mainly include:

(1) Basic data in the reservoir basin such as topography, geology, vegetation, soil distribution, water system, precipitation, evaporation, temperature, wind direction and strength, and other characteristic parameters like reservoir basin area, river length, river slope, etc.;
(2) Basic data in the reservoir basin such as storm, flood and ice condition, rainfall data collected by related precipitation stations, yearly flood data collected by related hydrometric or gauging stations, and data of human activities' impact on hydrological parameters [29];
(3) Basic data regarding other water control projects in reservoir upstream and downstream.

2. **Data on dam engineering features**

The data of general information, table of engineering features, past engineering drawings shall be collected. Data in this area include:

(1) General engineering data on dam construction, follow-up upgrading, reconstruction, reinforcement, and dam features (the project grade and categories of structures, flood control standard, characteristic water level and storage capacity, basic information of the hydraulic structures, and information regarding gate, hoist facility and power supply);
(2) Plan of the current engineering hub, and plan and typical cross-section views of main hydraulic structures;
(3) Current storage capacity curve, water curve of level-flow relation at important control stations downstream;
(4) Protection targets of flood control measures, and function indicators like irrigation, water supply, power generation, and navigation;
(5) Social, economic and population data in downstream areas both under flood protection and with possible flood risks from dam breaks.

3. **Data on engineering geology**

Data on geological geology at different phases shall be collected and analyzed. Targeted surveys, tests and experiments shall be conducted as required. Engineering geological survey reports and inspection and testing reports shall be collected to get the following data and information:

(1) Particle composition and permeability coefficient of soil layers for earth-rock fill dam and non-rock dam foundation, compaction, index of strength, compressibility, contact relation and permissive seepage gradient of earth and rock materials, and distribution and physico-mechanical index of liquefiable soil, soft clay and collapsible loess in dam foundation;
(2) Attitude, width, composition, length and location of such geologic structures as fault and fractured zone and weak interlayer in dam foundation, and the impacts on the dam foundation and body by seepage, piping, erosion and sliding;
(3) Basic seismic data:

① Basic seismic data shall include parameters of seismic risk and ground motion. Historical seismic data of the site and adjacent areas and characteristics of ground motion during the biggest earthquake shall be collected. The following data should be obtained for Grade I and II dams.

(a) Seismic activity data: Time, epicenter location, magnitude, hypocenter depth, epicenter intensity, site intensity, damage records, and comments on seismic activity, etc.
(b) Ground motion parameters: Characteristic index of hypocenter during typical earthquakes, records of vibration process monitored and measured at different locations, vibration duration, amplitude of maximum vibration (acceleration, velocity, displacement), characteristic period of site, residual displacement, response spectrum of structure acceleration.

(c) Special analysis and report of seismic risk.

② Test data of dynamics characteristics (including deformation and strength) for earth-rock fill dam of Grade I with designed seismic intensity of 8 and 9°; Dynamic strength index of earth-rock fill dam of Grade II.

③ Data regarding seismic damages and geological disasters in history.

(a) Data on past earthquakes and their damages to reservoir dams;
(b) Data on geological disasters in reservoir area and stability of dam-adjacent banks [30];
(c) Data of reservoir-induced earthquakes [31].

4. **Data on dam design and construction**

Data in this area mainly include:

(1) Design reports and drawings for construction, upgrading and expansion and reinforcement of the dam shall be collected.

① Data on flood standard, designed flood calculation, flood regulation calculation, recheck calculation of dam crest elevation;

② Data on excavation and treatment of foundation and bank slope, including excavation and removal of useless rock mass, shape control of foundation surface, consolidation and contact grouting of foundation, foundation seepage control and drainage, concrete replacement for weak layers and zones, prestressed anchorage, etc.;

③ Data on seepage design, seepage control, detailed structure of anti-seepage and drainage body and its contact transition with neighboring material, designed seepage pressure distribution, seepage quantity, and the permissive seepage gradients and seepage lines of different materials;

④ Data on structure safety design, including analysis of stability, deformation and stress and corresponding measures;

⑤ Data on seismic safety design, including seismic calculation and anti-seismic measures [32]. The seismic calculation mainly include designed seismic load and combination, cross sections of seismic sensitive structures (interior gallery, cavity, pipelines, drainage, filter, body shape) and the location of seismic sensitive materials (like liquefiable material), and calculation models, related parameters, judgment criteria and calculation results used for dynamic reaction or liquefaction analysis;

⑥ Data on hydraulic design of water release and conveyance structures and its structure, seismic and seepage control;

⑦ Data on design and test of metal structures;

⑧ Data of self-inspection on water storage safety and technical appraisal on construction completion.

(2) Documents of all previous design inspection, comment, and approval shall be collected.
(3) Construction data shall be collected.

① Data of relevant drawings, reports and documents regarding construction methods, quality control and inspection, supervision and acceptance for dam body, seepage control work, drainage facility, foundation (including bank slope) excavation and treatment, water release and conveyance structures, etc.;
② Data on material, manufacture, transportation, installation and completion acceptance of metal structures;
③ Data of important design changes;
④ Data of quality defects found during construction and operation and related solutions;
⑤ Data of project as-built drawings, including plan layout of the project, plan views and typical cross section views of main hydraulic structures.

(4) Acceptance documents shall be collected.

① Data of acceptance in different project phases;
② Data of water storage safety appraisal and technical appraisal on construction completion.

5. **Data on dam safety monitoring**

Data in these areas mainly include:

(1) Data of design and installation of dam safety monitoring system, including plan views, cross section layout, and installed checklist;
(2) Data of monitoring regarding seepage, deformation, pressure (stress), strain, hydraulics and environmental variables covering downstream and upstream water level, water temperature in reservoir, precipitation, temperature, sedimentation and downstream scouring [33];
(3) Data of all previous compilation and analysis of dam safety monitoring.

6. **Data on dam operation and maintenance**

Data in these areas mainly include:

(1) Data on organization, team and system of reservoir management;
(2) Data of reform plan of reservoir management mechanism, and fund for personnel and maintenance;
(3) Data on reservoir operation plan, the highest and lowest water levels and reservoir benefits;
(4) Data on reservoir emergency management, including emergency plan for flood control and disaster rescues, dam safety, and layout of traffic, communication, power and alarm facilities for flood control;
(5) Data on chronicle of reservoir operation events, including all previous records of operation, monitoring, maintenance, repair, overhaul, and technical upgrading of hydraulic and metal structures;
(6) Data on engineering and management safety and reservoir benefits since its operation.

7. **Data on dam safety condition**

Data in these areas mainly include:

(1) Data of all previous dam safety evaluations, including dam safety analysis and evaluation reports [20];
(2) Data on action plan for the conclusions of previous dam safety evaluations and its effect;
(3) Data of document regarding quality defects, potential risks (like crack, slide, abnormal seepage) and accidents (like gate failure) revealed during operation and corresponding treatment. The document shall include time, locations, nature, external conditions, whole development process and damage of the accident.

1.4 Assessment Contents and Conclusions

Comprehensive evaluation on dam safety means to carry out evaluation in an overall way on dam safety and to further assess on the safety class of a dam, based on on-site inspection and monitoring data, in accordance with the evaluation results of specific items of flood control ability, seepage safety, structural safety, seismic safety and metal structure safety and with reference to evaluation conclusions on construction quality and dam operation management evaluation. By comprehensive evaluation on dam safety, it is to carry out a comprehensive analysis and evaluation on a dam as one system. In the final evaluation on the class of dam safety, quantitative evaluation is combined with qualitative evaluation, that is to say great importance is attached to project safety on the one hand, and the result of quantitative evaluation is taken as the main basis and focus on the other hand, to encourage people to do good job in dam safety management (Fig. 1.7).

The safety assessment of dams shall be carried out on the basis of site inspection and analysis of safety monitoring data. In accordance with the provisions and requirements of the current relevant specifications, project scale, the categories of structures, flood control standards and seismic fortification standard shall be reviewed, the construction quality, present state and actual operating conditions of the a dam reviewed, and the flood control capability, seepage safety, structural safety, seismic safety, metal structure safety and operation management of a dam reviewed and evaluated. In addition, a comprehensive assessment of dam safety shall be made by summarizing the above assessment results. The results of the latest flood routing calculation, monitoring, testing and detection shall be used to calculate loads and parameters.

Specific reports should be worked out for dam safety evaluation with reference to contents required. Then a report of comprehensive evaluation on dam safety shall be prepared by integrating these specific reports. The technical route of dam safety assessment shown in Fig. 1.8.

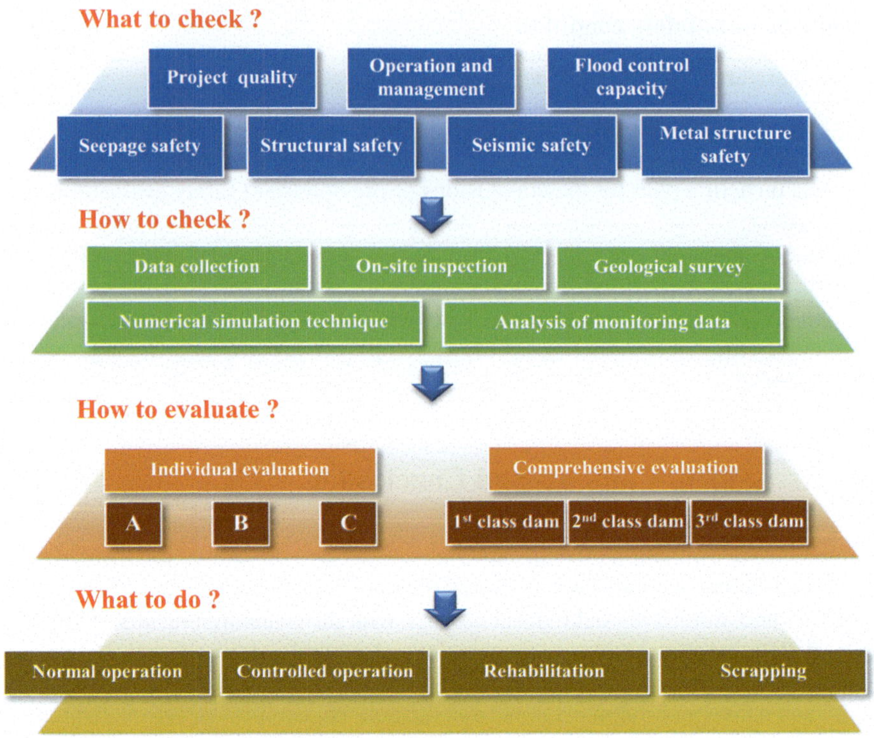

Fig. 1.7 The safety appraisal measures for dams

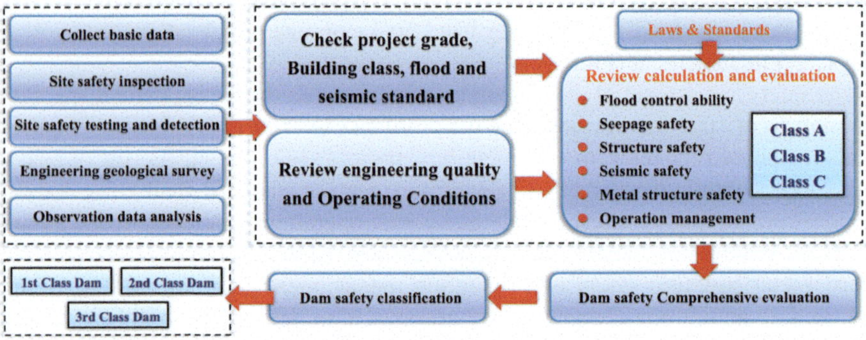

Fig. 1.8 Technical route of dam safety assessment

In conclusion of the evaluation on flood control ability, seepage safety, structure safety, seismic safety and metal structure safety, dam safety is classified into three classes, i.e. Class A, B and C, indicating safe dam, basically safe dam with defects and unsafe dam, respectively [34]. The grading of various evaluation conclusions is detailed in each chapter. Principles for rating evaluation results are described below:

- **Class A**: Safe and reliable: The results of review meet the requirements of specifications, operation behavior is normal, and there are no obvious cracks, sinks, slumps or leakage.
- **Class B**: Fairly safe with defects: The result of review meets the requirements of specifications, but there exist construction quality defects and local cracks, sinks, slumps or leakage occurred during operation.
- **Class C**: Unsafe: The result of review does not meet the requirements of specifications, or there exist serous construction quality defects and local cracks, sinks, slide (shift), inclination or seepage occurred during operation.

In conclusion of construction quality evaluation, construction quality is classified as "Accepted", "Basically accepted" and "Not accepted". In conclusion of operation management evaluation, operation management is classified as "Standardized", "Basically standardized" and "Not standardized". These will be taken as reference basis for comprehensive evaluation on dam safety.

1.5 Dam Safety Classification

This book pays more attention to management and plays an important role in securing dam safety by management ability, safety monitoring, scientific regulation, emergency preparedness plan, maintenance and repair and other non-engineering measures. This evaluation standard keeps up with current development trends of dam safety management and is more operable. Principles and standards for the classification of dam safety are described as below.

The safety performance of a dam shall be classified in the report of comprehensive evaluation on dam safety. It is classified into dams of Grade I, II and III, which are rated as below:

- **1st Class Dam**: The current flood control ability of a dam meets requirements. The dam is free of obvious quality defects. All the results of review and calculation satisfy the requirements of specifications. Management facilities like safety monitoring facilities are sound and complete. Maintenance and repair are ascertained and in place. Management is standardized and the dam can normally operate in accordance with design standards.
- **2nd Class Dam**: The current flood control ability of a dam fails to meet requirements, but does meet standard for a short-term and emergency operation of flood for reinforcement of water control project issued. The holistic safety of dam structures, seepage safety and seismic safety meet specifications and operation state is basically normal. However, construction quality defects exist or management facilities like safety monitoring facilities are not sound and complete. Maintenance and repair are not ascertained. Management is not standardized, but the

dam can normally operate in accordance with design standards. The dam can only operate safely under the given controlled operation conditions.

- **3rd Class Dam**: The current flood control ability of a dam fails to meet standard for a short-term and emergency operation of flood for reinforcement of water control project issued, or serious quality effects and potential risks exist and the dam cannot operate normally as per the design.

1st Class Dam is reliably safe and can be operated properly as design. 2nd Class Dam is basically safe and can be operated under enhanced monitoring and control. 3rd Class Dam is not safe, belonging to unsafe dam. For 2nd Class Dam or 3rd Class Dam where construction quality defects and hidden troubles exist, counter measures for disposal include controlled operation, enhanced management, reinforcement, degradation or abandonment.

A dam can be rated as 1st Class Dam when the evaluation result of all specific items in respects of flood control ability, seepage safety, structural safety, seismic safety and metal structure safety is Class A, and construction quality is accepted and operation management standardized. If the evaluation result of more than one (including one) specific items is Class B, the dam shall be rated as 2nd Class Dam. To guarantee dam safety, one-vote veto is applied for Grade III dam. That is, a dam will be rated as Grade III dam as long as the evaluation result of one of specific items (including flood control safety, seepage safety, structure safety, seismic safety and metal structure safety) is rated as Class C.

In the meantime, to strengthen dam safety management, a dam can also be rated as 2nd Class Dam when management facilities such as safety monitoring is not sound and complete, maintenance and repair not ascertained and management not standardized even though the evaluation results of all the specific items are rated as Class A. As to a 2nd Class Dam with accepted engineering quality and standardized operation management but having one or two specific items rated as Class B, the specific items of Class B can be upgraded to that of Class A in a given time limit and then the dam can be upgraded to Grade I dam.

A dam rated as 1st Class Dam or 3rd Class Dam shows that quality defects and latent defects exist in the dam. To ensure dam safety, the dam shall be operated with limited water level before necessary measures are taken. In addition, monitoring and walk-around inspection shall be enhanced and emergency preparedness plan worked out.

For a dam rated as 3rd Class Dam with remarkable functions and beneficial results, reinforcement measures shall be taken to deal with dangers and potential safety problems, which are feasible in technology and rational in economy. On the contrary, for a 3rd Class Dam with serious dangers, very high risks and functions decrease, where reinforcement is not feasible in technology and not rational in economy, they shall be degraded or abandoned.

1.6 Consider Safety Evaluation Standards for Functional Assessment

One to three types of risk-based criteria and functional assessment standards shall be established on the basis of the existing "three types of dams" classification, taking into account the risk-based system safety standards, reservoir dam design function, safety status, the current stage of demand, the future demand for local economic development, funding, public and environmental safety, management capacity and other key elements. For reservoir dam function assessment results, corresponding classification disposal plan shall be developed, mainly including normal operation, rectification and upgrading, and demotion and retirement:

1. **Normal operation**: The status quo shall be maintained for reservoir dams operated with matched functions and needs, good ecological environment, clear ownership rights, appropriate management agency and personnel, daily management and protection, and funding.
2. **Rectification and upgrading**: Rectification and upgrading shall be carried out for reservoir dams operated with unmatched demand, diseases and risks, or inadequate supporting facilities, imperfect management system and mechanism.

    (1) **Adjustment**. Adjustments shall be made if the actual demand does not match the original design function, such as decreasing role irrigation, power generation function, and increasing role of urban and rural water supply, downstream flood control, and eco-environment improvement. Relevant government water agency shall organize technical discussion, consult industry authorities, reservoir management units and stakeholders, and then propose a program for function adjustment of reservoir dams. The adjustment procedure of reservoir function shall be made in accordance with the provisions of reservoir dam registration methods.

    (2) **Rectification**. Rectification shall be carried out if reservoir dam management does not meet the requirements of regulations, including unclear property right ownership, defective management system and mechanism, inadequate personnel, funding, and support policy for management agencies. The property owner shall develop a rectification program, set a deadline, and supervise the implementation.

    (3) **Reinforcement**. Reinforcement shall be made for reservoir dams with poor safety condition (concerning diseases and risks) and inadequate supporting facilities. The property owner shall to propose a reinforcement and upgrading program and participate the implementation.

3. **Demotion and retirement**: For reservoir dams with reduced scale or functionality, they should be operated and managed after demoting their original design grade by one or more levels. Some reservoir dams have severe diseases and risks, and their reinforcement is technically unfeasible or economically unreasonable. Some have lost basic functions. These reservoir dams should be retired. The work of demotion and retirement should be entrusted to a qualified agency. The

property owner shall submit a demonstration report written by the agency to the relevant government department for approval before implementation. After the implementation, the property owner shall submit an application for acceptance by the government department. After the acceptance, the registration of the reservoir dams shall be changed or canceled in accordance with relevant provisions.

References

1. International seminar on extreme flood and drought disaster risk management successfully held (2023) J Yangtze River Acad Sci 40(03):67
2. Guan X, Zang Y, Meng Y et al (2021) Study on spatiotemporal distribution characteristics of flood and drought disaster impacts on agriculture in China. Int J Disaster Risk Reduct 64:102504. https://doi.org/10.1016/j.ijdrr.2021.102504
3. Chen S (2020) Issues and countermeasures of safety management of reservoir dams under new situation in China. China Water Resour 904(22):1–3
4. Boulange J, Hanasaki N, Yamazaki D, Pokhrel Y (2021) Role of dams in reducing global flood exposure under climate change. Nat Commun 12:417. https://doi.org/10.1038/s41467-020-20704-0
5. Chen J, Shi H, Sivakumar B, Peart MR (2016) Population, water, food, energy and dams. Renew Sustain Energy Rev 56:18–28. https://doi.org/10.1016/j.rser.2015.11.043
6. Yang X, Lu X (2014) Drastic change in China's lakes and reservoirs over the past decades. Sci Rep 4. https://doi.org/10.1038/srep06041
7. Zhang H, Dou Y, Ye L et al (2022) Realizing the full reservoir operation potential during the 2020 Yangtze river floods. Sci Rep 12. https://doi.org/10.1038/s41598-022-06801-8
8. He X (2023) Dams, cropland productivity, and economic development in China. China Econ Rev
9. Sheng JB, Xiang Y et al (2022) Key technologies and applications for safety diagnosis and intelligent management of reservoir dams. J Geotech Eng 44(07):1351–1366
10. Song C, Fan C, Zhu J et al (2022) A comprehensive geospatial database of nearly 100 000 reservoirs in China. Earth Syst Sci Data 14:4017–4034. https://doi.org/10.5194/essd-14-4017-2022
11. Ministry of Water Resources of the People's Republic of China (2022) 2020 statistic bulletin on China water activities. China Water Power Press
12. Hydropower and Water Resources Planning and Design Institute (2023) 2022 China renewable energy development report. China Water Resources and Hydropower Publishing House, Beijing
13. Jun X, Jin C et al (2023) Opportunities and challenges in the construction of modern water networks in China under changing environments. J Geogr 78(07):1608–1617
14. Zhao DX, Wang ZQ et al (2023) Exploration and application of green comprehensive prevention and control technology system for termites in earth-rock dams in the Xiaolangdi project. China Water Resour 1–4
15. Tan JX, Ren X et al (2021) On the safety of reservoir dams in the new era. People's Yangtze River 52(05):149–153
16. Yang Z, Liu C et al (2019) Leading-edge technologies in hydropower development in China. Glob Energy Interconnect 2(03):244–253
17. Ministry of Water Resources of the People's Republic of China (2017) SL 252-2017, classification and flood control standards for hydropower projects. China Water Resources and Electric Power Press, Beijing
18. Regulations on the safety management of reservoir dams: State Council of the People's Republic of China No. 78 (1991)

19. Measures for safety appraisal of reservoir dams (2003) Bull State Counc People's Repub China 32:22–24
20. Jiang C, Wang LJ (2021) Typical problems and work suggestions for safety assessment of small reservoir dams. China Rural Water Resour Hydropower 12:171–173, 180
21. Chen H, Tian BR et al (2022) Study on the safe operation management model of small reservoirs. China Rural Water Resour Hydropower 02:174–178, 183
22. Ministry of Water Resources of the People's Republic of China (2012) Technical specification for safety monitoring of earth-rock dams: SL 551-2012. China Water Resources and Hydropower Publishing House, Beijing, p 2012
23. Zhang JY, Sheng JB et al (2015) Key technologies for safety assurance of reservoir dams. Water Resour Hydropower Technol 46(01):1–10
24. Ministry of Water Resources of the People's Republic of China (2013) Technical specification for safety monitoring of concrete dams: SL 601-2013. China Water Resources and Hydropower Publishing House, Beijing, p 2013
25. Wu ZR (2003) Theory of safety monitoring of hydraulic buildings and its application. Higher Education Press, 7-04-011306-6
26. Ministry of Water Resources of the People's Republic of China (2017) Guidelines for safety assessment of reservoir dams: SL 258-2017. China Water Resources and Hydropower Publishing House, Beijing, p 2017
27. Zheng S, Gu C, Hu Y et al (2022) A novel outlier detection method for monitoring data in dam engineering. Expert Syst Appl 193:116476. https://doi.org/10.1016/j.eswa.2021.116476
28. Gabriel-Martin I, Sordo-Ward A et al (2019) Hydrological risk analysis of dams: the influence of initial reservoir level conditions. Water 11:461. https://doi.org/10.3390/w11030461
29. Fluixá-Sanmartín J, Altarejos-García L et al (2018) Review article: climate change impacts on dam safety. Nat Hazards Earth Syst Sci 18:2471–2488. https://doi.org/10.5194/nhess-18-2471-2018
30. Zhang Y, Yin ZH et al (2019) Research on the stability of ancient landslides near dam banks and their impact on dam safety. Hydroelectr Power 45(08):23–26, 75
31. Martin W (2020) Effect of reservoir-triggered seismicity on safety of large dam projects. INCOLD J 9(2)
32. Sheng JB, Wang ZS (2010) Lecture 14 on water conservancy forum: safety evaluation of reservoir dams. China Water Conserv 04:63–66
33. Wang SJ, Gu YC et al (2019) Evaluation system for dam safety monitoring system. J Water Resour Transp Eng 04:63–67
34. Sun JH (2018) Achievements and challenges in safety management of reservoir dams in China. China Water Resour 20:1–6

Chapter 2
Dam Safety On-site Inspection and Test

2.1 Introduction

It is the key to carry out effective dam health operation inspection to identify the daily operation safety problems of the dam, and then carry out targeted safety maintenance [1]. There are three general types of dams health and operational condition inspections: **regular technical inspection, regular maintenance inspection and daily inspection**. The technical inspection is carried out by entrusting experts to carry out dam safety assessment, that is, dam safety appraisal; maintenance inspection is more frequent than technical inspection by assessing the operational capacity and structural stability of each component of the dam, so as to identify the adverse factors of dam development in the early stage; daily inspection is carried out by the daily dam management personnel (including dam construction personnel, hydropower plant operators, maintenance personnel and reservoir responsible persons) within their areas of responsibility. Some hydraulic structures are with hidden safety defects. The safe and normal operation of these structures can be ensured if the defects were found timely during the daily inspection and correct maintenance, repair, reinforcement and de-risking measures were taken thereafter. Therefore, daily inspection and regular maintenance is necessary for the safe operation of reservoir dams [2, 3].

2.2 Regular Visual Inspection

Regular visual inspection of reservoir dam is the most economical way that managers can use to ensure the safety of the dam and its surrounding environment, and ensure the long-term safe and stable operation of the dam [4].

Y. Xiang et al., *On-site Inspection and Dam Safety Evaluation*, Hydroscience and Engineering, https://doi.org/10.1007/978-981-97-3987-5_2

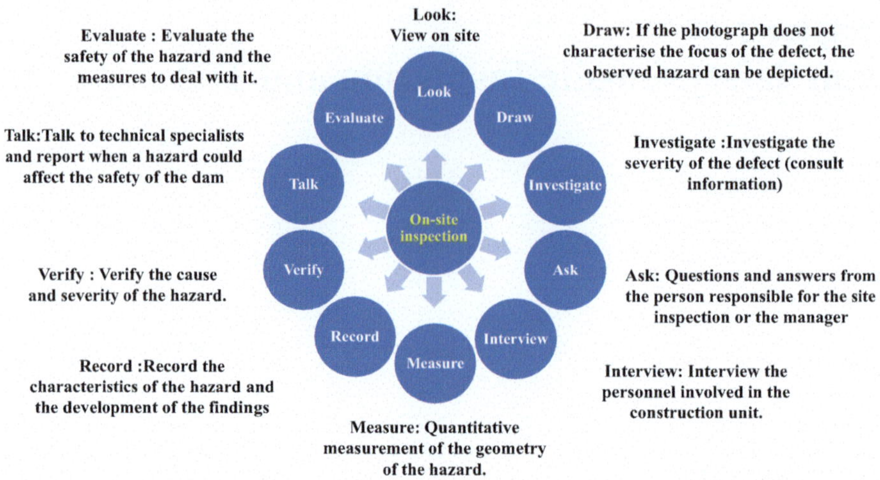

Fig. 2.1 The "10-step" principle for on-site inspection

The visual inspection is a simple procedure, strong operability and operable method that can be used by any properly trained practitioner, and are inexpensive and time-consuming, with inspection programs that can be fully completed in a short period of time. It mainly involves careful observation of all structural surfaces of the dam and its various components, including the surrounding environment of the dam, in order to make a reasonable and accurate assessment of the dam's operational behavior. The "10-step" principle for regular visual inspection can be applied to identify a potential hazard (Fig. 2.1).

2.2.1 Inspection Methods and Equipment

All inspections should be organized and systematic, and there should be a dedicated person in charge (Table 2.1). The inspectors should use appropriate methods and equipment to accurately record the results of the inspection, and comprehensively examine the structural state of the dam system [5].

2.2.2 Recording Inspection Results

Record all measurements and observed details to obtain the current condition of the dam and possible problems, and accurately describe what was observed during each inspection in detail, so as to compare the conditions observed during different periods. The information contains the following three elements:

Table 2.1 Inspection equipment and its purpose

No.	Check equipment	Purpose
1	Task list	Make a list before the inspection starts, and specify the relevant items for this inspection to improve work efficiency
2	Notebook, ballpoint pen	Use paper records, voice records, picture records and other means to record the examination results in detail to reduce the error caused by memory bias and avoid repeated examination. Using picture records can compare the differences between the past and the current situation
3	Recording equipment	
4	Phones, cameras and other devices	
5	Hand-held level	It can be used to determine the height and slope of the levee
6	Probe	It can be used to determine the depth and softness of saturated areas below the surface
7	Safety helmet hat	It should be worn throughout the inspection process to ensure the personal safety of the inspectors
8	Steel tape measure	Accurate measurement of dimensions, including deformations, cracks, falling blocks, missing, etc.
9	Flashlight	For checking areas with low light and low visibility
10	Shovel	It helps to clean up the drain, remove debris, and locate monitoring points
11	Stone hammer	It can be used to check the stability of problematic slope protection or concrete. But it should not cause unnecessary damage to the structure in the process
12	Knocker	It can be used to determine the condition of the supporting material in a concrete structure by forcefully striking the surface of the facing material. There is a noticeable sound difference between concrete that is completely supported by the filling material and the surface material that has voids or cavities
13	Volume container and timer	For accurate measurement of leakage. Depending on the flow rate, different sizes of containers may be required
14	Items to use for marking	Items such as markers, chalk or paint, etc. are used to mark the areas that need to be paid attention to in the future, and to mark the boundaries of existing conditions such as cracks and wet areas for future comparative analysis
15	Waterproof boots	For checking areas of the dam site where standing water exists
16	Insect repellent	Prevent mosquito bites and ensure the efficiency and effectiveness of the inspectors
17	First aid kit	Keep yourself safe during the examination

1. **Defect location**: The location of any defect areas or anomalies must be accurately described, with photographic evidence retained if necessary, so that changes in these conditions can be analyzed over time or to facilitate re-inspection by professionals.
2. **Defect extent or area**: This includes the length, width, height, and depth of the defect area or anomaly.
3. **Detailed description of defects**: A brief but detailed description of the defect area or anomaly should be provided [6]. The contents include: the amount of drainage outflow, the amount of seepage from point and surface sources, the color and quantity of sediment in water, the depth of concrete deterioration, the length, displacement and depth of cracks, the area of wet, moist, the adequacy of the protection surface, the adequacy of surface drainage, the steepness and structure of the slope, the proportion of obvious deterioration, the different variable conditions, etc.

2.2.3 Considerations

(1) **Inspection route**: The specific inspection route should be determined before the inspection procedure begins. The reasonable formulation of the inspection route can ensure the inspection efficiency while minimizing unnecessary energy consumption, so as to reduce mistakes in the inspection process. Systematic full coverage inspection of the entire dam site should be in the following order: **upstream dam slope → dam top → downstream dam slope → seepage area → outlet → spillway → other water drainage buildings**. Inspections in this continuous order can reduce important conditions that have been overlooked. At the same time, when carrying out inspection work on the downstream dam slope, a "Z" shape should be used to check the path to ensure that all cracks are detected. It is recommended to write the inspection results report in the same order to ensure the consistency of records [7].

(2) **Preservation of inspection results**: Each inspection result should be dated and archived with all inspection results such as images, audio and paper reports (photos should also be dated). In addition to the results observed during the inspection, the data and weather conditions measured during the inspection (especially recent rainfall, prolonged drought and snow cover) should also be recorded in the inspection report.

Immediately after the inspection, the observations should be compared to previous records to see if there are any trends that indicate problems with the dam. If changes or trends are noticed but the problem is not urgent, dam managers can consult relevant experts to quickly improve the situation that needs attention to ensure dam safety and to keep the dam safe and stable, while preventing the need for costly repairs in the future [8].

(3) **Critical inspection events**: Regardless of the normal schedule, an emergency inspection should be carried out immediately when the following situations occur.

- Check the condition of conveyance and drainage structures such as dams and spillways before heavy rain or snow melt is forecast.
- Check the condition of conveyance and drainage structures such as dams and spillways during or after a rainstorm.
- Immediately after an earthquake, inspect the area; over the next few months, carry out multiple inspection efforts to check for delayed effects.
- Before the first storage of the reservoir and during the initial storage operation; Before the reservoir is filled for the first time, a safe and regular comprehensive inspection can ensure that the dam design and site conditions meet the requirements of water storage. In the initial water storage operation process of the dam, the safe and regular comprehensive inspection can find the problems existing in the reservoir dam in time to ensure the safe and stable operation of the dam.

2.2.4 The Inspection Parts of the Earth-Rock Dams

The main inspection parts of the earth-rock dams include: upstream dam slope, downstream dam slope, dam crest and seepage condition. Many of the principles and guidelines mentioned in this section also apply to concrete structures.

2.2.4.1 Upstream Dam Slope

In general, the main problems that occur on upstream dam slopes are: **cracks, landslides, cave traps and severe erosion** [9]. The first three problems are mainly found on levees, but severe erosion can damage the structure.

Upstream revetments and high water levels can hide problems. (When walking on riprap, caution should be used to avoid injury.) When there is no water in the reservoir, the exposed slope should be thoroughly inspected for settling areas, rat holes, landslides or sliding conditions, and when there are no empty reservoir inspection conditions, a third party should be commissioned to carry out diver or underwater drone inspection [10]. The bottom of the reservoir should also be inspected for landslides or sinkholes. It is of utmost importance to use crossing paths so that cracks and landslides can be easily detected. In many cases, a change in slope uniformity can be seen by looking along the axis of the dam [11].

The inspector should stand at one end of the dam and check the straightness and uniformity of the dam axis. If a crack is seen, then the dam crest and downstream dam slope in its immediate vicinity should be carefully inspected. Cracks may indicate a shift in the dam foundation, the failure of the dam, or sliding on the surface. The crack may be less than 1 cm wide, but it may be several meters deep.

Slips can be just as hard to detect as cracks. When building a dam, the slope may not be graded uniformly. Understanding the slope structure can help better detect slope movement after construction is complete. Slopes may slide over very small distances, for example, they may only settle or bulge by about 1 m over distances of 100 m or more, but this is still a considerable amount of subsidence. When checking for these changes, previous inspection records can provide a contrast.

The cause of the caves and depressions is erosion within the dam—a serious problem for earth dams [12]. Internal erosion or piping erosion can be reflected by turbid seepage at the outlet. The surface soil of the dam slope may be eroded by wave action, rainwater runoff, and animal burrowing activity, which may result in a reduction in the thickness of the embankment and a weakening of the embankment structure [12].

2.2.4.2 Downstream Dam Slope

In order to adequately inspect the downstream dam slope, the area should be kept free from the shade of weeds, shrubs, or trees, and the downstream area should be cleaned and remedied regularly [13]. If the downstream dam slope is covered by thick brush or vegetation, a more serious search must be conducted. The main problems that occur on the downstream dam slope are: **cracks, landslides, or seepage**. When these conditions occur, the designated dam safety authority should be notified immediately.

Cracks may occur as a result of **soil subsidence, drying, and shrinkage**. Regardless of the cause, cracks should be detected and changes in their length and width should be noted. Cracks due to soil dryness may appear and disappear seasonally, and usually do not show changes in vertical displacement like subsidence cracks or sliding cracks. The occurrence of landslides requires a detailed assessment. Early warning signs include a bulge near a dam site or a vertical displacement in the upper part of a levee. Seepage issues will be discussed separately later. If any of these three conditions arise or there are indications that the problem may fall under any of these three conditions, the dam safety management agency should be notified immediately.

2.2.4.3 Dam Crest

The crest area is the main access for inspection and maintenance, and dam managers need to re-level the crest on a regular basis. Problems that occur at the crest of a dam should not be simply graded or covered up, and the dam safety management agency should be notified immediately when a problem is discovered. The main problems that occur at the crest of a dam are: **longitudinal cracking**, **transverse cracking or misalignment** [14].

Longitudinal cracking indicates the local instability of the embankment, differential settlement, or displacement between adjacent sections. Longitudinal cracks are typically characterized by a single crack or a dense system of cracks along the crest of the dam and more or less parallel to the axis of the dam. Typically, these cracks are continuous in length and can be distinguished from dry cracks. Dry cracks are usually not continuous, irregularly shaped, shallow in depth, narrow in width, and have a large number of cracks.

Transverse cracking indicates differential settlement or movement in adjacent parts of the dam. A transverse crack is usually a single crack or a dense, parallel system of cracks oriented more or less perpendicular to the dam across the crest. Transverse cracks can compromise the safety and integrity of the dam. If the crack develops below water level, seepage can flow out along the crack and through the dam, causing severe erosion, and if not repaired, a dam failure.

Dislocation indicates movement between adjacent parts of a dam, and usually occurs in a direction perpendicular to the axis of the dam. Excessive settlement of dam materials and dam foundations can also cause misalignment. Most problems can be detected upon close inspection. However, it is only when viewing the dam from its base that misalignment can be detected. If the structure at the top of the dam is straight in length when examined up close, you can stand on the shoulder of the dam at a distance and observe along the upstream and downstream edges of the dam top to further check for misalignment. For a curved dam, you can stand at either end of the dam and observe along the upstream and downstream edges of the dam top, noting all curvature and dislocation in that section to check whether the dam is aligned.

2.2.4.4 Seepage Condition

Seepage occurs in all dams, but seepage in any area of the dam or in an area adjacent to the dam should be considered a potential problem. If the area downstream of the dam is wet, it is usually not a natural spring, but a seepage area. It is important to control the flow rate and seepage volume. Water flowing through the dam at high speed will erode and eventually cause the dam to break. Massive landslides can occur in the saturated areas of the dam or the shoulder of the dam, which can lead to other problems [15].

Seepage can occur anywhere in the dam, including downstream of the dam, at the toe of the dam, or on the downstream shoulder of the dam below normal reservoir level [16]. One potentially dangerous situation is water seepage on the downstream side above the toe of the dam. Water seepage on the downstream side can cause landslides or internal erosion (piping, etc.). The result of water seepage may be a soft, moist area or a flowing spring. Balsam, reeds, mosses, and other swamp vegetation often grow in the seepage area, so it may initially be just a dark green colored area with heavy vegetation. The area downstream of the abutment should always be carefully inspected for signs of seepage, as well as areas in contact with it such as levees, spillways, and other ancillary structures. Slippage in dikes and dam bases can be caused by saturated soil as well as seepage from high pore water pressure [17].

Water seepage is not usually noticed immediately, so all possible seepage areas should be searched intensively. Even in areas where the grass is short, seepage may not be visible and must be stepped on. Ideally, a leak check should be performed when the reservoir is full.

2.2.5 The Inspection Parts of the Concrete Dams

2.2.5.1 General Inspection

Although concrete dams are in the minority compared to earth-rock dams, they are usually higher and have a greater capacity to hold water. Therefore, the failure of concrete dams usually causes greater harm to people's life and property safety. It is important for concrete dam managers to understand all the failure modes of concrete dams and identify situations that threaten the safety of the dam and require repair. From a safety point of view, the main advantage of concrete dams over earth dams is that they are relatively immune to the erosion failure of the overflow process, the sliding of the dam slope and the internal erosion failure [18].

The cause of failure of concrete dams is very different from that of earth dams. The main reasons include: **structural cracks**, **weakness of dam foundation and abutment and** deterioration **due to alkali aggregate reaction** [19]. If these conditions are found during an inspection, the dam managers should immediately seek help from the relevant personnel.

Cracks occur when the dam is under too much pressure under part of the load due to improper design, improper construction, or poor material quality. Structural cracks are usually irregular and may be the same Angle as the main shaft of the dam, or they may suddenly change direction. These cracks may also be accompanied by significant radial, lateral, or vertical displacements [20].

Concrete dams transfer large loads to the abutments and foundations. Natural abutments and foundations may crack, collapse or slip over large areas under the action of extreme loads [21, 22]. If this happens, the dam will lose its support and collapse. The initial movement is generally small, so it can be difficult to spot an imminent problem with the abutment or foundation.

A chemical reaction occurs between the alkali [23] in the cement and some forms of silica present in the mixture, which may lead to severe deterioration of the material and the production of by-products of silica, resulting in concrete expansion and strength of the concrete reduction. Alkali reactions [24] are usually accompanied by observable characteristics such as cracking (usually large in scale and random in pattern) and excessive expansion both internally and overall. Other signs include the presence of gel-like exudates or white amorphous deposits on the surface, and the highly contrasting appearance of newly cracked concrete.

The alkali aggregate reaction only occurs in the presence of water. Surfaces that have water or are wet due to water seepage will deteriorate rapidly. If this is suspected, it can be confirmed by drilling a hole. Although the deterioration of the alkali aggregate reaction is gradual, there is currently no way to effectively deal with the continued deterioration brought about by the alkali aggregate reaction, which may eventually lead to structural reconstruction.

2.2.5.2 Additions

The inspection of concrete dams is similar to that of earth-rock dams, but the following elements need to be added: **inspection and safety measures, monitoring instrument layout, expansion of building cracks and construction joints, shrinkage cracks, deterioration caused by spalling, and minor leakage**.

Because the surface of concrete dams is usually vertical, and the site is usually steep rock canyon, so the use of reasonable **inspection and safety measures** for concrete dam inspection is very important. Access to the downstream face, toe, and abutment of such dams may be difficult for inspectors and requires special safety equipment such as safety ropes or specialized scaffolding for working at heights. With the continuous advancement of drone technology, the use of drones to carry out concrete dam inspection has become the first choice.

Monitoring instrument layout [25] helps to detect structural problems in concrete dams, such as cracks in the dam, dam abutment, or dam foundation. If structural problems are found, the relevant monitoring instruments can be installed to monitor the changes in width, length, and offset of the development of structural problems, and the data can be read regularly for analysis [26].

Since concrete dams are built in sections, construction joints will appear, while expansion joints are built in the dam to adapt to the volume change of the structure after concrete pouring. The latter are referred to as **"design" cracks**. These joints are usually constructed without any bonding or reinforcement other than unbonded stops and pins extending across the joint.

Shrinkage cracks [27] usually occur during construction, and irregular parts of the abutment contact surface or gaps filled with concrete that cannot be fully cured before pouring. Subsequent consolidation shrinkage of the concrete may result in irregular cracks at the dam abutment or near the joints; temperature changes can also cause the appearance of contractile cracks [28, 29]. During the winter months, the upper part of the dam can get much colder than the part that comes into direct contact with the reservoir water [30]. This temperature difference can cause cracks to extend some distance down from the top of the dam, which may appear in construction joints or expansion joints, if any. Shrinkage cracks may indicate that some part of the dam cannot withstand the amount of load it is designed to carry. In this case, the total compressive load must be borne by a small proportion of the structure, and it may be necessary to restore the load-bearing capacity of the affected area by grouting, which requires the assistance of the design department [31].

Spalling is the process by which concrete breaks apart due to freezing and thawing. Under the influence of cold weather, almost every concrete dam will continue to experience slight deterioration due to spalling [32]. Usually spalling only affects the surface of the structure, so it is not usually considered dangerous. However, if spalling is allowed to continue, it can cause structural damage, especially if the dam has a thin cross-section. Alternatively, when spalling causes the rebar to become exposed, repair is necessary. The method of repair for the spalling area depends on the depth of the damage. If the damage is severe, engineering assistance will be required.

Minor leakage in concrete dams, although difficult to detect, is usually not dangerous if it is not accompanied by structural cracking, the effects of which may be exacerbated by freezing and thawing. Dam managers should be aware, however, that increased seepage may indicate that material in the dam has been chemically washed away by running water. Water seepage will also be reduced, due to the formation of mineral deposits in some leakage channels. Both of these conditions are not inherently dangerous, and both need to be studied in detail to determine whether repairs are needed for reasons other than aesthetics [33].

2.2.6 The Inspection Parts of the Spillway

2.2.6.1 General Problems

The function of the spillway is to provide a safe passage for the drainage of the reservoir. If the spillway does not have enough capacity, overtopping or dam failure can occur. Similarly, rapid erosion can cause defects in the spillway, which in turn can lead to dam failure. The spillway should always be kept open and resistant to erosion, and prevented from deteriorating. Reservoir hubs are expensive to build and spillways account for most of the cost of reservoir hub construction, so maintenance programs should be carefully implemented not only to protect the public, but also to minimize costs.

The main problems of spillway include **insufficient discharge capacity, occupied flood passage, erosion damage, structural deterioration, cracks, and damage to the spillway outlet** [34].

The **insufficient discharge capacity of the spillway** is caused by the following factors, such as the area of drainage area, the intensity of rainstorm in the basin, the storage capacity of the reservoir, and the speed of rainwater flowing into the reservoir and filling the reservoir. Insufficient discharge capacity can lead to reservoir overtopping.

Overgrowth of aquatic plants and weeds, dense undergrowth, trees, debris, or landslide sediments that **occupy flood channels** will greatly reduce the discharge capacity and result in overtopping. Grass is usually not considered an obstacle, but the spillway should be regularly cleared of tall weeds, bushes, and small trees. Similarly, soil and large rocks due to caving, landslides, or sediment transport should also be cleared immediately, otherwise, they will hinder the flow of water and cause insufficient flood capacity.

When the spillway operates at high intensity for a long time, it can cause **water erosion of the spillway**. If the spillway cannot resist the erosion action of the water flow, it will cause serious damage to the structure or complete erosion. Erosion failure usually does not occur if the spillway is built on a rock foundation or lined with concrete. However, erosion protection measures will be very important if the spillway sits on sand, metamorphic granite, clay, or silt deposits. Generally speaking, if the slope of the spillway is relatively gentle, or if it is covered with a layer of grass or riprap with bedding material, its erosion resistance can be increased.

If the spillway has **structural deterioration**, it will not be able to continue normal operation. Deteriorating conditions include: slope collapse, riprap collapse, failure of concrete lining, break of approach section, collapse of steep chute, failure of stilling pool, damage to drainage channel or protective grass cover, etc. These problems can cause water to flow under and around the protective material and serious erosion damage to occur. Remedial action must be taken as soon as any signs of structural deterioration are noticed.

Dry **cracks** in earthen spillways are not usually considered a functional issue. However, missing stones in riprap linings ("cracks" in the protective layer) should be patched immediately. Cracks in spillway concrete linings due to uneven settlement of dam foundations, shrinkage, plate displacement, and excessive water pressure or earth pressure are common problems. Larger cracks can cause the water to rush out of the fine material beneath the concrete base plate, causing erosion, which can lead to more cracks, or even severe plate displacement, causing the plate to be washed away by the current. Severely cracked concrete spillways should be inspected and repaired under the supervision of industry experts.

One of the common problems with spillways is damage to spillways or other flood discharge and energy dissipation facilities that result in **damage to the spillway outlet**. Severe damage to the spillway outlet can cause displacement of some flood discharge facilities, scouring the downstream toe of the slope and causing landslides downstream of the dam, ultimately leading to dam failure. Upstream water must be safely transported from the reservoir to the designated location downstream of the dam without damaging the spillway and dam. Often, spillway outlets are adequately protected under normal flow conditions, but not under extreme flow conditions. The factors that cause erosion to occur at spillway outlets are as follows:

(1) The flow of water discharged from the spillway outlet is higher than the channel. If the discharge flow is at the correct height, the spillway's tailrace canal can absorb a considerable amount of high-speed water flow, and the water energy can be eliminated by the energy dissipation tank according to the design.
(2) The water flowing out of the spillway is usually free of sediment and therefore has a large sediment-carrying capacity. In the process of acquiring sediment, the flow scours soil material from the channel and leaves behind areas that have

been eroded. This erosion is difficult to design out, so the outlet needs to be protected from a safe distance downstream from the dam.

(3) The flow of water leaving the spillway at high speed creates negative pressure, which in turn "cavitation" occurs, causing material to loosen and detach from the floor and side walls.

(4) Water leakage from the joint of the flood passage or flowing along the flood passage may weaken the soil structure around the flood passage. During the construction process, insufficient compaction of adjacent flood passage will exacerbate this problem.

2.2.6.2 Inspection Procedures

Spillway inspection is an important part of dam safety procedures. The primary purpose of spillway inspection is to detect any signs of **obstruction, erosion, structural deterioration, misalignment, or cracking** [35].

(1) **Soil spillway inspection**

When inspecting soil spillways, determine if there are any landslides on the side slopes, if there is excessive vegetation in the channel, if there are signs of erosion, and if there are signs of rodent activity. The hardness and moisture content of the soil should also be determined, and when testing the hardness and moisture content of the soil, pay attention to locations that are particularly wet or soft and to see if the stilling poor or settling structure is properly protected with rock or riprap.

Since some erosion is inevitable, dam managers should also determine whether such erosion could endanger the structure itself. If sills are installed on the spillway, the dam manager should also determine if there are any cracks or misalignments in the sills, and if there are any conditions of erosion below or downstream of the sills.

(2) **Concrete spillway inspection**

Usually, finer cracks are harmless. Large cracks should be carefully examined and noted for their location, width, length, and orientation. Determine the degree of deterioration of the rebar and pay attention to the exposure of the rebar.

The spillway is subject to freeze–thaw cycles, and surface spalling often occurs. At the same time, chemical action, pollution, and poor-quality aggregates can also cause spalling. If the spalling is severe, a sketch of the spalled area should be made or photographs taken to record the length, width and depth of the area. Determine the degree of further deterioration of the concrete. The concrete should be struck with a "knocker" or a stone hammer to determine if there are gaps in the concrete. Deeper areas of spalling should be patched as soon as possible.

The spillway will have drain holes in the side wall and the bottom plate of the discharge chute, and the blockage of the drain holes increases the likelihood of failure of the retaining wall or bottom plate of the discharge chute. The working condition of the drain holes should be checked regularly to examine whether there is blockage by mud or mineral deposits, determine the cause of the hole blockage, and take appropriate measures to prevent the problem from happening again.

The spillway retaining wall and the bottom plate of the discharge chute are usually constructed in sections. Gaps or joints between adjacent sections must be tightly sealed with a flexible material such as asphalt, epoxy resin, or another compound. Sometimes rubber or plastic diaphragm material or copper foil is used to obtain watertightness. During the inspection, note the location, length, and depth of any missing sealant, and probe open gaps to determine if the soil behind the wall or under the board has been damaged.

Foundation settlement, soil pressure, or water pressure can cause the spillway retaining wall or the bottom plate of the discharge chute to be misaligned. Inspectors should look closely at the wall near the upstream or downstream end of the spillway to determine if the wall has developed an inward or outward tilt. Relative displacements or offsets between adjacent sections can be easily determined at the joints. The fence at the top of the retaining wall is usually erected in a straight line during construction; therefore, any deformation of the fence may indicate that the wall is deformed.

The entire chute will form a smooth surface at the time of construction. Therefore, measuring the relative motion between the bottom plates of the adjacent discharge chute at the joint can clearly reveal whether the bottom plate has shifted. The misalignment or displacement of the wall or floor is usually accompanied by the appearance of cracks. The shape of the crack should be clearly recorded or photographed to help understand the nature of the displacement.

2.2.7 The Inspection Parts of the Discharge and Conveyance Structures

The drainage project of a dam is essential to the operation of the dam [36]. Inspection and special precautions include **reservoir water level, drainage outlet, rust, trash rack on pipeline spillway, cavitation, and metal structure inspection**.

(1) **Reservoir water level**

The reservoir water level is controlled by the sluice gate or valve of the drainage structure. For slopes made of clay or mud and sand materials, the rate of water level decline should not exceed 0.3 m per week, except in emergencies. Very flat slopes or upstream with freely draining soil can withstand a faster rate of water level decline. Situations that cause or require temporary or permanent adjustments to the water level of a reservoir include:

- When a problem arises that requires lowering the reservoir water level, temporarily lower the reservoir water level until the problem is resolved.
- Release water downstream to replenish the river's water volume during dry weather.
- Fluctuating demand for water in downstream service areas.
- Winter dock maintenance and coastal aquatic plant growth.
- Management requirements for recreation, hydropower, waterfowl, or fish.

(2) **Drainage outlet**

The drainage outlet should always be operational so that the reservoir water level can be lowered in case of emergency or if necessary repairs need to be made [37]. Drainage valves or gates are a particular problem for dam managers if they are not used for a long period of time. If the valve cannot be closed after it has been opened, it may drain the reservoir of water. A rapid and uncontrolled drop in the water level at this point can cause more serious problems, such as landslides along the upstream slope of saturated dams or flooding downstream. Therefore, when using valves or gates, check, lubricate, and repair all appropriate components. Residents downstream should be informed if heavy and prolonged drainage is carried out.

The drainage inlet upstream of the valve or gate must be firmly blocked if the valve or gate is being tested to see if it affects drainage. Drainage facilities are equipped with double valves or gates, or slots upstream of the valve for placing stop logs (sometimes called partitions). Or hire a diver or drone to inspect the drainage outlet. Sediments in the reservoir can build up and block the drainage outlet, or debris can get into the valve chamber and cause the valve to fail.

(3) **Rust**

Rust is a common problem in metal pipelines. Exposure of pipelines to moisture, acidity, or salt can accelerate corrosion. In particular, acidic runoff from open-pit mining areas can cause rapid corrosion of steel pipes. In these areas, pipelines made of non-rusting materials such as concrete or plastic should be used. Metal pipes coated with resistance to accelerated rust can also be used. This paint can be epoxy resin, aluminum, zinc (galvanized), asbestos, or mortar. Because it is difficult to establish a bond, using paint on pipes is usually not very effective. Similarly, asphalt paint is only effective on pipes for one to two years. Rust on the metal parts of the operating mechanism can be effectively treated and prevented by the use of grease or paint. Corrosion can also be prevented by installing cathodic protection in the machine.

(4) **Trash rack**

Tunnels (culverts) and metal pipe inlets blocked by debris or garbage can reduce flood discharge capacity. The possibility of overtopping is greatly increased. If the dam has an emergency spillway, the blockage of the main spillway can cause the emergency spillway to flow more frequently and more heavily than normal. Since emergency spillways are usually designed for infrequent flows over a short period of time, serious damage can be caused. For these reasons, the trash rack should be installed at the entrance to the buildings that drain water. A well-designed trash rack will stop large pieces of trash that may clog pipes, but water and smaller trash will still be able to pass through the drain grating. If the grate is too small, it can hold back small debris such as twigs and leaves, but will cause larger objects to build up and eventually clog the intake entirely. Also the sewage grate needs to be strong enough to withstand the weight of fast-moving rubble, heavy objects, and ice.

Check the trash rack regularly for rusty or broken parts and make repairs as needed. During and after a storm, the trash rack should be inspected to ensure that it is functioning properly and to remove accumulated dirt.

(5) **Cavitation**

When the local pressure in the liquid decreases, the process of formation, development, collapse of vapor or gas cavities (bubbles) inside the liquid or at the liquid–solid interface is called cavitation. After the object is affected by cavitation in the liquid movement, the deformation and material erosion phenomenon will appear on the surface, which will damage the outlet pipe and control valve. This process can be a serious problem for dams with high discharge velocities.

(6) **Testing the outlet/drain system**

All valves should be fully open or fully closed at least once a year. This will not only stop corrosion building up on the control rod or valve guide rod, but also check whether the system is running smoothly. Jitter or erratic operation can be a signal that something is wrong, indicating the need for more detailed inspection.

The full range of gate settings should be checked. The inspector should slowly open the gate to check for noise and vibration—certain gate settings may cause greater turbulence. The inspector should also listen for any noise that sounds like gravel moving quickly through the valve. The presence of this sound indicates cavitation and should be avoided. The operation of all electromechanical systems, backup motors, generators, and outlets related to power and lighting wiring should also be inspected.

(7) **Checking the outlet/drain system**

The accessible part at the outlet/drain can be easily inspected on a regular basis. However, serious problems are usually related to the deterioration or failure of parts buried in the dam or normally in water. Checking includes an inspection of the inner wall of the pipe. If necessary, underwater drones can be used to perform the inspection, or divers can be hired to conduct underwater inspection.

2.2.8 The Inspection Parts of Other Areas

Other areas to be inspected are **electromechanical systems, near-dam reservoir banks, upstream basins and catchment surfaces, and downstream floodplains**.

Mechanical equipment includes gates or valves, stoplogs, blowdown pumps, sluice plates, relief wells, emergency power supplies, siphon pipes, etc. of drainage structures. All mechanical and electrical equipment should be operated at least once a year (or more frequently). The inspection of the electromechanical system shall cover the full operating range of the equipment under actual operating conditions. The operating equipment should be marked and placed in an easy-to-operate location. Check that all electromechanical control equipment is safe and protected from destruction. Make sure the operating instructions are clear and keep them in a safe and easily accessible location.

It is necessary to inspect the near-dam reservoir banks to ensure that there are no issues with the actual structure. Severe landslides can cause water to overflow the dam.

The flood originates in the upper river basin. Thus, the characteristics of the upstream watershed (such as impervious areas) are directly related to the magnitude of the flood. Urban development within the watershed will increase the size of the flood peak and the runoff, which will lead to the inadequacy of the flood capacity of the drainage structures obtained in the design stage. Understanding upstream development and other factors that may affect reservoir flow can help predict potential problems and make necessary dam modifications.

The development of downstream floodplains is also very important to dam owners, and the degree of development and flood protection is directly related to the loss of life and economic and social losses after dam failure.

2.2.9 Different Types of Dam Failure Conditions

This section gives detailed guidelines for earth and rock dams, concrete dams, spillways, tunnels and other areas. This section lists common dam failure conditions, possible causes, possible consequences and remedial measures, providing guidance for understanding various problems of dams (See Tables 2.2, 2.3 and 2.4). There are three types of dam failures, namely seepage, deformation, and operational and maintenance problems. However, if necessary, relevant industry experts are also required to assess the defects and recommend remedial measures.

Dam safety managers should make their best efforts to identify all the abnormal behaviors mentioned in this section that may cause safety problems in dam operation, and respond to the abnormal behaviors quickly, which can extend the service life of the dam and prevent costly repairs in the future [38].

Table 2.2 Seepage problem

Problem detail	Internal erosion or piping	Concentrated leakage	Water flows out of cracks in dam crest	Leakage water flows out of the dam foundation
	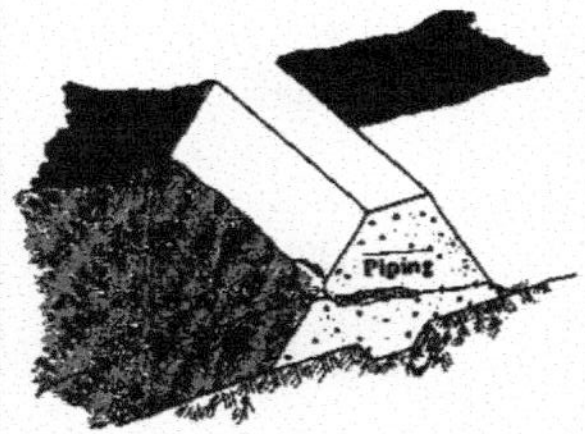	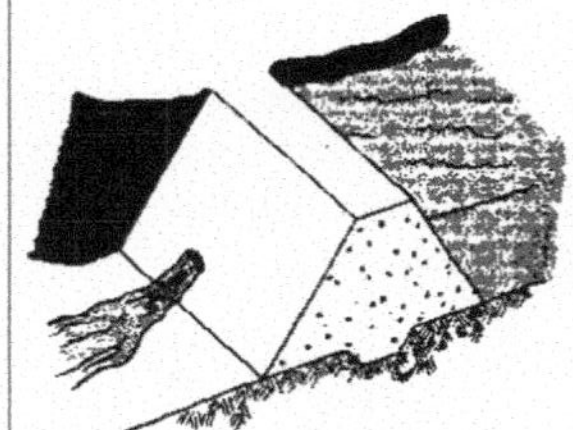	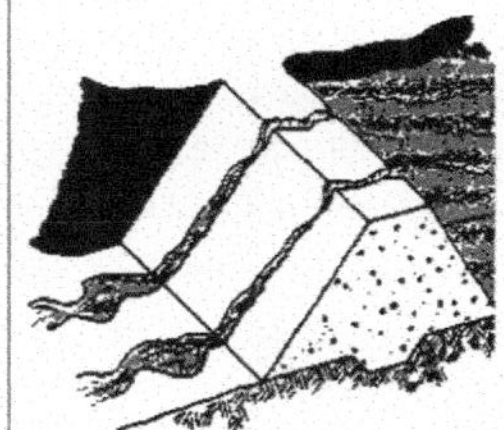	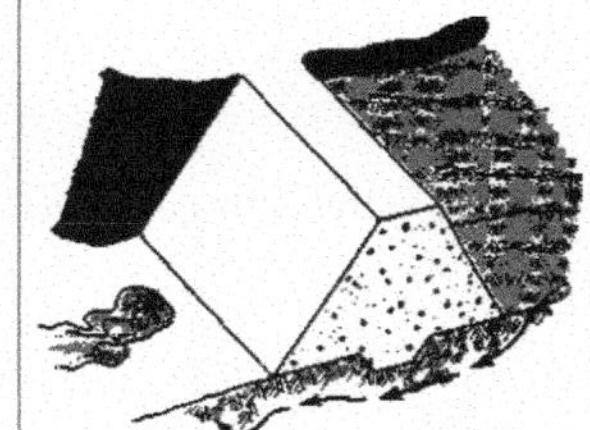
Probable cause	1. Seepage pits caused by internal erosion or piping of embankment materials and foundation 2. Karst cave collapses after being eroded 3. The turbidity downstream seepage indicates that the dam has been eroded	1. A large amount of water accumulates at the downstream slope. The current carries the dike material out. The churning of the current may cause muddy water 2. Water erodes embankment materials and washes them away, or rodents, freeze–thaw action or poor construction make water form a through passage on the dam	1. Severe drying causes shrinkage of embankment materials 2. Lateral cracks caused by embankment or dam foundation settlement	Some parts of the dam foundation (sand or gravel in the dam foundation) form flow channels. leakage water flows out of the dam foundation (sand boiling)
Consequences	Piping may cause the water in the reservoir to be drained, or cause dam break by erosion	The continuous flow of water will further erode the dam materials, leading to dam break	Cause dam failure	The increase of flow will lead to dam foundation erosion and dam break

(continued)

Table 2.2 (continued)

Recommended actions	1. Check other parts of the dam for leakage or seepage pits 2. Find out the exact cause of the seepage pit 3. Check whether the water seepage is turbid	1. Measure the flow rate to determine whether the water becomes turbid, or remains unchanged, or becomes clear 2. If the flow increases, the water level of the reservoir shall be lowered until the flow stabilizes 3. Check whether there are cracks on the upstream side, and plug them as much as possible	1. Block the upstream side of the crack 2. Lower the reservoir water level until it is below the crack position	1. If soil particles move downstream, sandbags or earth shall be used to build berms around the leakage. The pressure generated by the water level can control the flow velocity and temporarily prevent further erosion 2. If the erosion is more and more serious, the reservoir water level should be reduced
Problem detail	Seepage exiting at abutment contact	Large area wet or producing flow	Marked change in vegetation	Bulge in large wet area
	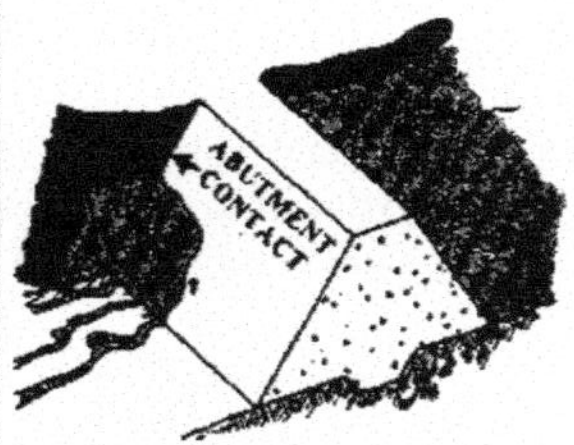			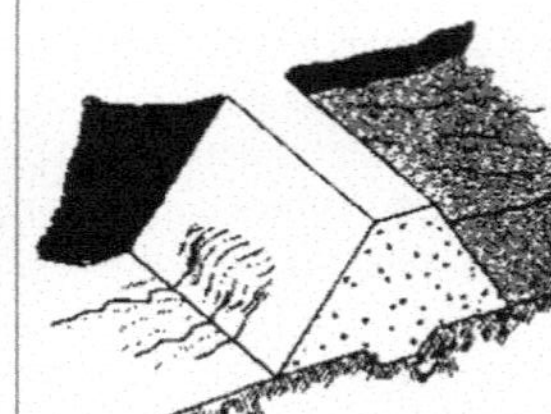
Probable cause	1. Water flowing through pathways in the abutment 2. Water flowing through the embankment	A seepage path has developed through the abutment or embankment materials and failure of the dam can occur	1. Embankment material are supplying flows paths 2. Natural seeding by wind 3. Change in seed type during early post construction seeding	Downstream embankment materials have begun to move

(continued)

Table 2.2 (continued)

Consequences	Can lead to erosion of embankment materials and failure of the dam	1. Increased flows could lead to erosion of embankment material and failure of the dam 2. Saturation of the embankment can lead to local slides which could cause failure of the dam	Can show a saturated area	Failure of the embankment result from massive sliding can follow these early movements
Recommended actions	1. Study leakage area to determine quantity of flow and extent of saturation. Inspect daily for developing slides 2. Water level in reservoir may need to be lowered to assure the safety of the embankment 3. A qualified engineer should inspect the conditions and recommend further actions to be taken	1. Focus on observing the saturated area and monitoring the increase or decrease of its area. Measure any flow as accurately as possible 2. If at a fixed pool level, the wet area increases, or the flow increases, the reservoir water level may need to be reduced 3. Relevant personnel shall timely check the situation and take further measures	1. Use probe and shovel to establish if the materials in this area are wetter than surrounding areas 2. If areas shows wetness, when surrounding areas do not, a qualified engineer should inspect the condition and recommend further actions to be taken	1. Compare embankment cross section to the end of construction condition to see if observed condition may reflect end of construction 2. Stake out affected area and accurately measure outflow 3. A qualified engineer should inspect the condition and recommend further actions to be taken
Problem detail	Trampoline effect in large soggy area	Dam abutment leakage	Wet zone of horizontal zone	Large increase in flow or sediment in drain outfall
		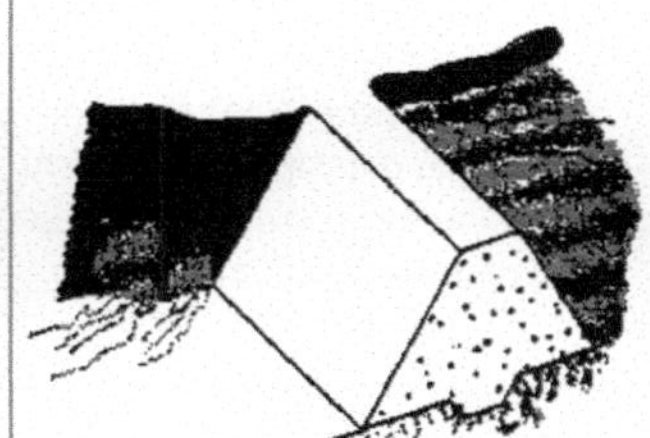	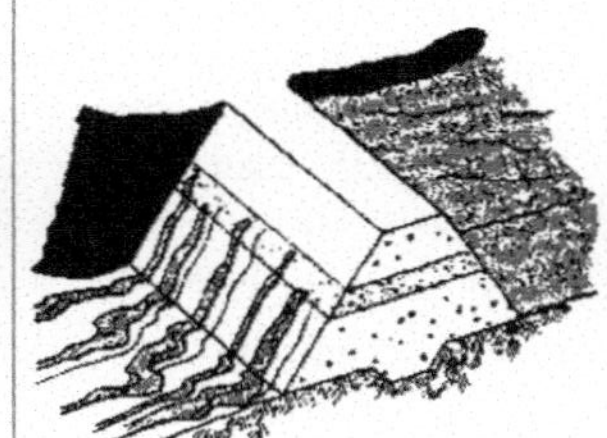	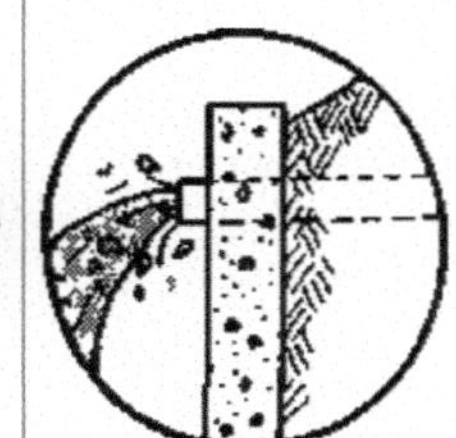

(continued)

Table 2.2 (continued)

Probable cause	Water flow (controlled or restricted by the perfect lawn root system) rapidly passes through the dam or foundation	Water seeps out in the cracks at the dam abutment	Frost layer or sandy material layer in the original building	A shortened seepage path or increased storage levels
Consequences	1. This situation indicates that there is too much water seepage in this area 2. If the control layer of turf is damaged, rapid erosion of dam materials may lead to dam break	Can lead to rapid erosion of abutment and evacuation of the reservoir. Can lead to massive slides near or downstream from the dam	1. The dampness below the excessive seepage area will lead to local instability of the dam (landslide) 2. Excessive flow will lead to accelerated erosion of local materials and dam failure	1. Higher velocity flows can cause erosion of drain then embankment materials 2. Can lead to piping failure
Recommended actions	1. Carefully check the flow and all materials transported in the area 2. The qualified engineer shall inspect the condition and recommend further action	1. Carefully inspect the area to determine quantity of flow and amount of transported material 2. A qualified engineer or geologist should inspect the condition and recommend further actions to be taken	1. Determine as closely as possible the flow being produced 2. If flow increases, reservoir level should be reduced until flow stabilizes or stops 3. Stake out the exact area involved using hand tools, try to identify the material allowing the flow	1. Accurately measures outflow quantity and determine amount of increase over previous flow 2. Collect jar samples to compare turbidity 3. If either quantity or turbidity has increased by 25%, a qualified engineer should evaluate the condition and recommend further actions

Problem detail	Erosion	Cave-in on crest	Gully on crest	Outlet releases eroding toe of dam
	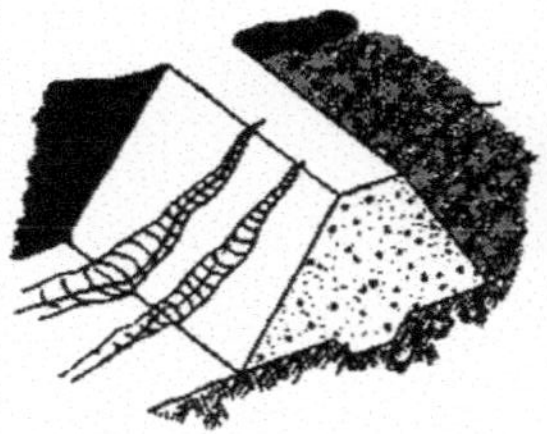		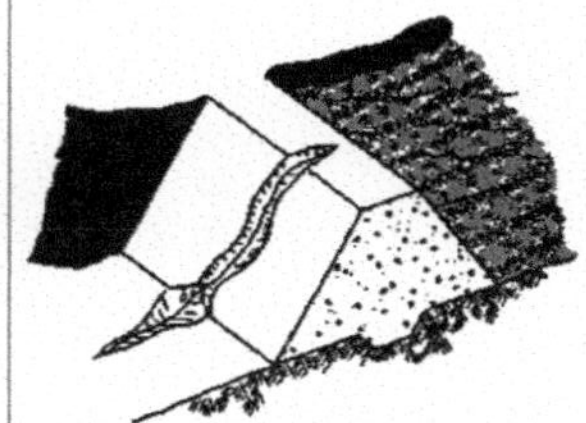	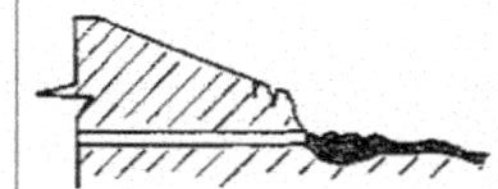
Probable cause	Water from intense rainstorms or snow-melt carries surface material down the slope, resulting in continuous troughs	1. Rodent activity. Hole in outlet conduit is causing erosion of embankment material 2. Internal erosion or piping of embankment material by seepage 3. Breakdown of dispersive clays within embankment by seepage waters	1. Poor grading and improper drainage of crest. Improper drainage causes surface runoff to collect and drain of crest at low point in upstream or downstream shoulder 2. Inadequate spillway capacity which has caused dam to overtop	1. Tunnel too short 2. There is no stilling basin at the downstream end of the tunnel

(continued)

(continued)

Consequences	Can be hazardous if allowed to continue. Erosion can lead to eventual deterioration of the downstream slope and failure of the structure	1. Void within dam could cause localized caving, sloughing, instability, or reduced dam cross section 2. Entrance point for surface water	1. The cross section of the dam is changed, resulting in the reduction of the cross section area of the dam 2. Inhibits access to all parts of the crest and dam 3. Can result in a hazardous condition if due to overtopping	Dam toe erosion makes the downstream slope too steep, resulting in progressive landslide
Recommended actions	1. The use of stones or riprap is the best way to protect eroded areas 2. If the problem is found in time, it is OK to re-plant protective grass seeds	1. Carefully inspect and record location and physical characteristics (depth, width, length) of cave in 2. Engineer should determine cause of cave in and supervise all steps necessary to reduce threat to dam and correct condition 3. Excavate cave in, slope sides of excavation, and backfill hole with competent material using proper construction techniques	1. Restore freeboard to dam by adding fill material in low area, using proper construction techniques 2. Regrading crest to provide proper drainage of surface runoff 3. If gully was caused by overtopping, provide adequate spillway which meets current design standards. This should be done by engineer 4. Re-establish protective cover	1. Extend pipe beyond toe (use a pipe of same size and material, and form watertight connection to existing conduit) 2. Protect dam with riprap over suitable bedding

Problem detail	Erosion behind poorly graded riprap	Excessive erosion in earth-slide causes concentrated flows	Poor surface drainage
	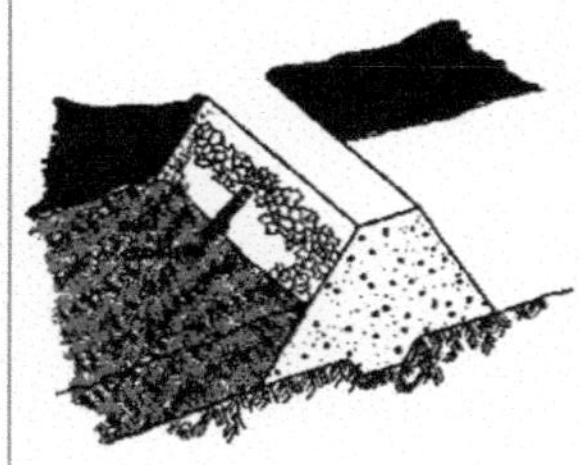	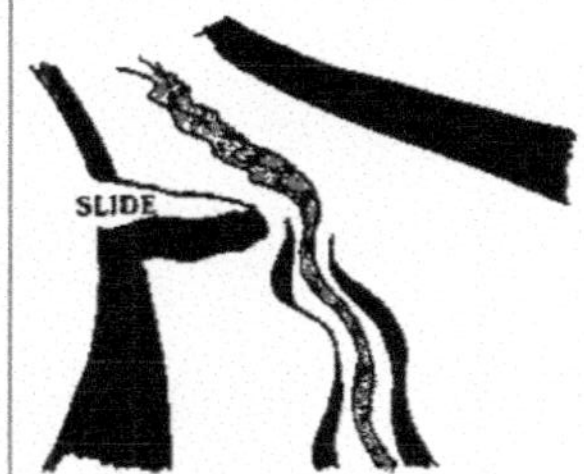	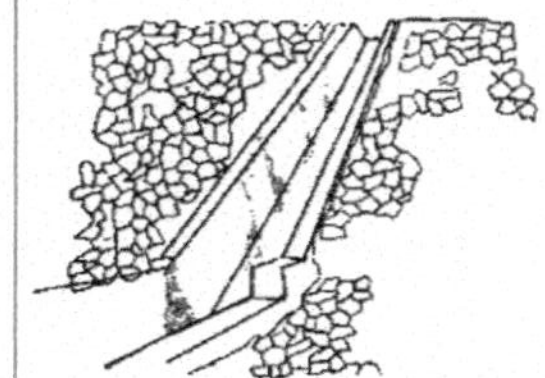
Probable cause	Similar-sized rocks allow waves to pass between them and erode small gravel particles and soil	Discharge velocity too high; bottom and slope material loose or deteriorated; channel and bank slopes too steep; bare soil unprotected; poor construction protective surface failed	1. No drain hole 2. No drainage facilities The drain pipe is blocked
Consequences	Soil is eroded away from behind the riprap. This allows riprap to settle, providing less protection and decreased embankment width	1. Material loss and sediment deposition load increase downstream 2. Dam collapse 3. Spillway damage. Severe erosion of spillway will lead to rapid emptying of reservoir	Wet foundation has lower supporting capacity; uplift pressure resulting from seepage water may cause damage to spillway chute; accumulation of water may also increase total pressure on spillway walls and cause damage
Recommended actions	Re-establish effective slope protection. Place bedding material. Engineer required for design for gradation and size for rock for bedding and riprap. A qualified engineer should inspect the conditions and recommend further actions to be taken	Minimize flow velocity by proper design, use sound material. Keep channel and bank slopes mild. Encourage growth of grass on soil surface. Construct smooth and well compacted surfaces. Protect surface with riprap, asphalt, or concrete. Repair eroded part using sound construction practices	Install weep holes on spillway walls. Inner end of hole should be surrounded and packed with graded filtering material. Install drain system under spillway near downstream end. Clean out existing weep holes. Backflush and rehabilitate drain system under the supervision of an engineer

Problem detail	Erosion channels	End of spillway chute undercut	Erosion, abrasion and fracture of concrete
	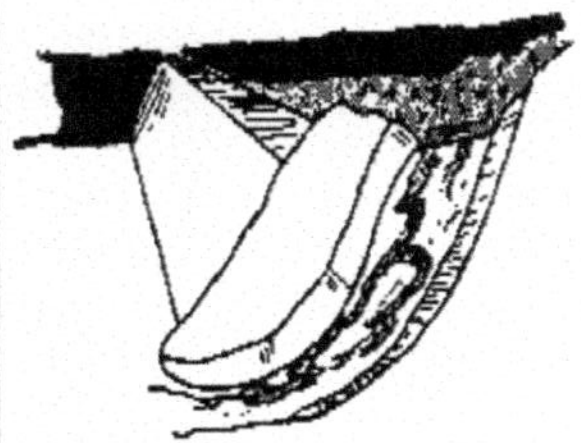	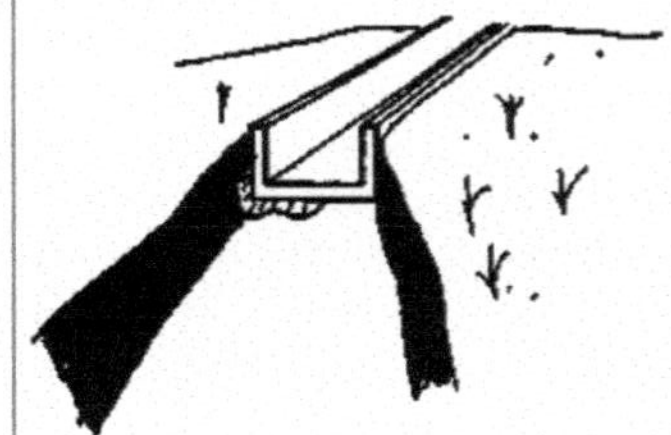	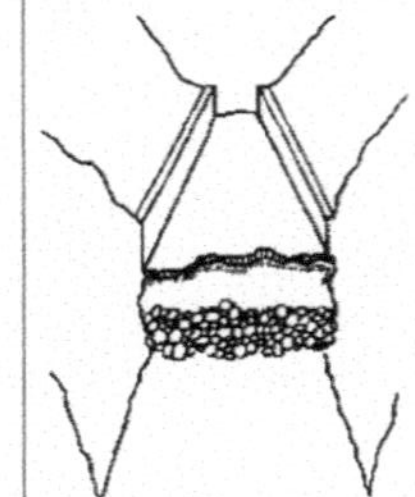
Probable cause	Surface runoff from intense rainstorms or flow from spillway carries surface material down the slope, resulting in continuous troughs. Livestock traffic create gullies where flow concentrates varies	Poor configuration of stilling basin area. Highly erodible materials. Absence of cutoff wall at end of chute	1. The flow velocity is too fast (usually at the end of the discharge chute of the high dam) 2. Gravel and rock rolling along the chute 3. There are voids behind or under the concrete slab

(continued)

(continued)

Consequences	Unabated erosion can lead to slides, slumps or slips which can result in reduced spillway capacity. Inadequate spillway capacity can lead to embankment overtopping and result in dam failure	Structural damage to spillway structure; collapse of slab and wall lead to costly repair	1. Voids and spalls on the concrete surface may deteriorate gradually 2. It may cause foundation subsidence and structural collapse
Recommended actions	Photograph condition. Repair damaged areas by replacing eroded material with compacted fill. Protect areas against future erosion by installing suitable rock riprap. Revegetate area if appropriate. Bring condition to the attention of the engineer during next inspection	1. Clear the eroded area and backfill properly 2. Improve the water flow channel under the slideway; riprap of appropriate size shall be laid in the stilling basin area 3. Install river closure measures	1. Remove the sand and gravel on the chute before the flood season 2. Raise the water level of stilling basin 3. The concrete quality meets the requirements 4. Ensure that the concrete surface is smooth

Problem detail	Leakage in or around spillway	Too much leakage from spillway under drains	Seepage from a construction joint or crack	Seepage water exiting from a point adjacent to the outlet
	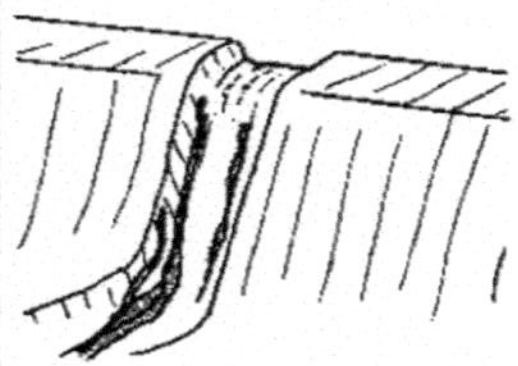	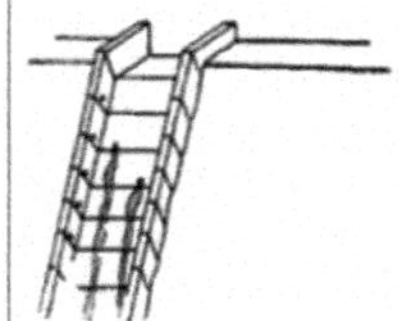	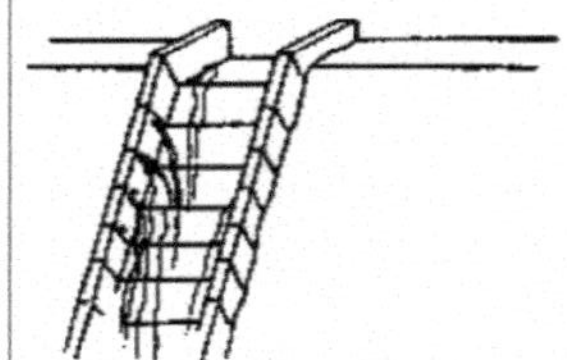	
Probable cause	1. Cracks and joints in geologic formation at spillway are permitting seepage 2. Gravel or sand layers at spillway are permitting seepage	Water is collecting behind structure because of insufficient drainage or clogged weep holes	Insufficient drainage or clogged seepage hole	1. A break in the outlet pipe 2. A path for flow has developed along the outside of the outlet pipe

(continued)

(continued)

Consequences	1. Could lead to excessive loss of stored water 2. Could lead to a progressive failure if velocities are high enough to cause erosion of natural materials	1. Excessive discharge from the spillway may erode the foundation materials and cause part of the spillway to collapse 2. Uncontrolled flow may result in loss of stored water	1. Can cause walls to tip in and over. Flows through concrete can lead to rapid deterioration from weathering 2. If the spillway is located within the embankment, rapid erosion can lead to failure of the dam	Continued flows can lead to raid erosion of embankment materials and failure of the dam
Recommended actions	1. Examine exit area to see if type of material can explain leakage. Measure flow quantity and check for erosion of natural materials 2. If flow rate or amount of eroded materials increases rapidly, reservoir level should be lowered until flow stabilizes or stops	The recommendation is the same as LEAKAGE IN OR AROUND SPILLWAY	1. Check area behind wall for puddling of surface water 2. Check and clean as needed; drain outfalls, flush lines, and weep holes 3. If condition persists a qualified engineer should inspect the condition and recommend further actions to be taken	1. Thoroughly investigate the area by probing and/or shovelling to see if the cause can be determined 2. Determine if leakage water is carrying soil particles 3. Determine quantity of flow 4. If flow increases, or is carrying embankment materials, reservoir level should be lowered until leakage stops

Table 2.3 Deformation problem

Problem detail	Large cracks	Landslide, collapse or sliding	Scarps, benches, oversteep areas	Broken down missing riprap
				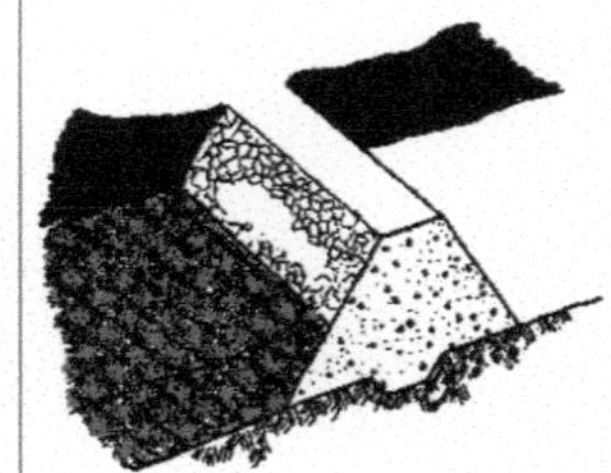
Probable cause	Some structures of the dam are unstable and move Or dam instability caused by dam foundation movement	Earth or rocks move down the slope along a slippage surface because of too steep a slope, or the foundation moves. Also, look for slides movement in reservoir basin	Wave action, local settlement, or ice action cause soil and rock to erode and slide to the lower part of the slope forming a bench	Poor quality riprap has deteriorated wave action or ice action has displaced riprap. Round and similar-sized rocks have rolled downhill
Consequences	Or will induce large-scale landslide or settlement	A series of slides can lead to obstruction of the outlet or failure of the dam	Erosion lessens the width and possible height of the dam and could lead to increased seepage or overtopping of the dam	The cross-section width of the dam will be reduced if the area without protective measures is subjected to wave action
Recommended actions	According to the situation of the dam, the water level of the reservoir will be reduced A qualified engineer should inspect the conditions and recommend further actions to be taken	1. Assess landslide risk and monitor landslide 2. If the dam safety is threatened, it is necessary to reduce the reservoir water level for operation 3. Relevant personnel shall timely check and take further measures	Determine exact cause of scarps. Do necessary earthwork, restore embankment to original slope and provide adequate protection (bedding and riprap)	Re-establish normal slope. Place bedding and competent riprap

Problem detail	Landslide/slough	Transverse cracking	Collapse	Longitudinal cracking
		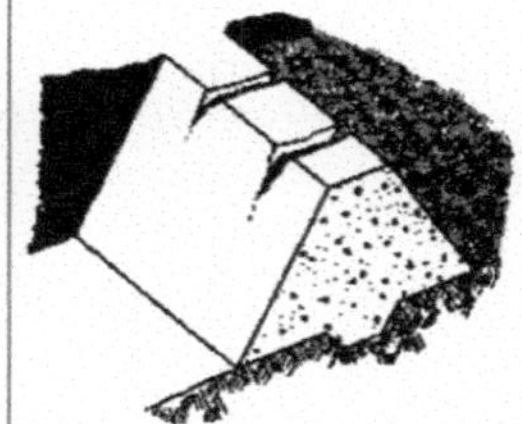		
Probable cause	1. Lack of or loss of strength of dam material strength 2. Loss of strength can be attributed to infiltration of water into the dam or loss of support by the foundation	The differential settlement of the dam body will lead to transverse cracking (the settlement of the center part is greater than that of the edge part)	1. Lack of adequate compaction 2. Rodent hole below 3. Piping through dam or foundation	1. Drying and shrinkage of surface material 2. Downstream movement of settlement of dam

(continued)

(continued)

Consequences	1. Massive slide cuts through crest or upstream slope reducing freeboard and cross section 2. Cause structural collapse or overtopping	Settlement or shrinkage cracks can lead to seepage of reservoir water through the dam. Shrinkage cracks allow water to enter the dam. This promotes saturation and increases freeze–thaw action	Dam structure is damaged or washed out	1. Early warning of potential landslide 2. The shrinkage crack makes the reservoir water enter the dam body, and the freeze–thaw action further cracks the dam 3. Settlement or slide showing loss of strength in embankment can lead to failure
Recommended actions	1. Measure extent and displacement of slide 2. If continued movement is seen, begin lowering water level until movement stops 3. Have a qualified engineer inspect the condition and recommend further action	1. If necessary, plug upstream end of crack to prevent flows from the reservoir 2. A qualified engineer should inspect the conditions and recommend further actions to be taken	1. Inspect for and immediately repair rodent holes. Control rodents to prevent future damage 2. Please check and take further measures in time	1. If cracks are from drying, dress area with well-compacted material to keep surface water out and natural moisture in 2. The settlement or landslide indicates that the strength of the dam is reduced and damaged

Problem detail	Slump (localized)	Transverse cracking	Longitudinal crack	Vertical displacement
		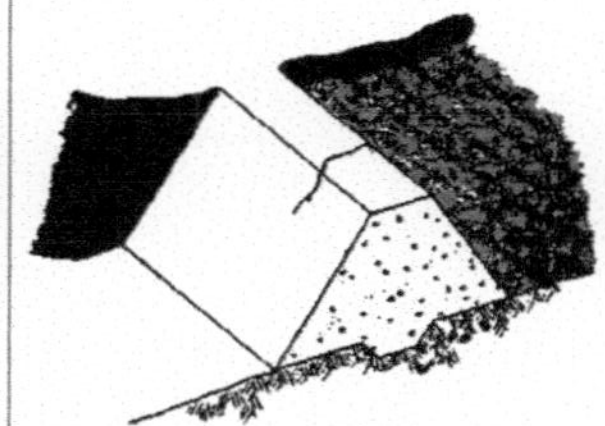	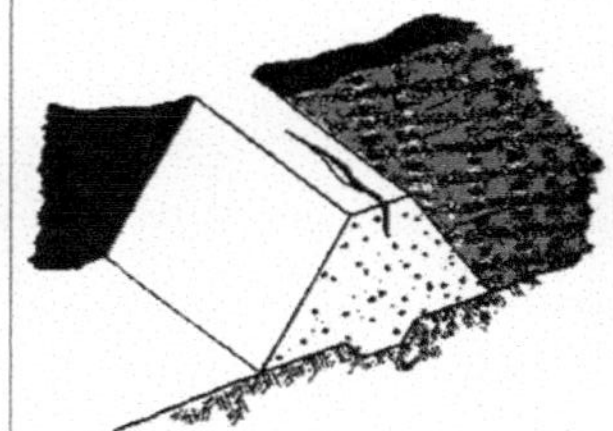	
Probable cause	Erosion weakens part of bank slope structure. This can also happen on steep slopes	1. Uneven movement between adjacent segments of the dam 2. Deformation caused by structural stress or instability	1. Uneven settlement between adjacent sections or zones within the dam 2. Dam instability caused by dam foundation failure 3. Initial stages of landslide	Differential settlement between adjacent parts of the dam. Structural deformation or failure caused by structural stress or instability, or by failure of the foundation

(continued)

(continued)

Consequences	The impervious area is exposed and eroded, leading to further landslides	1. Form leakage channel 2. Forming local low intensity areas. Future structural movement, deformation or failure could begin 3. Provide access for surface runoff to flow into the dam body	1. The local strength of the dam is low. It may be the precursor of partial structural deformation or damage 2. As a channel for surface runoff to enter the dam, it will damp the adjacent embankment area and may lead to local landslide	The formation of local low strength areas within the dam may lead to uneven displacement. Provide leakage channels for surface water. Cause structural instability or damage. The available embankment profiles are reduced
Recommended actions	1. Inspect area for seepage 2. Monitor for progressive failure 3. Please check and take further measures in time	1. Inspect crack and carefully record location, length, depth, width, alignment, and other pertinent physical features. Immediately stake out limits of cracking. Monitor frequently 2. Engineer should determine cause of cracking (displacement) and supervise steps necessary to reduce danger to dam and correct condition 3. Effectively seal the cracks at the crest's surface to prevent infiltration by surface water 4. Excavate area to the bottom of the displacement. Backfill excavation using competent material and correct construction techniques and under supervision of engineer 5. Continue to routinely monitor crest for evidence of further cracking		

Problem detail	Crest misalignment	Low area in crest of dam	Cracked deteriorated concrete face	Cracks due to drying
	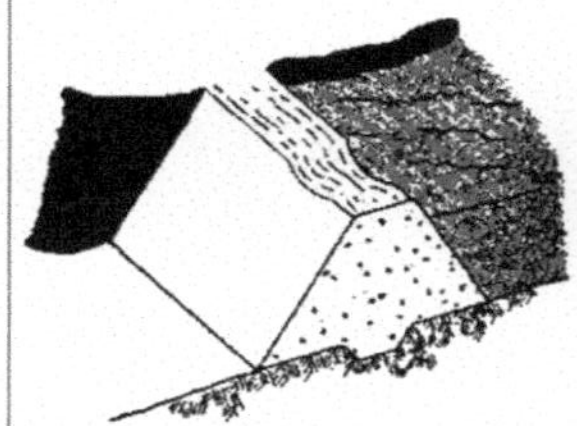	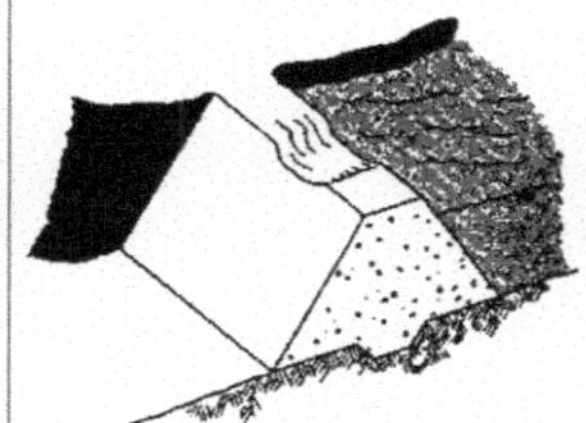		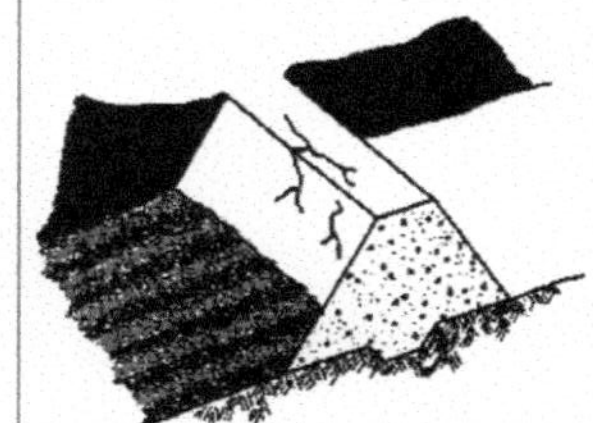
Probable cause	1. Movement between adjacent parts of the structure of the dam 2. Uneven deflection of dam under loading by reservoir 3. Structural deformation or failure near area of misalignment	1. Excessive settlement in the embankment or foundation directly beneath the low area in the crest 2. Internal erosion of dam crest material. Foundation spreading to upstream and/or downstream direction 3. Prolonged wind erosion of crest area. Unreasonable grading	Concrete deteriorated resulting from weathering. Joint filler deteriorated or displaced	The soil loses its moisture and shrinks, causing cracks. NOTE: Usually seen on crest and downstream slope mostly

(continued)

(continued)

Consequences	1. The low-lying area of the dam crest is formed, resulting in the reduction of the dam height 2. Local structural strength will be reduced or even destroyed	Reduces freeboard available to pass flood flows safely through spillway	Soil is eroded behind the face and caverns can be formed Unsupported sections of concrete crack. Ice action may displace concrete	Heavy rains can fill up cracks and cause small parts of embankment to move along internal slip surface
Recommended actions	1. Determine the exact location, number, and extent of misalignment (and dam crest settlement) 2. Relevant personnel shall determine the causes of the deviation (and dam crest settlement) to reduce the risk of the dam 3. After the repair measures are taken, the dam crest deformation is regularly monitored to monitor the possible movement 4. Fill the earth in low-lying areas and restore the dam crest under the guidance of relevant personnel. Daily monitoring of dam crest settlement		1. Determine cause. Either patch with grout or contact engineer for permanent repair method 2. If damage is extensive, a qualified engineer should inspect the conditions and recommend further actions to be taken	1. Monitor cracks for increases in width, depth, or length 2. A qualified engineer should inspect the condition and recommend further actions to be taken

Problem detail	Open or displaced joints	Breakdown/loss of riprap	Wall displacement	Large cracks
	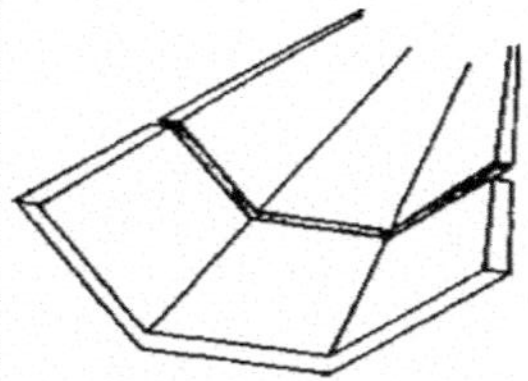	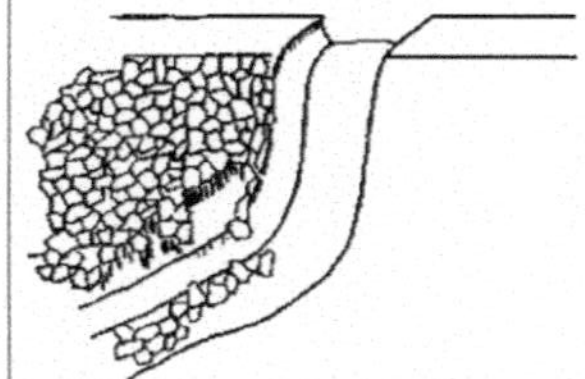	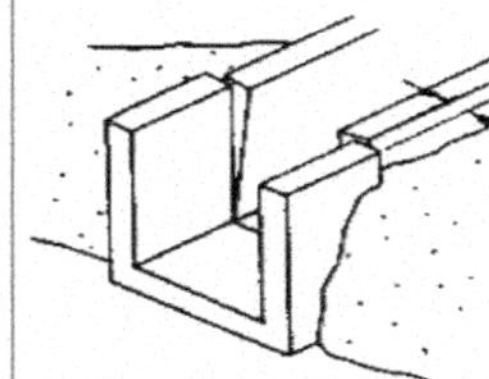	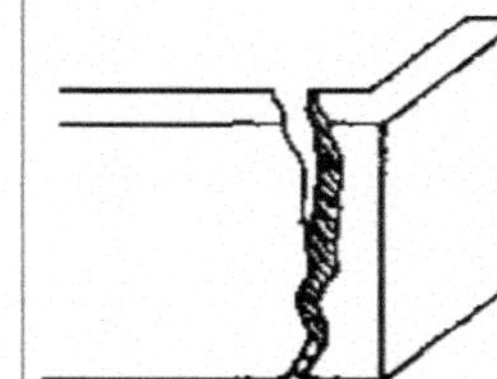
Probable cause	Excessive and uneven settlement of foundation; sliding of concrete slab; construction joint too wide and left unsealed sealant deteriorated and washed away	Slope too steep; material poorly graded; failure of subgrade; flow velocity too high; improper placement of material; bedding material or foundation washed away	Poor process quality; uneven settlement of foundation; excessive earth pressure and water pressure; insufficient reinforcement in concrete	Construction defect; local concentrated stress; local material deterioration; foundation failure, excessive backfill pressure

(continued)

(continued)

Consequences	Erosion of foundation material may weaken support and cause further cracks; pressure induced by water flowing over displaced joints may wash away wall or slab, or cause extensive undermining	The bottom of the channel and the river bank are eroded and the spillway is damaged	Slight displacement will produce eddy current and turbulence in the water flow, resulting in soil erosion behind the wall. Serious displacement will lead to serious cracks and eventually structural damage	Disturbance in flow patterns; erosion of foundation and backfill; eventual collapse of structure
Recommended actions	All joints should be sealed with asphalt or other flexible materials. Waterstops should be used where feasible. Clean the joint, replace eroded materials, and seal the joint. Foundations should be properly drained and prepared. Underside of chute slabs should have ribs of enough depth to prevent sliding. Avoid steep chute slope	Design a stable slope for channel bottom and banks. Riprap material should be well graded. Sub-grade should be properly prepared before placement of riprap. Install filter fabric if necessary. Control flow velocity in the spillway by proper design. Riprap should be placed according to specification	Carry out risk removal and reinforcement work, and the foundation should be carefully constructed. Install sufficient drain holes to relieve the water pressure on the side wall. Sufficient reinforcement shall be used in the concrete to anchor the wall to prevent further displacement. Props need to be installed between the spillway walls. Ensure that the drain pipe is in normal operation	Large cracks without large displacement should be repaired by patching. Surrounding areas should be cleaned or cut out before patching material is applied. Installation of weep holes or other actions may be needed

Problem detail	Crack	Hole	Joint offset	Failure of concrete outfall structure
	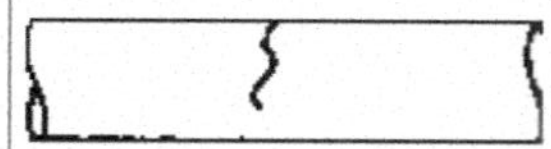		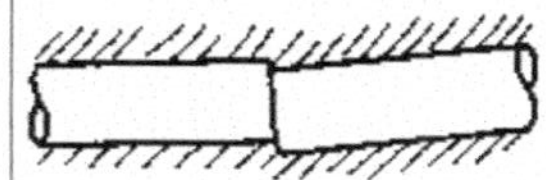	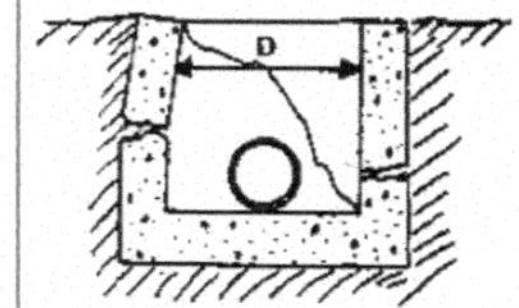
Probable cause	Settlement; impact	Rust (steel pipe) Erosion (concrete pipe) Cavitation	Settlement or poor construction practice	Excessive side pressures on nonreinforced concrete structure. Poor concrete quality
Consequences	Excessive seepage, possible internal erosion	Excessive seepage, possible internal erosion	Provides passageway for water to exit or enter pipe, resulting in erosion of internal materials of the dam	Loss of outfall structure exposes embankment to erosion by outlet releases
Recommended actions	Check for evidence of water either entering or exiting pipe at crack/hole/etc.	Tap pipe in vicinity of damaged area, listening for hollow sound which shows a void has formed along the outside of the conduit	If a progressive failure is suspected, request engineering advice	1. Check for progressive failure by monitoring typical dimension, such as "D" shown in figure 2. Repair by patching cracks and supplying drainage around concrete structure. Total replacement of outfall structure may be needed

Problem detail	Cracked gate leaf	Bearing block rupture	Control rod bent/ broken, rusted	Control rod damaged/missing, rusted
	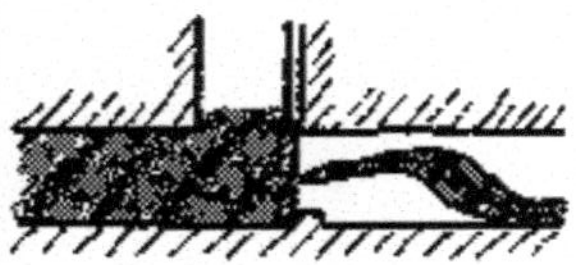	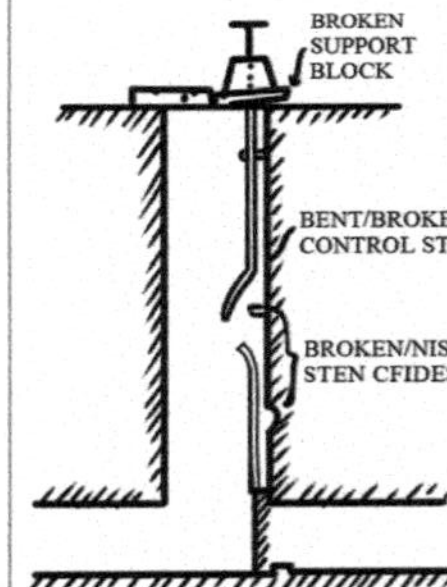		
Probable cause	Ice action, rust, affect vibration, or stress resulting from forcing gate closed when it is jammed	Concrete deterioration. Excessive force exerted on control stem by trying to open gate when it was jammed	Excessive force to open or close the gate. Insufficient or damaged stem guides	Insufficient lubrication. When the gate is stuck, the force for opening or closing is too large

(continued)

(continued)

Consequences	Gate-leaf main fail completely, evacuating reservoir	Cause the support block to tilt: the control lever may get stuck. There may be problems with the control work The gate may not open all the way The support block may fail completely, leaving the outlet inoperable	Outlet is inoperable	Loss of support for control stem. Stem may buckle and break under even normal use
Recommended actions	Use valve only in fully open or closed position. Minimize use of valve until leaf can be repair or replaced	1. It means that the control device cannot operate, or it can only operate partially at most 2. The use of the system should be minimized or discontinued 3. If the outlet system has a second control valve, consider using it to adjust the release until maintenance is carried out 4. The qualified engineer shall check the situation and recommend further measures		

Table 2.4 Operation and maintenance problems

Problem detail	Ruts along crest	Puddling on crest	Drying cracks
	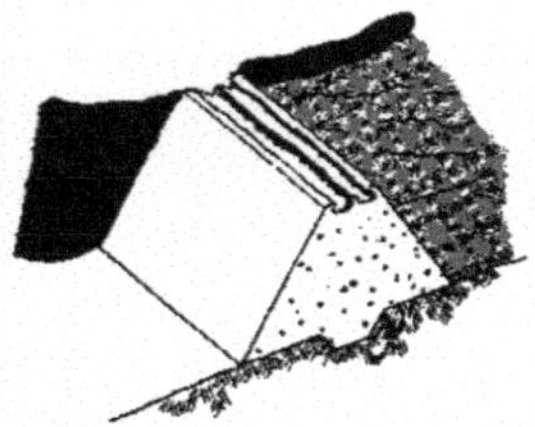		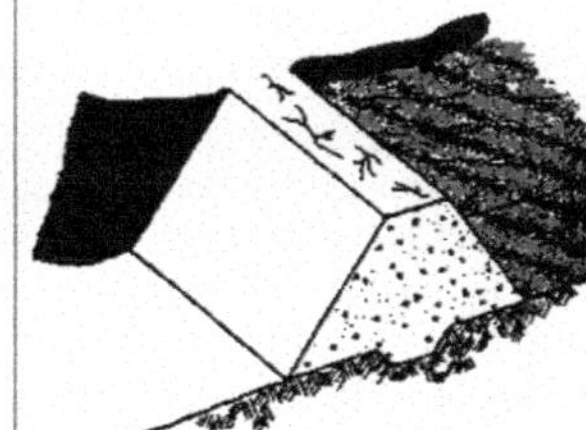
Probable cause	Heavy vehicle traffic without adequate or proper maintenance or proper crest surfacing	1. Poor grading and improper drainage of crest 2. Localized consolidation or settlement on crest allows puddles to develop	Material on the crest of dam expands and contracts with alternate wetting and drying of weather cycles. Drying cracks are usually short, shallow, narrow, and many

(continued)

Table 2.4 (continued)

Consequences	1. Inhibits easy access to all parts of crest 2. Allows continued development of rutting 3. Allows standing water to collect and saturate crest of dam 4. Operating and maintenance vehicles can get stuck	1. Cause localized saturation of the crest 2. Inhibits access to all parts of the dam and crest 3. Becomes progressively worse if not corrected	Provides point of entrance for surface runoff and surface moisture, causing saturation of adjacent embankment areas. This saturation, and later drying of the dam, could cause further cracking
Recommended actions	1. Drain standing water from ruts 2. Regrade and recompact crest to restore integrity and provide proper drainage to upstream slope 3. Provide gravel or roadbase material to accommodate traffic 4. Do periodic maintenance and regrading to prevent reformation of ruts	1. Drain standing water from puddles 2. Regrade and recompact crest to restore integrity and provide proper drainage to upstream slope 3. Provide gravel or roadbase material to accommodate traffic 4. Do periodic maintenance and regrading to prevent reformation of low areas	1. Seal surface of cracks with a tight, impervious material 2. Routinely grade crest to provide proper drainage and fill cracks—OR 3. Cover crest with non-plastic (not clay) material to prevent large moisture content variations

Problem detail	Rodent activity	Livestock/cattle	Obscuring vegetation	Trees/obscuring brush
	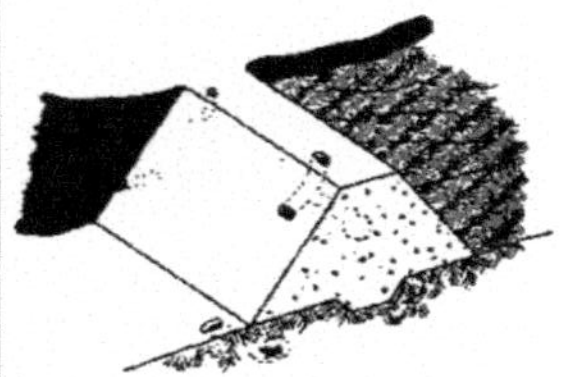		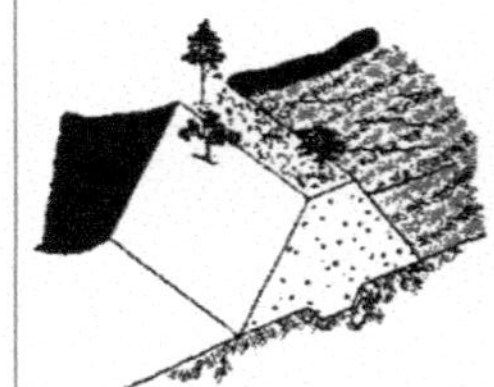	
Probable cause	Animal burrows create pits, tunnels and caves. Some places near the reservoir (such as Typha plants and trees) are conducive to the survival of animals	Excessive travel by livestock especially harmful to slope when wet	Neglect of dam and lack of proper maintenance procedures Natural vegetation in area	
Consequences	Can reduce length of seepage path, and lead to piping failure. If tunnel exists through most of the dam, it can lead to failure of the dam	Creates areas bare of erosion protection and causes erosion channels. Allows water to stand. Area is susceptible to drying cracks	1. Obscures large parts of the dam, preventing adequate, accurate visual inspection of all parts of the dam. Problems which threaten the integrity of the dam can develop and remain undetected until they progress to a point that threatens the dam's safety 2. Associated root systems develop and penetrate into the dam's cross section. When the vegetation dies, the decaying root systems can provide paths for seepage. This reduces the effective seepage path through the embankment and could lead to possible piping situations 3. Prevents easy access to all parts of the dam for operation, maintenance, and inspection. Provides habitat for rodents	

(continued)

(continued)

Recommended actions	1. Control rodents to prevent more damage 2. Backfill existing rodent holes 3. Remove rodents. Determine exact location of digging and extent of tunneling remove habitat and repair damages	1. Fence livestock outside embankment area 2. Repair erosion protection	1. Clear all vegetation on the dam. Grass planting should be encouraged in all areas of the dam to prevent erosion by surface runoff. Plant roots shall also be removed. The voids shall be backfilled with good and compacted materials 2. Destructive vegetation should be removed and planned into the annual maintenance plan 3. The rubbish generated by the removal of all vegetation on the dam shall be immediately cleared and properly disposed of outside the reservoir

Problem detail	Excessive vegetation or debris in channel	Material deterioration—spalling/disintegration of riprap, concrete, etc.	Debris stuck under gate	Damage gate seat or guides
	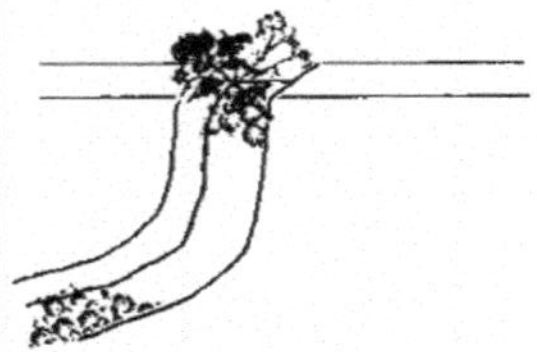		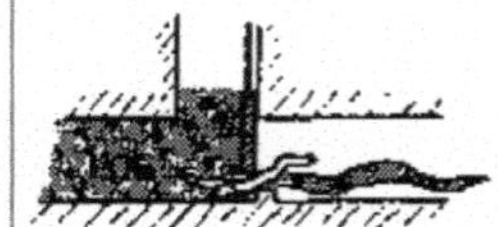	
Probable cause	Accumulation of slide materials, dead trees, excessive vegetative growth, etc., in spillway channel	Use of defective or defective materials; the structure is susceptible to freeze–thaw cycles; improper maintenance measures; chemical contamination	Trash rack missing or damaged	Rust, erosion, cavitation, vibration, or wear
Consequences	1. Drainage capacity decreases 2. Overflow of spillway 3. Overtopping of dam 4. Prolonged overtopping can cause failure of the dam	Structure life will be shortened; premature failure	Gate will not close. Gate or stem may be damaged in effort to close gate	Leakage and loss of support for gate leaf. Gate may bind in guides and become inoperable
Recommended actions	1. Clean out debris periodically 2. Control vegetative growth in spillway channel 3. Install trash rack in front of spillway entrance to intercept debris	Avoid using shale or sandstone for riprap. Add air entraining agent when mixing concrete. Only clean high-quality aggregate shall be used in concrete. The concrete protective layer meets the requirements. Concrete shall be kept moist during curing and protected from freezing	1. Slowly raise and lower the gate until debris is loose and floats through the gate 2. When the reservoir is lowered, repair or replace the trash rack	Minimize use of valve until guides/seats can be repaired. If cavitation is the cause, check to see if air vent pipe exist, and is unobstructed

2.3 The On-site Safety Inspection in Dam Safety Evaluation

The on-site safety inspection in dam safety evaluation means an overall inspection and evaluation on the dam appearance, operation performance, equipment, management facility. The inspection team shall fill in checklist form, write inspection report and make proposals and recommendations. This inspection work shall be conducted by the evaluation organizer, management agency, competent authority, or property owner. The inspection expert group shall be composed of experts from various disciplines including hydrology, geology, metal structure, dam construction and management, and government authorities. The on-site safety inspection report shall be written by the evaluation organizer [39].

Overall inspection shall be carried out on the appearance and operation of dam, water conveyance and discharge structures, bank slope near the dam, metal structure and electrical, monitoring, flood control, traffic and communication facilities through discussion, data review, and on-site observation and measurement, including underwater inspection if necessary. See Sect. 3.4 for details of underwater inspection. The inspection purpose is to find the potential changes in reservoir function and operating condition, quality defects and safety risks. Simple measurements, tests and excavation are also useful means for the inspection. Unclear threats shall be reported immediately to technical personnel. The on-site inspection checklist of dam safety is in Appendix. Some addition and deletion are suggested in accordance with actual situations [40].

Safety tests shall be carried out on the dam, dam foundation, soil structure, latent defects, concrete, masonry, and metal structures when abnormal deformation, seepage, aging, uneven settlement, collapse, cracking, sliding and inclination occur. The tests should meet the requirements of relevant regulations and minimize disturbances and adverse effects on the object structure. The results of the safety test shall be compared with historical and monitoring data to provide comprehensive parameters for dam safety assessment. The aim of safety test is to figure out the current quality status and provide latest calculation parameters for dam safety assessment. See Sects. 2.5–2.8 for details of dam safety test.

2.4 Drilling Test and Hidden Danger Detection

Geological survey and drilling tests shall be carried out when the existing geological data or dam body filling data are insufficient. In the case of suspected quality defects or serious operational hazards, drilling tests and hidden danger detection shall be necessary if the existing data are inadequate for safety assessment [41]. Drilling test (See Fig.2.2) includes tunnel and trench exploration. Holes and pits shall be drilled on weak and key parts (including junction of building across dam, anti-seepage body, and foundation surface). Field water injection and pump-in test, geotechnical test, indoor sample geotechnical test shall be carried out [42].

Fig. 2.2 Diagram of drilling test

Drilling tests shall be carried out in accordance with related regulations, and their impacts on the reservoir dams shall be minimized. For example, damage of boreholes to anti-seepage body shall be reduced and blocking of the downstream drainage system shall be avoided (Fig. 2.3).

In recent years, being widely used to detect hidden dangers, non-destructive test has become one of the most important auxiliary means to find out quality defects of dams [43]. Comprehensive analysis shall be carried out on the test results, geological conditions and dam construction quality. By using the information carried by physical fields such as sound waves, electromagnetic waves, electric fields, and magnetic fields, the internal structure or defects of the inspected object can be determined, and hidden dangers such as cracks, voids, leaks, and damages within the dam and anti-seepage system can be detected [44, 45]. Table 2.5 introduces several commonly used non-destructive testing methods in the water conservancy industry.

1. **Ground penetrating radar**

Ground penetrating radar [46] uses high frequency electromagnetic waves (main frequency ranges from tens of megahertz to even gigahertz) in the form of short

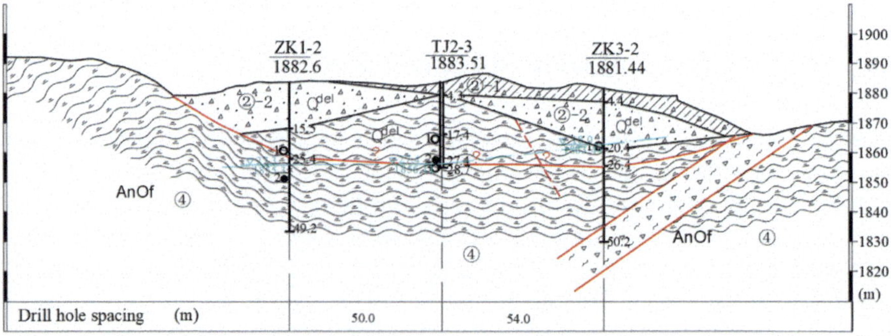

Fig. 2.3 Drilling layout diagram

Table 2.5 The non-destructive detection methods used in the water conservancy industry

Non-destructive means	Dam type							
	Concrete dam		Earth dam			Earth-rock dam		Concrete face rockfill dam
	Crack	Concrete aging	Leakage	Landslide	Cavity	Leakage	Strength	Separation
Resistivity method	×	×	√	√	×	√	×	×
Flow potential method	×	×	√	√	×	√	×	×
Magnetoelectric resistivity method	×	×	√	×	×	√	×	×
Ground penetrating radar	√	√	√	√	√	√	×	√
Ultrasonic reflection method	×	×	×	×	×	×	×	√
Electromagnetic wave CT	√	√	√	×	√	√	×	×
Seismic wave CT	√	√	√	×	√	×	√	×

broadband pulses (See Fig 2.4). The waves are emitted from the transmitting antenna *T* and go into the underground through the ground surface. The receiving antenna *R* catch the waves on the ground after they are reflected by the underground strata or target body. Pulse wave travel time is shown in Eq. 2.4.1.

$$t = \sqrt{4z^2 + x^2}/v \tag{2.4.1}$$

If the wave velocity v in the underground medium is known, the depth of the reflector (m) can be calculated using the above equation based on the measured time t value (ns). The value of x (m) in the equation is fixed in the profile detection. V value (m/ns) can be measured directly with a wide-angle way or calculated approximately according to Eq. 2.4.2.

$$v \approx c/\sqrt{\varepsilon} \tag{2.4.2}$$

Radar patterns are often recorded in the form of pulsed reflected waves. Figure 2.5 shows the basic principle of ground-penetrating radar. The waveform record graph is plotted against a simple geological model. Waveforms are recorded at each measurement point on the record graph in the perpendicular direction of the measurement line, forming a radar profile (GPR profile). The specific operating principles are as follows. After a radar system uses an antenna to transmit broadband high-frequency electromagnetic waves into the subsurface, reflection, transmission and refraction will occur as the electromagnetic wave signal propagates through a medium and encounters a medium interface with a large dielectric difference. The difference between the dielectric constants of the two mediums are in direct proportion to the reflected electromagnetic wave energy. The radar host accurately record the movement of the reflected electromagnetic waves monitored by the receiving antenna which moves in tandem with the transmitting antenna. The movement characteristics are then processed by signal technology to form a full cross-section scan.

Feasibility Analysis for Defect Detection

The ground-penetrating radar test results are shown in the Fig. 2.6. On the geological radar image profile, based on the comparison of reflected waves on adjacent channels, the contrast line connecting the same reflected wave with the same phase on different channels is called the phase axis. Under the scenario of dense, zero-defect dam structure, on the radar image, there will be obvious and clear concrete layer, bedding layer

Fig. 2.4 Dam inspection using ground-penetrating radar

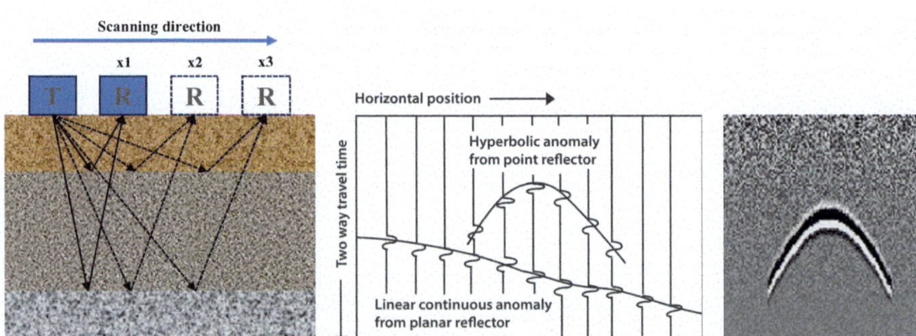

Fig. 2.5 Basic principle diagram of ground-penetrating radar

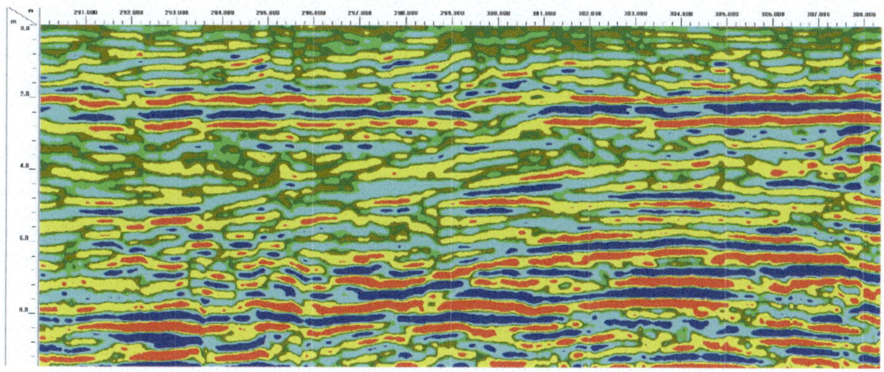

Fig. 2.6 Ground-penetrating radar test results

and interface layer of transition zone, evenly and horizontally distributed medium layers, continuous radar waves phase axis, stable waveform and consistent amplitude. When the dam structure contains large water volume or becomes loose, the electromagnetic wave phase axis will emerge with other miscellaneous waves, scattering and bypassing, chaotic and disorderly. At the same time, the amplitude of the electromagnetic wave in the disturbance area becomes larger, and the energy relative to the surrounding dense area stronger. The abnormal area appears as a sudden change on the radar with interrupted phase axis is, disordered waveform. The reflection coefficient of the electromagnetic wave in the abnormal area becomes larger, forming a coarse reflected wave phase axis.

The geo-radar waveform over a large void usually shows a parabolic arc. The top arc center acts as of the void center, and the top depth the void depth.

2. **Multi-electrode resistivity method**

The multi-electrode resistivity method [47] is an array exploration method based on the difference in electrical conductivity between rock and soil and investigates the distribution of conductive currents in the ground under the influence of an artificially applied steady current field (See Fig. 2.7). The field measurements are carried out by simply placing all the electrodes (dozens to hundreds) at each measurement point in the observation section, and then using a programmable electrode conversion device and a microcomputer engineering electrode meter to achieve rapid and automatic data acquisition, and when the measurement results are fed into the microcomputer, the data can be processed and various graphical results about the distribution of the geoelectric section can be given [48].

The physical premise of the high density resistivity method is the difference in electrical conductivity between subsurface media. As with the conventional resistivity method, a current I is applied to the subsurface through electrodes A and B, and the potential difference ΔV between electrodes M and N is then measured to determine the apparent resistivity value at that recording point. Based on the measured apparent resistivity profile (Fig. 2.8), calculations, processing and analysis can be carried out to obtain the resistivity distribution in the stratum, and the change in the corresponding measured apparent resistivity value reflects the change in electrical properties along the vertical direction below the point. delineate stratigraphy, trap anomalies, etc.

The high density electrical method has the following advantages over conventional resistivity methods [49].

1. Electrode placement is done in a single pass, which not only reduces errors and disturbances caused by electrode placement, but also provides the basis for rapid and automated field data acquisition.
2. The ability to effectively scan and measure multiple electrode arrays provides a wealth of geological information on the structural characteristics of the geoelectric section.
3. Field data acquisition is automated or semi-automated, which is not only fast (approximately 2–5 s per point), but also avoids errors due to manual operation.
4. The data can be pre-processed to show the shape of the profile, and off-line processing can also automatically draw and print a variety of results.
5. Compared with the traditional resistivity method, the cost is low, the efficiency is high, the information is rich and the interpretation is convenient.

Fig. 2.7 Multi-electrode resistivity method

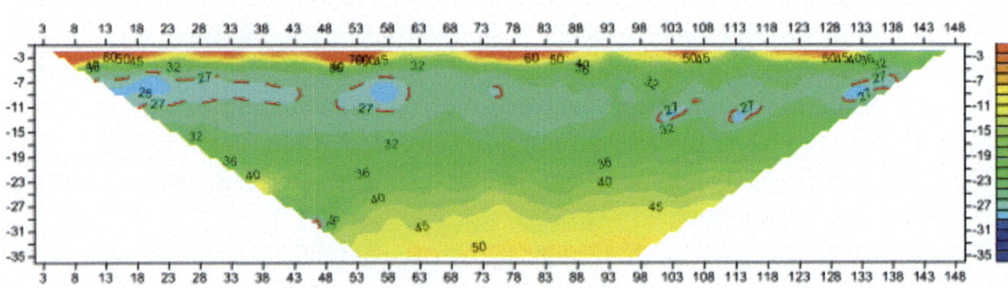

Fig. 2.8 Multi-electrode resistivity method test results graph

2.5 Underwater Inspection

The research and development of underwater detection, operation, and repair technologies originated and applied to the field of marine engineering. Some reservoir dam structures work underwater, including slope, dam body, energy dissipation facility, guide wall. Erosion and cracks are inevitable for these structures which will threaten the safe operation of reservoir dams if there is no regular underwater inspection. For large and medium-sized reservoir dams, traditional underwater inspection means such as diving and camera shooting, with high cost and low efficiency, cannot fulfil the inspection mission [50, 51].

2.5.1 *Classification of Underwater Detection Technologies*

Underwater robot, also known as submersible, is an intelligent equipment that operates under water and independently completes specific missions. It has become a primary tool in human exploration of the deep water environment. Playing an important role in marine resource development and deep sea exploration, it has caught attention globally. Submersibles are classified into three main types, namely human occupied vehicles (HOVs), remotely operated vehicles (ROVs), and autonomous underwater vehicles (AUVs), as shown in Fig. 2.9 [52].

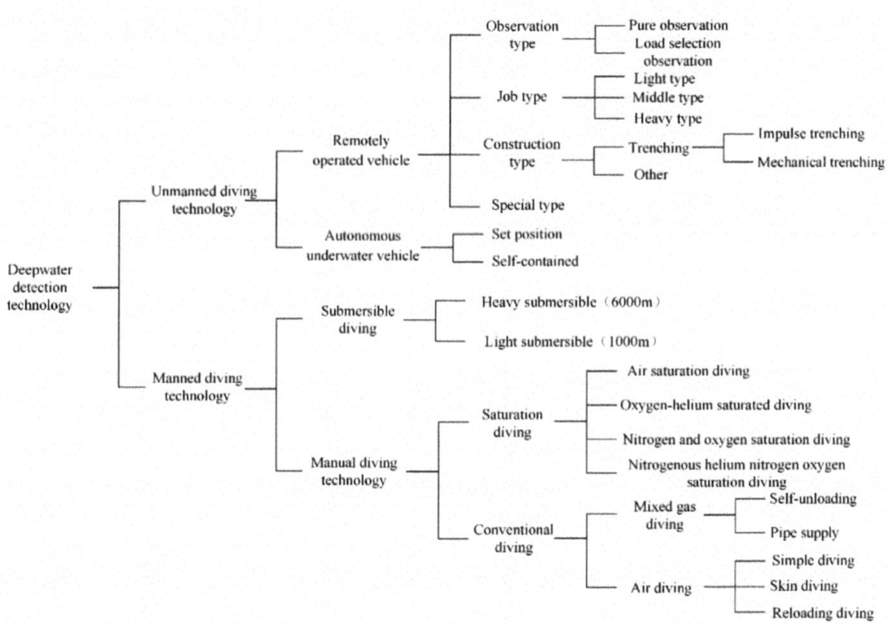

Fig. 2.9 Classification of underwater detection technologies

Fig. 2.10 Underwater unmanned underwater vehicle types

Underwater inspection, exploration, and operations are carried out using devices equipped with observation-level or working-level remotely operated vehicle (ROVs) depending on the purpose of the work or the water environment. Working-level ROVs can provide more powerful propulsion and a platform for carrying the equipment in underwater areas [53, 54] (Fig. 2.10).

Nanjing Hydraulic Research Institute has led the independent research and development of the "Yulong" deep-water detection submersible for dams [55], achieving a breakthrough from 0 to 1 in the equipment of manned submersibles for deep-water complex environments in the water conservancy and hydropower industry for the first time, and forming experience in deep-water detection and repair and reinforcement. Solving many key technologies in deep-water inspection of manned submersibles for dams: hovering operation, attachment cleaning, leakage tracing, laser ranging, etc. New technology means and equipment for high-dam and large reservoir deep-water detection and repair reinforcement were provided (Fig. 2.11).

2.5.2 Working Principle of ROV

ROV is an unmanned underwater vehicle equipped with underwater observation and detection, navigation equipment, and underwater manipulators, which connects power, video, data, and control signals to surface control equipment through cables or umbilicals, and is remotely operated by surface personnel to accomplish specific underwater tasks [56]. Based on its motion patterns in water, it can be divided into types such as floating, towed, crawling, and structure-dependent, and is mainly

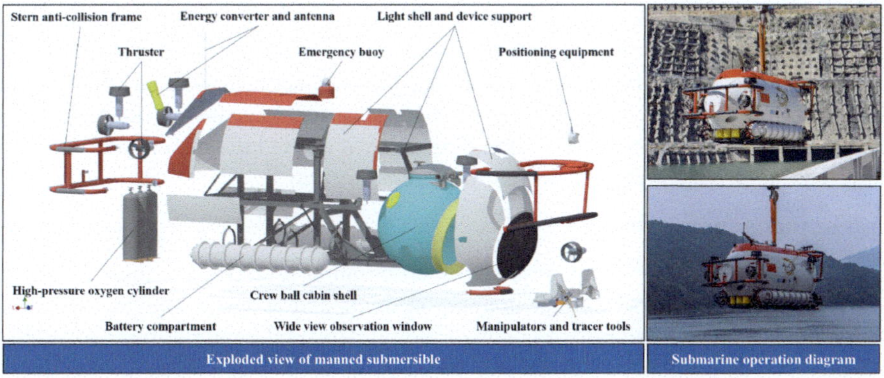

Fig. 2.11 The "Yulong" manned submersible used in high dam reservoir inspection

composed of a submersible, an imaging system, a navigation and positioning system, and a water surface control system [57–59] (Fig. 2.12).

1. **The submersible**

Submersibles are typically constructed from materials known for their corrosion resistance and high strength. They provide a platform for the installation of ROV underwater equipment and buoyancy. The propulsion of the submersible is generated by multiple sets of thrusters driven by electric motors. These thrusters usually consist of propulsion motors, controllers, and complementary propellers. These thrusters

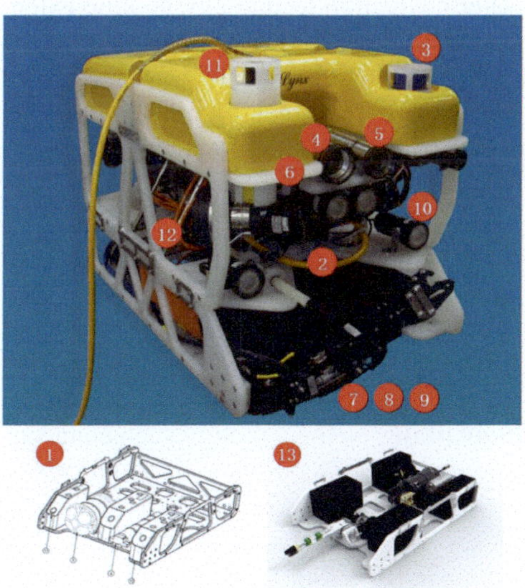

	component name	Type
1	Inertial navigation system	Acoustic equipment
2	Image sonar	Acoustic equipment
3	Mechanical scanning sonar	Acoustic equipment
4	Low light level camera	Optical equipment
5	Optical wide angle camera	Optical equipment
6	Vertical mechanical PTZ	PTZ
7	3D real-time multibeam sonar	Acoustic equipment
8	Multifunctional manipulator	Operate device
9	Side scan sonar	Acoustic equipment
10	Underwater high current lamp	Artificial light source
11	Ultra Short Base Line	Acoustic equipment
12	Propeller	Dynamical system
13	high pressure cleaning system	Cleaning system

Fig. 2.12 Working-level ROV working components

enable the ROV to perform precise movements in the water [60]. Each submersible is equipped with an electronic control system, which is the core component of the ROV and consists of computers and interface circuits, serving as the ROV's brain. The electronic control system can receive and process real-time instructions from the surface control system, while also handling its own status information and various collected image data, allowing for real-time remote control.

2. **The imaging system**

According to the usage characteristics of observation-level ROVs, their imaging systems mainly consist of optical imaging systems and acoustic imaging systems [60]. Optical imaging equipment is one of the fundamental configurations of ROVs and includes color cameras, black and white cameras, and photography cameras, among others. The effectiveness of optical imaging is significantly influenced by factors such as light sources and water quality, and it is typically equipped with LED light sources or halogen light sources.

Acoustic imaging systems are used when optical imaging conditions are poor or for searching large-area targets. Based on different working principles, sonar technology can be categorized into various techniques, including multibeam detection technology, side-scan sonar detection technology, shallow subsurface profiling technology, dual-frequency identification sonar detection technology, synthetic aperture sonar detection technology, and underwater 3D scanning sonar detection technology, among others.

3. **The navigation and positioning system**

Navigation and positioning technology are the primary guarantees for the safe operation of remotely operated vehicles (ROVs) underwater, and this constitutes a technically challenging system [61]. Underwater navigation technology is quite extensive, and the current main technical systems include gravity navigation, geomagnetic field navigation, and terrain-based navigation, among others. The advantages and disadvantages of different technologies are also apparent, making the selection and development of technology a significant challenge within the ROV industry [62].

4. **The water surface control system**

This includes the system control computer, the controller, power modules, and related display and recording equipment, among other components. The umbilical cable connects the submersible to the surface control unit, allowing the operators to monitor the basic status of the ROV in real-time. Through the controller, operators can send control commands to the submersible, controlling both the submersible itself and its onboard sensors, imaging systems, and more [63].

2.5.3 *Analysis of Different ROV Applications in Underwater Environments*

Underwater detection technologies, mainly underwater robots or divers, are suitable for a wide range of application scenarios, including resource development, military security and rescue, and marine environment protection [64, 65]. These technologies have been used to inspect damages and defects of under-water structures, sediment, pipelines, cables, metal structures, and monitor the underwater environment. They have also been used for underwater construction, repair and maintenance, scientific research, and emergency rescue. Unmanned underwater vehicle, supported by side-scan sonar, multibeam sonar, stratigraphic profiler and other equipment, can be used for inspection, maintenance, and reinforcement of underwater structures of reservoir dams [66] (Fig. 2.13).

2.5.3.1 Underwater Inspection in Clearwater Environments

In cases where the water quality within a reservoir is good, characterized as a clearwater environment, underwater inspections are typically conducted using ROVs equipped with high-definition cameras. High-definition cameras provide clear imaging, but due to the rapid attenuation of light in water, their identifiable range is relatively close. ROVs originating from marine engineering often have lower pixel requirements for high-definition imaging and instead focus more on the technology for identifying targets in sonar images [67].

However, in the context of hydraulic engineering inspections, where reservoir water quality is generally good, and the search range is limited, optical imaging camera systems are more widely applied. On the other hand, ROVs equipped with

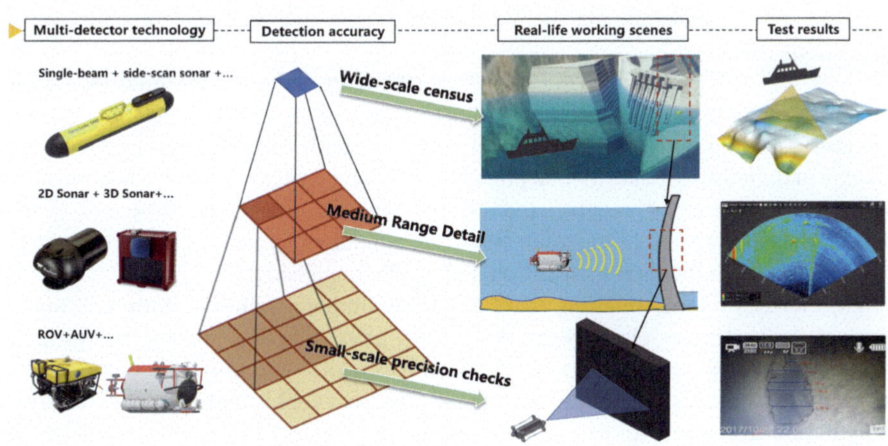

Fig. 2.13 Multi-scale collaborative inspection program

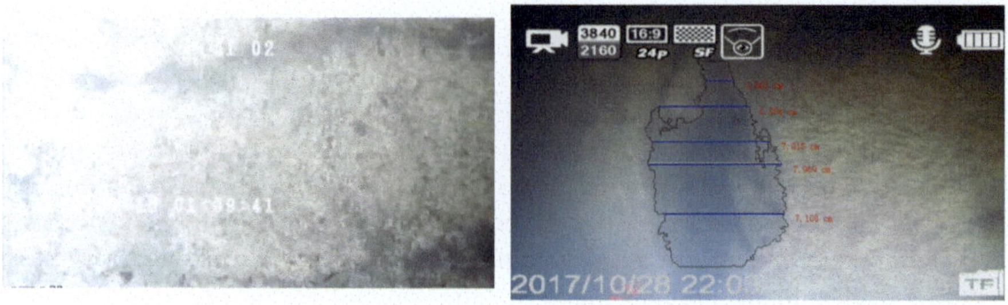

(a) Low resolution imaging (b) High resolution imaging

Fig. 2.14 Imaging comparison of different resolutions

high-resolution cameras place higher demands on data storage and signal transmission, which can have some impact on endurance and inspection efficiency. When conducting underwater inspections in hydraulic engineering, it is often necessary to identify cracks at the millimeter level, which places higher requirements on the pixel quality of the ROV's onboard cameras (Fig. 2.14).

2.5.3.2 Underwater Detection in Muddy Water Environment

Unlike the clearwater environment typically found in reservoirs, certain hydraulic structures, such as tunnels or culverts, have low visibility due to factors like sediment, aquatic plants, and suspended organisms. They do not provide suitable conditions for optical imaging. Since sound waves experience much lower losses when propagating through water compared to electromagnetic waves and light waves, underwater acoustics technology is widely used for the positioning, navigation, observation, imaging, and information communication of underwater vehicles. Therefore, in murky water environments, ROVs typically use sonar systems for underwater inspections [68].

Sonar technology has a history of over 100 years, initially invented by Lewis Nixon of the British Navy in 1906 for detecting icebergs [69]. Sonar is a device that utilizes sound waves to detect underwater targets. It can be considered a device capable of radiating (in a horizontal plane) to search for targets. Sonar emits ultrasonic waves into the water and receives the reflected waves generated by these emissions. By utilizing the reflected ultrasonic waves, it can detect targets, as well as transmit information about the target's location and approximate shape.

The clarity of sonar imaging is related to its image resolution. Although many sonar imaging devices claim resolutions down to centimeters or even millimeters, the actual imaging quality depends on various factors, including the equipment's working environment and conditions. It may not always achieve the expected results. In hydraulic structures, for example, where the surrounding environment consists of rock, reinforced concrete, or steel pipes, sound waves emitted in the confined space produce multiple coherent reflections on the tunnel walls. This results in a complex

Fig. 2.15 Measured results of 3D sonar

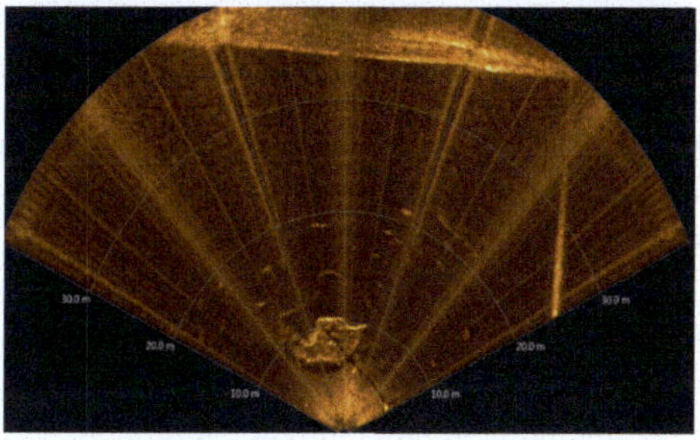

Fig. 2.16 Measured results of image sonar

reverberation background, reducing the signal-to-noise ratio of echo signals and significantly increasing the difficulty of analysis. While sonar imaging technology has high-frequency sound waves and strong anti-interference capabilities, interference from multiple reflected waves can lower the signal-to-noise ratio of echo signals, affecting the image quality.

When visibility is too low and optical imaging is not feasible, high-resolution image-scanning sonar is used to perform sonar image scanning and obstacle avoidance for hydraulic structures. This device emits sound waves that, upon encountering a target, reflect back to the sonar's surface. By analyzing these echoes, multi-beam target sonar image signals are obtained [70]. The collected sonar data is transmitted to the computer in the ROV's surface control unit for processing, allowing the generation of two-dimensional sonar images [71]. This enables the ROV to identify and locate target objects in low-visibility environments, as shown in Figs. 2.15 and 2.16.

2.5.3.3 Underwater Detection in High Resistance Environment

When inspecting hydraulic tunnels using an ROV, it's common to encounter situations where the tunnel axis is not straight and may have bends in both plan and profile. In such cases, the ROV's umbilical cable makes contact with the tunnel walls at the bends, causing friction. This friction has two main effects: it increases the resistance to the vehicle's movement, and it may lead to cable breakage. In general, the greater the number of bends, the sharper the angles, the longer the distance between bends and the entrance or exit of the tunnel, and the longer the total length of the tunnel, the more challenging the inspection becomes [72].

Additionally, when using an ROV for underwater inspections, the vehicle must overcome greater cable gravity and cable resistance. Cable resistance comprises frictional resistance and viscous pressure drop resistance as the umbilical cable moves through water. Frictional resistance is related to the wetted surface area of the system, while viscous pressure drop resistance is related to the vehicle's shape. As the cable length increases, the surface area increases, resulting in higher frictional resistance.

To address these challenges in high-resistance environments, ROVs typically employ strategies such as zero-buoyancy cables, enhancing cable strength, increasing the vehicle's propulsion force, and improving power. Some examples include [73]:

- Designing new umbilical cables with reduced surface resistance, improved tensile strength, and buoyancy.
- Increasing the main propulsion power to ensure the ROV can pull the umbilical cable through.
- Using crawling robots that can adhere to tunnel walls and have greater cable-pulling capacity.
- Implementing specialized umbilical cable management devices in the entrance section.

These measures are taken to enhance the ROV's performance and ensure successful inspections in high-resistance environments.

2.6 Safety Test of Concrete Structure

2.6.1 Safety Test Contents of Concrete Structure

According to the relevant design codes, hydraulic structures should be designed to ensure the necessary serviceability, durability and safety during the design reference period and under normal conditions of use and maintenance. To understand the extent and cause of deterioration of the concrete structure of a hydraulic structure, the necessary tests are carried out and if the results of these tests indicate that the structure is in a dangerous condition, strengthening measures must be taken [74]. Figure 2.17 shows on-site inspection photos, and Table 2.6 introduces the relevant items and methods used for concrete inspection.

Safety test of concrete structure shall include the following contents which may be confirmed in accordance with requirements of dam safety evaluation and site test conditions.

(1) Test on appearance quality and defects of the concrete;
(2) Test on concrete strength of main structural members;

Fig. 2.17 Concrete structure safety test site photos

(3) Test on carbonation depth of concrete, thickness and degree of corrosion of steel protective layer;
(4) In case of crack, hole, cavity on the main structural members or on structures with anti-seepage requirement, its distribution, width, length and depth shall be measured and the cause of its occurrence be analyzed;
(5) When corrosion occurs due to the action of corrosive medium, the compositions and contents of the corrosive medium shall be determined and the degree of corrosion of the structure shall be measured.

2.6.2 Safety Inspection Methods for Concrete Structures

2.6.2.1 Test on Deformation of Concrete Structure

In the test, the displacement measurement instruments should be arranged according to the needs of deformation measurement of the specimen, and it should calculate the deformation parameters such as deflection and angle of rotation of the specimen from the measured displacement values. The instruments and meters for displacement measurement can be chosen according to the requirements of accuracy and data collection, such as electronic displacement meter, percentage meter, micrometer, level meter, latitude and longitude meter, inclinometer, total station, laser rangefinder, straightedge, etc.

2.6.2.2 Test on Concrete Internal Defects

Test on concrete internal defects should be done by using non-destructive methods like ultrasonic method, impact-echo method, etc. Partial destructive method can be used to verify the test results by non-destructive method when necessary [75, 76] (Fig. 2.18).

Table 2.6 On-the-spot check of concrete

Inspection items	Selection method	Possible local damage	Main restricting conditions
Strength and performance of components	Load test and deformation measurement	Components destroyed	Test members need to be isolated
Strength and compactness of concrete	1. Coring test 2. Rebound method, ultrasonic method and γ radiation measurement 3. Penetration resistance method 4. Pull-out method 5. Drilling internal fissure method	1. Left holes must be repaired 2. No damage 3. Left cone holes must be repaired 4. Left holes pulled out 5. Left cone holes must be repaired	1. Size, distribution of test pieces and safety protection of components 2. Smoother test surfaces are required 3. Minimum margin and member thickness 4. Careful planning is required in advance 5. Difficult drilling
Concrete cracks	Ultrasound, acoustic emission and holographic techniques		
Permeability	Absorption, flow and capillary water rise test		
Cement content, moisture content, mix ratio and composite materials	Neutron method, chemical analysis method, resistance method, microwave absorption method, coring method, micrometer method		
Rebar detection	Electromagnetic, X-ray and γ radiographic method		

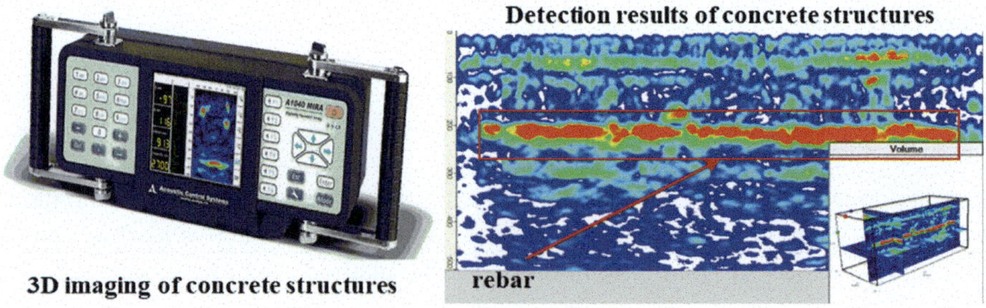

Fig. 2.18 Detection results of concrete structures

2.6.2.3 Testing of Compressive Strength of the Concrete of Structural Members

Rebound method, ultrasonic-rebound combined method, pin-shooting method, core drilling method, etc. can be used for **testing of compressive strength of the concrete of structural members** [77]. The method to be used shall be chosen depending on the site conditions. Core drilling method shall be used for correction of other methods if site condition permits. Tests on internal defects of concrete and compression strength of structural members shall be carried out by using as much as possible non-destructive or local destructive method for purpose of prevention or reduction of adverse effects brought to the structures [78].

1. **Rebound measurement**

A number of $200 \times 200\ \text{mm}^2$ rebound measurement zones are established on the structure and the loose layer on the surface is removed and smoothed with a wire brush and grinding wheel. If the rebound test angle is not horizontal and the test surface is an uncured face, the rebound value of the test area is corrected to give the final representative rebound value R_a.

2. **Sound velocity measurement**

The flat measurement method is used to measure the sound velocity value of the measurement area. Three measurement points are arranged on the concrete surface of the rebound measurement area (the line of measurement points is at an angle of 450° to the reinforcement axis), the measurement points are flat and free of defects such as cracks between the measurement points, and the wave velocity value v_p of the measurement area is measured using a probe with a distance of 15 cm between the transmitter and receiver of the ultrasonic instrument.

Select a certain number of measurement areas in the same structural section, in these measurement areas to 10, 20, 30, 40, 50 cm spacing arrangement of five measurement points to measure the above points distance concrete acoustic propagation time value t. From the above five points of the distance l and t series of values, through the regression can get the linear equation $l = a + bt$, the equation b that is the measurement area of the concrete wave velocity value v_d, to $\lambda = v_d/v_p$ value of other wave velocity values for correction, so as to get all the measurement area of the sound velocity representative value v_a.

3. **Area intensity conversion**

The following equation was used to convert the representative values of rebound and sound velocity in the measured area into measured area intensity values:

$$f_{cu,i}^{c} = 0.0056 v_{ai}^{1.439} R_{ai}^{1.769} \tag{2.6.1}$$

2.6.2.4 Test of Stress of Concrete Structure

Test of stress of concrete structure [79] shall include test of strain of concrete and rebar. Strain measurement instruments should be selected according to the purpose of the test and the requirements for strain measurement of concrete and reinforcing steel specimens. The strain of reinforcing steel and concrete should be measured by resistance strain gauge, vibrating string strain gauge, fiber optic light flute strain gauge, extensometer, etc. When resistance strain gauge is used to measure strain, there should be reliable temperature compensation measures. When mechanical strain gauges are used to measure strain in places where the temperature varies greatly, the temperature effect should be corrected.

2.6.2.5 Measurement of Concrete Carbonation Depth

Atmospheric CO_2 and concrete $Ca(OH)_2$ role in the surface of the component to form a neutral $CaCO_3$ layer, called the component of the concrete carbonation layer, its thickness is the depth of carbonation. Concrete before carbonation is alkaline, in which the reinforcement is protected by a layer of surface passivation film and does not rust, after the generation of $CaCO_3$ layer in the concrete, due to the pH value in the void decreases, destroying the highly alkaline medium in the concrete, so that the passivation film of the reinforcement is destroyed. If the carbonation depth exceeds the burial depth of the reinforcement in the member, the passivation film on the surface of the reinforcement is destroyed and rusts rapidly. Concrete is alkaline when not carbonated and weakly alkaline after carbonation, when exposed to alcoholic phenolphthalein solution, the former is red, the latter is colourless [80].

After the rebound value is measured, **the depth of carbonization** [81] shall be measured on representative measuring areas, and the measuring points shall not be less than 30% of the number of measuring areas of components. The average value shall be taken as the depth of carbonization of each measuring area of components. When the range of depth of carbonization is greater than 2 mm, the depth of carbonization should be measured separately in each measuring area.

The main points of the method for determining the depth of carbonation of the concrete component are as follows: in the part of the measuring area where the strength is determined by the rebound method, a hole is made in the surface of the component to be tested using an impact drill, the powder is removed from the hole, the reagent is sprayed into the hole and the thickness of the surface layer of undiscoloured concrete is measured using a caliper.

2.6.2.6 The Measurement of Thickness of Protective Layer of Rebar

The measurement of thickness of protective layer of rebar [82] should be conducted by using non-destructive electromagnetic induction method or radar method. Verification can be done by chiseling of the concrete when necessary.

The measurement of the protective layer thickness of the reinforcement is divided into two steps: (1) Detect the position of the rebar: use the electromagnetic induction probe to move slowly on the concrete surface, when the probe detection surface is close to the position of the rebar, the electrical signal output from the probe increases, when the probe is directly above the rebar, the electrical signal output is the largest, the electrical signal is amplified and read on the gauge, the position of the rebar can be determined. (2) Select the same measuring gear as the measured rebar diameter on the instrument and place the probe directly over the rebar position, then the protective layer thickness of the rebar can be measured and read on the instrument.

2.6.2.7 Corrosion Detection of Steel Reinforcement

In detection of steel reinforcement corrosion, chisel method or electrochemical determination method can be chosen depending on the test conditions and requirements [83]. In addition, the following provisions shall be adhered to:

1. By chisel method, the steel reinforcement shall be chiseled out to directly measure the remaining diameter of the steel.
2. The operation of electrochemical determination method shall be carried out in accordance with the provisions. Verification should be made in combination with the chisel method.

The process of rebar corrosion in reinforced concrete structures is divided into two stages [84]:

1. Uncorroded stage—from the time the concrete is poured until the carbonation layer of the concrete reaches the reinforcement and the reinforcement starts to rust.
2. Corroded stage—from the start of corrosion of the reinforcement to the development of severe corrosion so that the structure is damaged to the point where it is no longer safe to use. In the rusting stage, the ingress of environmental erosive media accelerates and rusting continues to intensify. This stage can be further divided into the following four periods:

    (1) The early stage—the individual protective layer on the smaller reinforcement begins to rust, a small amount of length and area are small by the bar expansion crack or spalling. At this stage, the average depth of concrete carbonation is still less than the thickness of the protective layer.
    (2) The middle stage—the concrete surface appears a number of lengths and areas of the larger along the reinforcement cracking or spalling, concrete carbonation depth average is close to or more than the protective layer thickness.
    (3) The late stage—concrete generally appears serious expansion cracking, spalling damage, that is, has reached an intolerable degree, must be a comprehensive overhaul.

(4) Advanced—the corrosion of the reinforcement has progressed to the point where it has damaged the structural area, rendering the structure unsafe for use.

Based on the depth of concrete carbonation, the thickness of the protective layer of reinforcement and the results of reinforcement corrosion tests, we can analyze and assess the current situation and development trend of corrosion of structural concrete reinforcement.

2.6.2.8 Detection of Cracks in Structural Elements

The cracks on the concrete structural member can be divided into stable crack and unstable crack depending on nature of its activity. To determine the reliability of the structure or to work out corrective plan, various factors in relation to it shall be fully considered, including cause of crack, stability of crack, etc. Cracks shall be monitored when necessary [85–87]. When necessary, oblique photography (Fig. 2.19) or 3D laser scanning (Fig. 2.20) can be used to reconstruct real scene data and monitor surface cracks in concrete. The following provisions shall be adhered to in **detection of cracks in structural members**:

1. The items for detection shall include the location, length, width, depth, form and number of the cracks. The records of detection can be in form of tables or graphs.
2. The depth of crack can be measured by ultra-sonic methods. Verification can be carried out by drilling core specimen.
3. Regular observations shall be carried out for those cracks that are still in development.

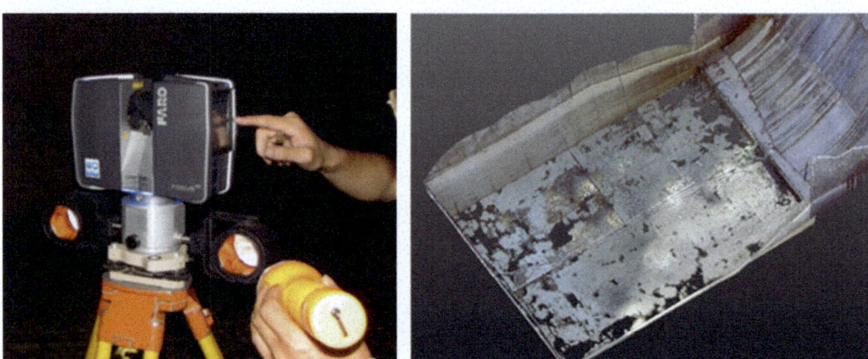

Fig. 2.19 3D laser scanning results

Fig. 2.20 The oblique photography results of drone

2.6.3 Defect Assessment Criteria and Defect Handling Determination Principles

The criteria for evaluating concrete defects are given in Table 2.7 and the principles for determining the treatment of concrete defects are given in Table 2.8.

2.7 Safety Test of Metal Structure

2.7.1 Steel Gates, Trash Racks and Hoists

The gate and hoist shall be subject to regular safety inspection. The safety inspection cycle can be determined according to the operation time and conditions of the gate and hoist, and should be divided into the first inspection, regular inspection and special case inspection. The first inspection shall be carried out within 5 years after the gate and hoist are put into operation. After the first inspection, the gate and hoist shall be inspected regularly every 6–10 years. The regular inspection items can be focused on according to the actual operation of the gate and hoist.

If the following conditions occur during the operation of the gate and hoist, special conditions shall be detected immediately. The inspection items can be determined according to the actual conditions of the gate and hoist.

- During the operation period, it has ever operated beyond the designed working conditions, has major accidents caused by misoperation, encountered irresistible natural disasters and other special circumstances.
- During the operation, it is found and confirmed that there are dangerous defects and major hidden dangers affecting the safety of the main structural members or main parts of the gate and hoist.

Table 2.7 Concrete defect assessment criteria

No.	Project	Type	Characteristic	Classification criteria	
				The slit width	The seam depth
1	Cracks in hydraulic mass concrete	A	Minor cracks	$\delta < 0.2$ mm	$h \le 0.3$ m
		B	Surface and shallow cracks	$0.2 \text{ mm} \le \delta < 0.3$ mm	$0.3 \text{ m} < h \le 1.0$ m
		C	Deep cracks	$0.3 \text{ mm} \le \delta < 0.5$ mm	$1.0 \text{ m} < h \le 5.0$ m
		D	Penetrating cracks	$\delta \ge 0.5$ mm	$h > 5.0$ m
	Cracks in hydraulic reinforced concrete	A	Minor cracks	$\delta < 0.2$ mm	$h \le 0.3$ m
		B	Surface and shallow cracks	$0.2 \text{ mm} \le \delta < 0.3$ mm	$0.3 \text{ m} < h \le 1.0$ m and not more than 1/4 of the thickness of the structure
		C	Deep cracks	$0.3 \text{ mm} \le \delta < 0.4$ mm	$1.0 \text{ m} < h \le 2.0$ m or more than 1/4 of the thickness of the structure
		D	Penetrating cracks	$\delta \ge 0.4$ mm	$h > 2.0$ m or greater than 2/3 of the thickness of the structure
2	Leakage	A	Minor	Minor spotting, surface leachate	
		B	Moderate	Localized concentrated leachate, dissolution	
		C	Severe	Presence of jet or inter-layer leakage	
3	Freeze–thaw erosion	A	Slight	$h \le 10$ mm	
		B	Moderate	$10 \text{ mm} < h \le 50$ mm	
		C	Severe	Freeze–thaw spalling depth $h > 50$ mm or spalling resulting in exposed reinforcement	
4	Reinforcing steel corrosion	A	Slight	The protective layer of the reinforcement is intact, but there is localized rust on the reinforcement	
		B	Moderate	No cracking or spalling of the concrete along the reinforcement, no extensive corrosion of the reinforcement and less than 10% loss of cross-section	

(continued)

Table 2.7 (continued)

No.	Project	Type	Characteristic	Classification criteria	
				The slit width	The seam depth
		C	Severe	Corrosion of most or all of the surface of the reinforcement, loss of more than 10% of the cross-section or failure of the bearing capacity, or cracking and spalling of the concrete along the reinforcement	
5	Abrasion and cavitation	A	Slight	Partial exposure of coarse concrete aggregates	
		B	Moderate	Extent and degree of concrete wear, localized loss of concrete coarse aggregate, formation of discontinuous wear surface (unexposed reinforcement)	
		C	Severe	Coarse aggregate exposed concrete forming a continuous wearing surface with exposed reinforcement	
6	Carbonization	A	Slight	Beginning to show signs of carbonization	
		B	Moderate	Carbonation of reinforced concrete to a depth less than the protective layer thickness of the reinforcement	
		C	Severe	Carbonation of reinforced concrete to a depth at or above the thickness of the protective layer of reinforcing steel	

- The operation of the gate and hoist is obviously abnormal, affecting the safe operation of the project.

Inspections and tests for steel gate, trash rack and hoist mainly include the following:

(1) Walkaround inspection;
(2) Gate appearance check;
(3) Test on performance state of hoist;
(4) Corrosion test;
(5) Material test;
(6) Non-destructive test;
(7) Stress test;
(8) Test on structure vibration;
(9) Test of hoisting force;
(10) Performance test of hoist;
(11) Special tests.

The above Items (1), (2) and (3) are the items for compulsory check and shall be carried out hole by hole while Item (4) to Item (9) are items for sampling check. Sampling check shall be conducted in proportion of the number of holes of the gates

Table 2.8 Principles for determining the treatment of concrete defects

No.	Project	Characteristic	Judgement principles	Remarks
1	Cracks in hydraulic mass concrete	A	Located in a Class I environment, no treatment can be carried out; in Classes II and III environments, treatment should be carried out	Class I environment: indoor or outdoor Class II environment: located in the upstream water surface, water level fluctuation zone or corrosive groundwater environment Class III environment: flow section, seawater or salt spray zone
		B		
		C	All should be treated	
		D		
	Cracks in hydraulic reinforced concrete	A	Located in a Class I environment, no treatment can be carried out; in Classes II and III environments, treatment should be carried out	
		B	All should be treated	
		C		
		D		
2	Leakage	A	Generally untreated, but should be treated if it affects operational safety	
		B	Should be treated and a structural safety analysis should be carried out for category C infiltration	
		C		
3	Freeze–thaw erosion	A	Outside the anti-friction zone it can be left untreated, inside the anti-friction zone it should be treated	
		B	Treatment is desirable	
		C	Should be treated and safety reviewed where freeze–thaw spalling is causing corrosion of reinforcing steel in reinforced concrete structures	

(continued)

Table 2.8 (continued)

No.	Project	Characteristic	Judgement principles	Remarks
4	Reinforcing steel corrosion	A	Surface protection treatment	
		B	Repairs and treatments	
		C	Full reinforcement treatment	
5	Abrasion and cavitation	A	No treatment	
		B	Repairs should be carried out and, in the case of Class C abrasions and cavitation, the design and safety analysis should be reviewed	
		C		
6	Carbonization	A	No treatment	
		B	Surface protection treatment	
		C	Treatment should be carried out by cutting out the carbonated concrete and replacing the protective layer of reinforcement	

and number of hoists of the same type. The operation state and the locations of arrangement of the gate and hoist shall be considered when samples are taken.

2.7.1.1 The Appearance Check

Before carrying out an inspection of appearance and condition, detailed information should be obtained on the completion and acceptance of the manufacture and installation of the sluice gates and gate lifts, major defects, operation, repairs, maintenance and various anomalies that have occurred during operation.

Describe and record, by visual inspection and using tools such as tape measures, calipers and digital cameras, the appearance of the main components of the metal structure, the ageing of coatings, the parts of the components that are corroded, the degree of corrosion, the number and parts of components that are deformed, damaged and defective, the ageing, wear and perfection of mechanical and electrical equipment, the operation of equipment, etc.

The appearance and status quo test results should be recorded in a timely manner, if necessary, can be used to record and describe the auxiliary methods such as video, photography (Figs. 2.21 and 2.22).

Fig. 2.21 On-site inspection results

Fig. 2.22 Steel gates, trash racks and hoists safety test site photos

2.7.1.2 Corrosion Detection of Metal Structures

(1) Coating thickness inspection

Metal structures are subject to erosion from water, gas, sunlight and aquatic life, as well as impact and friction from water, sediment and floating objects, and the uneven electrochemical properties of the structure material itself. Corrosion of the material reduces the load-bearing capacity of the structure and, in serious cases, threatens safe operation. To prevent and retard corrosion of steel doors, non-metallic (e.g. paint, coating) or metallic (e.g. zinc and zinc alloy coatings) coatings are often used for protection [88].

The coating thickness of the metal structure is sampled using a coating thickness gauge, with the sampling distribution being uniform and representative.

In the sampling point 1 dm^2 of the reference surface for three measurements, each measurement position 25 ~ 75 mm apart, take the arithmetic mean of the three measurements for the reference surface of the local thickness.

(2) Erosion thickness detection

The use of ultrasonic methods to detect the corrosion thickness of the main components of the steel structure: the ultrasonic probe placed on the surface of the measured body, the ultrasonic reflection to the steel propagation, in the encounter with the air, the sound waves are reflected and then received by the ultrasonic probe, measurement of sound waves from the reflection to be received by the time t, because the wave speed of steel v is a known quantity, then the thickness of the measured point $T = vt$.

The use of ultrasonic thickness gauge to detect the thickness of the main components of the metal structure of the steel plate (i.e. etch residual thickness), in the gate corrosion detection, through sampling statistical methods to determine the etch residual thickness of the gate components, the etch residual thickness of the components as a sample of the whole, with a number of measurement points as a sub-sample of the whole, with sub-samples subject to normal distribution for processing analysis, the components in the 95% confidence interval of the statistical value as the components of the thickness of the component within the 95% confidence interval is taken as the residual thickness of the component. The upper limit of the interval is the maximum etch thickness value and the lower limit is the minimum etch thickness value. The design value is subtracted from the residual thickness value to give the pit depth value.

2.7.1.3 Ultrasonic Weld Flaw Detection

Ultrasonic waves are able to penetrate deep into the metal material and reflect at the edges of the interface as they pass from one section to another to inspect metal components for flaws as a method of non-destructively detecting the size, shape and distribution of internal defects. Prior to flaw detection, the flaw location and quantitative method should be determined according to the material and thickness of the component being tested [89].

Weld flaw detection using ultrasonic flaw detector [90], when the ultrasonic beam from the surface of the component from the probe to the metal inside, encounter defects and the bottom surface of the component when the reflected waves, respectively, in the fluorescent screen to form a pulse waveform, according to these pulse waveform to determine the location and size of the defect [91].

2.7.2 Penstocks

The steel penstock shall be inspected regularly for safety. The inspection period can be determined according to the operation time and condition of steel penstock, and it should be divided into first inspection, regular inspection and special inspection. The penstock shall be fully inspected before capacity expansion of the unit.

The penstock shall be inspected for the first time within 5 years after it is put into operation. After the first test, the steel penstock shall be inspected periodically every 5–10 years. Items for periodic inspection can be selected according to the actual operation condition of steel penstock.

Special inspection shall be carried out first by patrol inspection and external inspection, and then by other items if necessary. Special inspection shall be carried out immediately when one of the following conditions occurs:

- The steel penstock is subjected to irresistible natural disasters during operation, operation under over-design conditions < quality and safety accidents, etc.
- Hazardous defects affecting safety are found and confirmed in steel penstock during operation. The operation condition of steel penstock is obviously abnormal, which may affect the safe operation of the project.

Inspections and tests for penstock mainly include:

(1) Walkaround inspection;
(2) Appearance check;
(3) Test of material quality;
(4) Non-destructive test;
(5) Stress test;
(6) Vibration test;
(7) Water quality test and geological test.

Items (1) and (2) are items for compulsory check.

Before the appearance test, understand the pressure steel pipe production, installation, acceptance, handling of major defects, operation, repair, maintenance, and operation of the various abnormalities that have occurred. During the appearance inspection, the inspection shall focus on the parts that have been repaired during the fabrication and installation of the penstock and have been abnormal during the operation [92].

The results of the appearance of the test should be recorded in a timely manner, if necessary, can be used to record and describe the auxiliary methods such as video, and photography. It is appropriate to use the level, straightedge, caliper, tape measure, feeler gauge, weld inspection ruler, and other necessary measurement tools and instruments, and equipment. Appearance inspection parts should mainly include the pipe body, support ring, and support, expansion joint, connection parts. If necessary, underwater robots can be used for the inspection of pressure steel pipes.

2.7.3 Other Metal Structures

Safety test of other metal structures like ship passage and ship lift shall be carried out with reference to the above provisions.

2.8 The Maintenance and Repair of Dam Operations

Maintenance means maintenance and protection measures carried out to guarantee normal operation of the dam. This maintenance is further divided into routine maintenance, regular maintenance and specialized maintenance. Repair means measures of remedy, treatment, reinforcement of various kinds that are taken to regain the original designed standard or restore the function of the dam in case of damage to dam, degradation and even failure of dam performance. It is divided into timely repair, annual repair, major repair and rush repair. Repair includes four working steps, investigation of damage; formulation, submission and approval of repair plan; implementation and inspection [93].

2.8.1 Key Points of Earth-Rock Dam Maintenance

Earth-rock dam maintenance mainly includes: Dam crest, dam end and dam slope maintenance; Maintenance of drainage facilities; Maintenance of observation facilities; Maintenance of dam foundation and dam area, etc. The evaluation of the maintenance effect of the reservoir can be carried out in combination with the on-site inspection of the reservoir. The main requirements are as follows:

(1) Whether trees and crops are planted on the dam surface, grass is shoveled, and banknote and stone materials are moved for slope protection and infiltration facilities.
(2) Whether there are activities endangering the safety of the dam, such as blasting, well drilling, quarrying, mining, sand digging, soil borrowing, grave repair, etc. within the scope of dam management and protection.
(3) Whether there are wharves and channels built on the dam body. If it is really necessary to build wharves and fish ponds within the scope of dam management and protection, it must be approved by the competent department of the dam, and keep a certain distance from the dam toe, water release and water conveyance structures, which shall not affect the dam safety, project management and emergency rescue work.

Earth-rock dam repair projects mainly include: Slope protection repair; Repair of dam cracks; Repair of dam leakage (including repair of dam body and sloping sand and stone inverted filter); Repair of dam foundation leakage and leakage around the dam; Repair of dam landslide; Repair of drainage facilities; Repair of concrete face slab dam.

The repair of earth-rock dam is divided into annual repair, overhaul and emergency repair. The annual repair is to carry out necessary repair and local improvement immediately according to the project damage and problems occurred in the operation of the dam and found in the inspection tour. The annual maintenance project shall be implemented after the management unit puts forward the annual maintenance plan according to the operation of the reservoir dam and submits it to the competent department for approval. Overhaul is a renovation project with a large amount of work, such as major damage to the reservoir hub project, large repair workload, more complex technical engineering problems, or engineering dangerous situations that have not been permanently treated after temporary emergency repair. The management unit shall timely put forward the feasibility study report or preliminary design report of the overhaul project, report to the superior competent department for approval, and then organize the design and construction according to the approved project. The design of the overhaul project shall be carried out by the design unit with corresponding qualification. Emergency repair means that when all kinds of dangerous situations endangering the dam safety suddenly occur, emergency repair must be carried out immediately.

2.8.2 Key Points of Concrete Dam Maintenance

Maintenance of concrete dams generally includes daily maintenance and protection of buildings. Repair of concrete dam includes repair of concrete cracks, leakage, erosion and underwater repair.

Maintenance and repair of concrete dam should be done well. Prevent the occurrence and development of damage defects. When damage is found during maintenance, it must be repaired in time to prevent expansion. The repair work procedure of concrete dam includes 4 work procedures: investigation of Engineering damage, design of repair scheme, construction and quality control, acceptance, etc. Approval for repair works: For large repair projects, the management unit shall propose a repair design scheme and submit it to the competent department at a higher level for approval and implementation; Major repair items affecting structural safety shall be designed by the original design unit or by a design unit with corresponding certificates and submitted to the competent authorities for approval for implementation.

For overhaul or repair works of reservoirs, the project leader must be clearly defined, and a quality and safety assurance system must be established, quality standards and process must be strictly implemented to ensure project quality. After completion, the acceptance must be presided over strictly by the engineering project approval department in accordance with relevant regulations. Generally, the procedures for general year-old repair of the project can be appropriately simplified.

2.8.3 Recording and Evaluation of Dam Maintenance

Detailed records and completion acceptance data shall be made for the overhaul and reinforcement works done on the dam. Appropriate acceptance data shall be provided to the evaluation unit during safety evaluation:

(1) The management unit shall submit the engineering hoarder's report, approval documents, all design documents and drawings.
(2) Construction units shall submit construction reports for reinforcement or overhaul, drawings for hoarding and final accounts for completion, original construction records and quality inspection records, etc.
(3) The supervision unit shall submit the project supervision report, the original record of the project supervision and the acceptance appraisal certificate at the project stage.
(4) All relevant units shall provide stage acceptance (or inspection) appraisal data and completion acceptance appraisal certificate for the concealed works in detail.

The evaluation of dam maintenance is mainly carried out through the relevant records of dam maintenance. Special attention should be paid to the periodic inspection and maintenance recorded by safety monitoring instruments.

References

1. Adamo N, Al-Ansari N, Sissakian V et al (2020) Dams safety: inspections, safety reviews, and legislations. J Earth Sci Geotech Eng 109–143. https://doi.org/10.47260/jesge/1114
2. Michalis P, Sentenac P (2021) Subsurface condition assessment of critical dam infrastructure with non-invasive geophysical sensing. Environ Earth Sci 80:556. https://doi.org/10.1007/s12665-021-09841-x
3. Niu X, Yang Q (2007) Safety evaluation of reservoir dams. China Water & Power Press. ISBN: 978-7-5084-4875-6
4. Xiang Y, Jing M (2021) Reservoir dam safety inspection guidelines. China Water & Power Press. ISBN: 978-7-5170-9477-7
5. United States (1991) Engineering guidelines for the evaluation of hydropower projects
6. Chow JK, Su Z, Wu J et al (2020) Anomaly detection of defects on concrete structures with the convolutional autoencoder. Adv Eng Inform 45:101105. https://doi.org/10.1016/j.aei.2020.101105

7. Froehlich D (2017) Guidelines for safety inspection of dams
8. U.S. Department of Homeland Security, Federal Emergency Management Agency (2013) Technical manual for dam owners. CreateSpace Independent Pub.
9. Upstream slope protection | Association of State Dam Safety. In: damsafety.org. https://damsafety.org/dam-owners/upstream-slope-protection
10. Tan J, Wang M, Tian J et al (2020) Research on dam leakage detection based on visual and acoustic integration: a case study of CFRD. IOP Conf Ser 525:012053. https://doi.org/10.1088/1755-1315/525/1/012053
11. Zhou JW, Lu PY, Yang YC (2017) Reservoir landslides and its hazard effects for the hydropower station: a case study. In: Mikos M, Tiwari B, Yin Y, Sassa K (eds) Advancing culture of living with landslides. WLF 2017. Springer, Cham. https://doi.org/10.1007/978-3-319-53498-5_81
12. Internal erosion of earth dams | Association of State Dam Safety. In: damsafety.org. https://damsafety.org/dam-owners/internal-erosion-of-earth-dams
13. Corcoran MK, Peters JF, Dunbar JB, Llopis JL, Tracy FT, Wibowo JL, Simms JE, Kees CE, McKay SK, Fischenich JC, Farthing MW, Glynn ME, Robbins BA, Strange RC, Schultz MT, Clarke JU, Berry TE, Little CD, Lee LT (2011) Initial research into the effects of woody vegetation on levees. US Army Corps of Engineers
14. He K, Song C, Fell R (2021) Numerical modelling of transverse cracking in embankment dams. Comput Geotech 132:104028. https://doi.org/10.1016/j.compgeo.2021.104028
15. Aldulaimi A, Alsaadi A (2021) Review about incidents in dams and dike behaviors induced by internal erosion. Int J Res Eng Technol 8:1507–1513
16. Malkawi AIH, Al-Sheriadeh M (2000) Evaluation and rehabilitation of dam seepage problems. A case study: Kafrein dam. Eng Geol 56:335–345. https://doi.org/10.1016/s0013-7952(99)00117-9
17. Groenier J (2012) Pocket safety guide for dams and impoundments. https://doi.org/10.13140/RG.2.1.4275.6968
18. Obernhuber P (2018) Richtlinien für den Nachweis der Tragsicherheit von Betonsperren. Oesterr Wasser Abfallw 70:388–396. https://doi.org/10.1007/s00506-018-0491-7
19. Zhang L, Peng M, Chang D, Xu Y (2016) Statistical analysis of failures of concrete dams, pp 53–56. https://doi.org/10.1002/9781118558522.ch4
20. Chen D, Huang B, Kang F (2023) A review of detection technologies for underwater cracks on concrete dam surfaces. Appl Sci 13:3564. https://doi.org/10.3390/app13063564
21. Shirato M, Fukui J, Koseki J (2006) Current status of ductility design of abutment foundations against large earthquakes. Soils Found 46:377–396. https://doi.org/10.3208/sandf.46.377
22. Saatcioglu M (2013) Structural damage caused by earthquakes. In: Encyclopedia of natural hazards, pp 947–959. https://doi.org/10.1007/978-1-4020-4399-4_346
23. Moreira KMV, Oliveira PVG, de Deus ÊP, Cabral AEB (2021) Alkali-silica reaction: understanding the phenomenon. J Build Pathol Rehabil 6. https://doi.org/10.1007/s41024-020-00100-3
24. Figueira RB, Sousa R, Coelho L et al (2019) Alkali-silica reaction in concrete: mechanisms, mitigation and test methods. Constr Build Mater 222:903–931. https://doi.org/10.1016/j.conbuildmat.2019.07.230
25. Zhao Z, Chen B, Wu Z, Zhang S (2021) Multi-sensing investigation of crack problems for concrete dams based on detection and monitoring data: a case study. Measurement 175:109137. https://doi.org/10.1016/j.measurement.2021.109137
26. de Rubertis K (2018) Monitoring dam performance. https://doi.org/10.1061/9780784414828
27. Lin P, Ning Z, Shi J et al (2021) Study on the gallery structure cracking mechanisms and cracking control in dam construction site. Eng Fail Anal 121:105135. https://doi.org/10.1016/j.engfailanal.2020.105135
28. Li M, Xu W, Wang Y et al (2020) Shrinkage crack inhibiting of cast in situ tunnel concrete by double regulation on temperature and deformation of concrete at early age. Constr Build Mater 240:117834. https://doi.org/10.1016/j.conbuildmat.2019.117834
29. Zhang X, Yan T, Liu Q et al (2022) Cracking analysis of induced joints in roller compacted concrete arch dam. Alex Eng J 61:3599–3612. https://doi.org/10.1016/j.aej.2021.08.082

30. He Z, Hu Y, Li Q, Ma R (2020) Restrained cracking failure behavior of concrete due to temperature and shrinkage. Constr Build Mater 244:118318. https://doi.org/10.1016/j.conbuildmat.2020.118318
31. Liu J, Tian Q, Wang Y et al (2021) Evaluation method and mitigation strategies for shrinkage cracking of modern concrete. Engineering. https://doi.org/10.1016/j.eng.2021.01.006
32. Nassif AY, Petrou MF (2013) Influence of cold weather during casting and curing on the stiffness and strength of concrete. Constr Build Mater 44:161–167. https://doi.org/10.1016/j.conbuildmat.2013.03.016
33. Dai Q, Lin F, Wang X et al (2017) Detection of concrete dam leakage using an integrated geophysical technique based on flow-field fitting method. J Appl Geophys 140:168–176. https://doi.org/10.1016/j.jappgeo.2017.03.013
34. Adamo N, Al-Ansari N, Sissakian V, Laue J, Knutsson S (2020) Dam safety and overtopping. J Earth Sci Geotech Eng 10:41–78
35. Lowe J, Regan RP, de S. Pinto NL, Scherich ET, Schrader EK (1988) Spillway design and construction. In: Jansen RB (eds) Advanced dam engineering for design, construction, and rehabilitation. Springer, Boston, MA. https://doi.org/10.1007/978-1-4613-0857-7_20
36. Ohio Department of Natural Resources (2019) Dam safety: lake drains, fact sheet 93-26. Ohio Department of Natural Resources, Division of Water, Dam Safety Engineering Program, Columbus
37. Morrison-Maierle, Inc. (2012) Guidelines for use of pumps and siphons for emergency reservoir drawdown
38. Unites States. Dam safety—an owners guidance manual
39. Zhu X, Bao T, Yeoh JKW et al (2021) Enhancing dam safety evaluation using dam digital twins. Struct Infrastruct Eng 1–17. https://doi.org/10.1080/15732479.2021.1991387
40. The People's Republic of China (2017) Guidelines on dam safety evaluation (SL271-2017) (in Chinese)
41. Sissakian VK, Adamo N, Al-Ansari N (2019) The role of geological investigations for dam siting: Mosul Dam a case study. Geotech Geol Eng 38:2085–2096. https://doi.org/10.1007/s10706-019-01150-2
42. Direct methods of exploration | Test pits, trenches, shafts and probing. https://www.civilengineeringterms.com/soil-mechanics-1/direct-methods-of-exploration-test-pitstrenchesshafts-and-probing/
43. Dwivedi SK, Vishwakarma M, Soni A (2018) Advances and researches on non destructive testing: a review. Mater Today Proc 5:3690–3698. https://doi.org/10.1016/j.matpr.2017.11.620
44. Lampman S, Mulherin M, Shipley R (2022) Nondestructive testing in failure analysis. J Fail Anal Prev 22:66–97. https://doi.org/10.1007/s11668-021-01325-1
45. Du W, Zhao Y, Roy R et al (2018) A review of miniaturised non-destructive testing technologies for in-situ inspections. Procedia Manuf 16:16–23. https://doi.org/10.1016/j.promfg.2018.10.152
46. Hui L, Haitao M (2011) Application of ground penetrating radar in dam body detection. Procedia Eng 26:1820–1826. https://doi.org/10.1016/j.proeng.2011.11.2372
47. Di Q, Wang M (2009) Determining areas of leakage in the Da Ye Dam using multi-electrode resistivity. Bull Eng Geol Environ 69:105–109. https://doi.org/10.1007/s10064-009-0222-1
48. Raji WO, Adedoyin AD (2020) Dam safety assessment using 2D electrical resistivity geophysical survey and geological mapping. J King Saud Univ Sci 32:1123–1129. https://doi.org/10.1016/j.jksus.2019.10.016
49. Zhou K, Zhang Q, Liu Y et al (2021) Internet-of-things-based four-dimensional high-density electrical instrument for geophysical prospecting. Geosci Instrum Methods Data Syst 10:141–151. https://doi.org/10.5194/gi-10-141-2021
50. Shi P, Fan X, Ni J, Wang G (2016) A detection and classification approach for underwater dam cracks. Struct Health Monit 15:541–554. https://doi.org/10.1177/1475921716651039
51. Wang L, Jiang L, Ma L et al (2019) The key technology of manned submersible in the field of high dam reservoir. China Water Transp 19:17–18, 110 (in Chinese)

52. Xiang Y, Sheng J, Wang L et al (2022) Research progresses on equipment technologies used in safety inspection, repair, and reinforcement for deepwater dams. Sci China Technol Sci 65:1059–1071. https://doi.org/10.1007/s11431-021-1958-y
53. Satria D, Wiryadinata R, Esiswitoyo DPA et al (2019) Hydrodynamic analysis of remotely operated vehicle (ROV) observation class using CFD. IOP Conf Ser Mater Sci Eng 645:012014. https://doi.org/10.1088/1757-899x/645/1/012014
54. Shimono S, Matsubara O, Toyama S et al (2015) Development of underwater inspection system for dam inspection. https://doi.org/10.23919/oceans.2015.7401929
55. Cai Y, Xiang Y, Sheng J, Meng Y (2023) Deep-water detection, monitoring, early warning and treatment of emergencies of major water conservancy projects: a review. Chin J Geotech Eng 45(3):441–458 (in Chinese)
56. Tan J, Wang M, Cai W et al (2015) Reservoir dam underwater reinforcement technology. Chang Jiang Press. ISBN: 978-7-5492-3545-2
57. Loupos K, Doulamis AD, Stentoumis C et al (2018) Autonomous robotic system for tunnel structural inspection and assessment. Int J Intell Robot Appl 2:43–66. https://doi.org/10.1007/s41315-017-0031-9
58. Seet G, Yeo SH, Law WC et al (2018) Design of tunnel inspection robot for large diameter sewers. Procedia Comput Sci 133:984–990 (in Chinese). https://doi.org/10.1016/j.procs.2018.07.078
59. Lynx | Saab Seaeye. In: www.saabseaeye.com. https://www.saabseaeye.com/solutions/underwater-vehicles/lynx. Accessed 15 Sept 2023
60. Huang Z, Sun H (2019) An application of remotely operated vehicle to underwater inspection of deep-buried long tunnel. J Yangtze River Sci Res Inst 36(7):170–174 (in Chinese)
61. Chen Y, Chen J, Wang H et al (2023) Key technology of underwater inspection robot system for large diameter and long headrace tunnel. J Tsinghua Univ (Sci Technol) 63(7):1015–1031 (in Chinese)
62. Zhang BN, Han B, Li NB et al (2021) Research progress on unmanned inspection technology and disease identification method of long-distance hydraulic tunnels in operation period. J Basic Sci Eng 29(5):1245–1264 (in Chinese)
63. Lai JT (2020) Research and application of underwater full coverage unmanned detection technology for large diameter and long diversion tunnel. Yangtze River 51(5):228–232 (in Chinese)
64. Li S, Liu J, Xu H et al (2018) Research status of autonomous underwater vehicles in China. Sci Sin Inf 48(9):1152–1164 (in Chinese)
65. Wang TW, Zhou HK, Yang F et al (2019) Engineering application of a special underwater vehicle for tunnel inspection of hydropower station. Water Resour Power 37(8):174–176 (in Chinese)
66. Xiong XH, Ke H, Fu YW et al (2020) Application of multi-element collaborative detection technology in hydropower engineering. Water Power 46(9):126–130 (in Chinese)
67. Lai JT, Nie Q, Li QD et al (2021) Application of underwater unmanned detection technology in operation and maintenance of hydropower stations in Yalong River Basin. Water Resour Power 39(11):144, 207–210 (in Chinese)
68. Shen Q (2018) Application of underwater robot technology in water conservancy project detection. China Strateg Emerg Ind 36:157–158, 160 (in Chinese)
69. Hu Q (2022) Sonar technology. In: Cui W, Fu S, Hu Z (eds) Encyclopedia of ocean engineering. Springer, Singapore. https://doi.org/10.1007/978-981-10-6946-8_296
70. Galceran E, Djapic V, Carreras M, Williams DP (2012) A real-time underwater object detection algorithm for multi-beam forward looking sonar. IFAC Proc Vol 45:306–311. https://doi.org/10.3182/20120410-3-pt-4028.00051
71. Tang Z, Lu J, Wang Z, Ma G (2020) Three dimensional height information reconstruction based on mobile active sonar detection. Appl Acoust 169:107459. https://doi.org/10.1016/j.apacoust.2020.107459
72. Zhao C, Thies PR, Johanning L (2022) Offshore inspection mission modelling for an ASV/ROV system. Ocean Eng 259:111899. https://doi.org/10.1016/j.oceaneng.2022.111899

73. Antonelli G (2014) Fault detection/tolerance strategies for AUVs and ROVs. In: Underwater robots. Springer tracts in advanced robotics, vol 96. Springer, Cham. https://doi.org/10.1007/978-3-319-02877-4_4
74. Chow JK, Liu K, Tan PS et al (2021) Automated defect inspection of concrete structures. Autom Constr 132:103959. https://doi.org/10.1016/j.autcon.2021.103959
75. Sugimoto K, Sugimoto T, Utagawa N et al (2018) Detection of internal defects of concrete structures based on statistical evaluation of healthy part of concrete by the noncontact acoustic inspection method. Jpn J Appl Phys 57:07LC13. https://doi.org/10.7567/jjap.57.07lc13
76. Cao S, Li J, Jie W et al (2021) Study on nondestructive test method for inner defects of mass concrete. IOP Conf Ser 668:012058. https://doi.org/10.1088/1755-1315/668/1/012058
77. Talaat A, Emad A, Tarek A, Masbouba M, Essam A, Kohail M (2020) Factors affecting the results of concrete compression testing: a review. Ain Shams Eng J. https://doi.org/10.1016/j.asej.2020.07.015
78. Sefrin R, Glock C (2022) Characteristic concrete compressive strength of existing structures—evaluation of EN 13791:2019 for small sample sizes. Struct Concr. https://doi.org/10.1002/suco.202100207
79. Nilimaa J, Nilforoush R (2023) A direct tensile strength testing method for concrete from existing structures. CivilEng 4:333–344. https://doi.org/10.3390/civileng4010019
80. Liu P, Yu Z, Chen Y (2020) Carbonation depth model and carbonated acceleration rate of concrete under different environment. Cem Concr Compos 114:103736. https://doi.org/10.1016/j.cemconcomp.2020.103736
81. Villain G, Thiery M, Platret G (2007) Measurement methods of carbonation profiles in concrete: thermogravimetry, chemical analysis and gammadensimetry. Cem Concr Res 37:1182–1192. https://doi.org/10.1016/j.cemconres.2007.04.015
82. Zhou F, Chen Z, Liu H et al (2018) Simultaneous estimation of rebar diameter and cover thickness by a GPR-EMI dual sensor. Sensors 18:2969. https://doi.org/10.3390/s18092969
83. Dixit M, Gupta AK (2021) A review of different assessment methods of corrosion of steel reinforcement in concrete. Iran J Sci Technol Trans Civ Eng 46:735–752. https://doi.org/10.1007/s40996-021-00644-5
84. Hu JY, Zhang SS, Chen E, Li WG (2022) A review on corrosion detection and protection of existing reinforced concrete (RC) structures. Constr Build Mater 325:126718. https://doi.org/10.1016/j.conbuildmat.2022.126718
85. Wu X, Tang N, Liu B, Long Z (2020) A novel high precise laser 3D profile scanning method with flexible calibration. Opt Lasers Eng 132:105938. https://doi.org/10.1016/j.optlaseng.2019.105938
86. Javaid M, Haleem A, Pratap Singh R, Suman R (2021) Industrial perspectives of 3D scanning: features, roles and it's analytical applications. Sens Int 2:100114. https://doi.org/10.1016/j.sintl.2021.100114
87. Zhao S, Kang F, Li J, Ma C (2021) Structural health monitoring and inspection of dams based on UAV photogrammetry with image 3D reconstruction. Autom Constr 130:103832. https://doi.org/10.1016/j.autcon.2021.103832
88. Reddy MSB, Ponnamma D, Sadasivuni KK et al (2021) Sensors in advancing the capabilities of corrosion detection: a review. Sens Actuators A Phys 332:113086. https://doi.org/10.1016/j.sna.2021.113086
89. Kim Y, Cho S, Park IK (2021) Analysis of flaw detection sensitivity of phased array ultrasonics in austenitic steel welds according to inspection conditions. Sensors 21:242. https://doi.org/10.3390/s21010242
90. Evident Corporation (2022) Ultrasonic flaw detection—theory, practice and applications. AZoM. Accessed 15 Sept 2023. https://www.azom.com/article.aspx?ArticleID=8021
91. Manjula K, Vijayarekh K, Venkatrama B (2014) Weld flaw detection using various ultrasonic techniques: a review. J Appl Sci 14:1529–1535. https://doi.org/10.3923/jas.2014.1529.1535
92. McStraw B, United States (1996) Inspection of steel penstocks and pressure conduits
93. Froehlich D (2018) Guidelines for preparing operation and maintenance manual for dams. https://doi.org/10.13140/RG.2.2.16558.97609

Chapter 3
Analysis on Safety Monitoring Data

3.1 Introduction

3.1.1 Concept of Monitoring

Dam safety monitoring is a comprehensive process involving the measurement and observation of various critical elements within a hydroelectric project [1, 2]. This monitoring is conducted through a combination of monitoring instruments and inspection visits, including observation and evaluation of the primary structures of the hydroelectric project, such as the dam itself, the foundation on which it's built, and any associated embankments along the riverbanks. Meanwhile, monitoring instruments must be installed at specific locations to collect data. These instruments measure parameters such as deformation, stress, seepage, temperature, and other factors that could affect the stability and safety of the dam [3].

Periodic visual inspections are conducted to visually assess the condition of the structures, dams, and surrounding area. Inspectors look for signs of damage, erosion, cracks, or other visible changes that may require attention. (See Chap. 2 for details.) In addition to visual inspections, instrumental probing involves the use of specialized equipment to assess various objects both on the surface and within structures. This helps to identify potential weaknesses that may not be visible through visual inspection alone. Monitoring extends to the overall operating environment, taking into account factors such as water levels, flow rates, and weather conditions. Changes in these conditions can affect the stability of the dam.

Monitoring data provides the basis for conducting periodic safety inspections and comprehensive assessments of the dam's condition. These inspections help identify potential problems and assess the overall safety status. Monitoring data helps assess suspected or existing hazards that could threaten the stability of the dam. This information guides decision-making regarding potential hazards and necessary mitigation

Y. Xiang et al., *On-site Inspection and Dam Safety Evaluation*, Hydroscience and Engineering, https://doi.org/10.1007/978-981-97-3987-5_3

measures. Monitoring data is critical in determining the need for and effectiveness of repair and strengthening measures. By analyzing the data, engineers can identify areas that need to be addressed and ensure that the actions taken are successful [4].

Effective dam safety monitoring requires accurate measurement of the relevant physical quantities. Appropriate instrumentation should be selected, maintained, and calibrated to provide data with acceptable accuracy. Monitoring should be systematic, following a predefined schedule and methodology. This will ensure consistent data collection and timely detection of changes. Monitoring should be conducted by personnel trained to accurately collect and record monitoring data. Their expertise will ensure reliable data collection. Monitoring data should be analyzed on an ongoing basis to identify trends, anomalies, or critical changes that may require immediate attention.

3.1.2 Monitoring Instruments and Equipment

Instrumental monitoring involves the utilization of monitoring instruments to gather measurements from engineered structures. These instruments typically encompass three key elements: a sensor for measurement, a communication medium for data transmission, and a readout or manual recording system. Various principles are employed for measurement, such as mechanical devices, optical, electrical, photographic, pneumatic, hydraulic methods, satellite positioning, or remote sensing.

There are two main categories of monitoring instruments: One type of monitoring instrument is capable of instantaneous response to obtain measurement data, while another type requires a period of time to obtain the required measurement data for the physical quantity being measured.

Monitoring instruments can exhibit different operational characteristics: One type of instrument is designed for continuous operation or measurement, known as "all-weather," while others measure data at regular intervals.

Monitoring instruments can also vary in terms of mobility and spatial coverage: Monitoring instruments can be either fixed or portable; they can provide measurements at a point, in a straight line, on an axis (vertical), or in a specific area.

Measurement data can be read remotely or locally. Methods of remote signal transmission include wired and wireless, wired being mainly in the form of cable and fibre optics, and wireless in the form of shortwave, ultra-shortwave, GPRS, Internet, satellite, etc. Data can be recorded manually, by chart recorders, mobile or fixed hard disks and other devices, and most automated monitoring systems monitor data directly from computer or server memory records [5, 6].

The choice between remote and local reading, as well as the methods used for data transmission and recording, depends on factors such as the complexity of the project, the desired frequency of data collection, the accessibility of the monitoring site, and the level of automation needed for efficient monitoring and analysis [7].

3.1.3 Purpose and Significance of Monitoring

The main purposes of safety monitoring are problem identification and analysis, design verification, early disease detection and risk prediction, and evaluation of the effectiveness of maintenance and strengthening actions.

(1) **Problem identification and analysis**. Monitoring data are often used to identify and analyze the presence of disease and hazard problems in reservoir dams [4]. For example, dam displacements caused by high reservoir levels need to be analyzed to determine whether the displacements are uniformly distributed along the dam; whether the displacements are caused by deformation of the dam body or the dam base, or both; and whether the displacements are increasing, decreasing, or remaining constant. This information is an important basis for developing maintenance and strengthening measures.
(2) **Design verification**. Monitoring instruments installed on the surface or inside the dam may rarely or never indicate that the dam is abnormal or dangerous, but the monitoring information obtained indicates that the dam is functioning as designed or that the project is operating normally, which is the value of dam safety monitoring and can verify the soundness of the design. Although problems may occur or are occurring with the dam, the monitoring data indicate that such "anomalies" are normal and were anticipated in the design of the dam, such as an increase in seepage caused by a rise in reservoir level.
(3) **Predictive warning**. Monitoring data can detect abnormal changes or trends that cannot be seen during inspections, such as fluctuations in seepage pressure, etc. [8]. Regular monitoring can also detect gradual changes that may not be apparent, such as a slow increase in seepage. Dam safety monitoring can provide early warning of serious problems [9].
(4) **Evaluate the effectiveness of maintenance and strengthening measures**. Dams are like living organisms, especially old dams, which inevitably become diseased and dangerous during operation. To eliminate diseases and hazards, improve flood resistance, and fully utilize the benefits of the reservoir, maintenance and strengthening or de-risking is required. The effectiveness of strengthening measures can be assessed by comparative analysis of monitoring data before and after the strengthening project [10].

3.1.4 Frequency of Dam Safety Monitoring

Determining an appropriate monitoring frequency is a critical aspect of evaluating a dam's performance and ensuring its safety. The frequency of monitoring should align with specific considerations related to code requirements, risk factors, engineering design, and other relevant characteristics [11]. Several factors contribute to deciding the monitoring frequency:

1. Engineering Design and Risk Factors: The level of risk associated with the dam's failure due to engineering design considerations must be taken into account. This includes factors such as the dam's structural complexity and the potential consequences of a failure.
2. Seismicity: The frequency and magnitude of seismic events in the region are crucial determinants. Areas prone to earthquakes may require more frequent monitoring due to the potential impact on the dam's stability.
3. Reservoir Operations: The frequency of reservoir level changes, including initial storage and significant water level fluctuations, can influence the monitoring frequency. Rapid changes in water levels can impact dam behavior.
4. Historical Anomalies: Any past anomalies or engineering issues in the dam's history may necessitate increased monitoring to ensure stability.
5. Environmental Risks: Environmental factors such as disease risks or extreme weather conditions may require more frequent monitoring to mitigate potential impacts.

The frequency of dam displacement monitoring should be determined according to the actual operation of the project and requirements, and displacement should be measured at least quarterly to determine the periodic effect of temperature on the concrete structure (the frequency of monitoring for earth-rock dams may not need to be as frequent). However, some conditions require more frequent measurements. These conditions include:

- Reservoir Initial Storage: The start of reservoir filling can lead to significant changes in dam behavior.
- Reservoir Water Level Changes: Frequent monitoring is crucial when water levels in the reservoir change notably.
- High Water Level Events: Innovative high water levels or extreme conditions demand heightened monitoring.
- Signs of Instability: Any signs of displacement or instability detected during routine inspections or patrols should trigger more frequent measurements.
- Recent Seismic Activity: Following seismic events, closer monitoring is needed due to potential impact on the dam's integrity.
- Signs of High Uplift or Pore Pressure: Elevated uplift or pore water pressure signals may necessitate increased monitoring.

3.2 Types of Instrumentation

The main monitoring quantities designed for the dam monitoring instrumentation system include seepage pressure, seepage (leakage), displacement, upstream and downstream reservoir water levels, etc. These basic physical quantities are of great importance in evaluating the safety and performance of concrete and rockfill dams. Other important monitoring quantities include earthquake response, pressure, stress and strain, internal concrete temperature, ambient temperature, and rainfall.

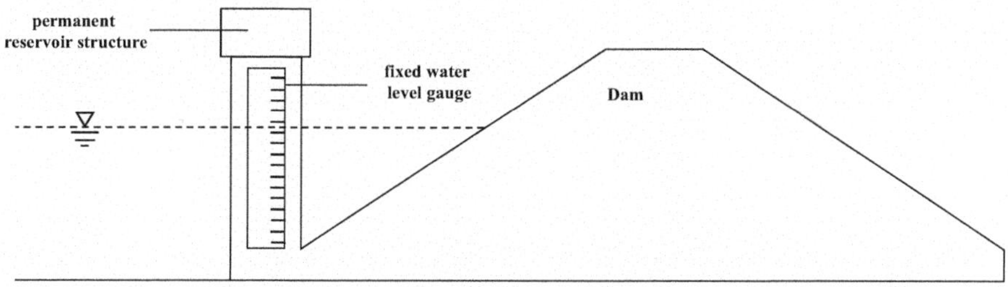

Fig. 3.1 Manual observation of reservoir water level

3.2.1 *Water Level*

Regular monitoring and recording of reservoir water levels are crucial components of dam safety inspections. Additionally, downstream water levels should also be documented. These water level measurements play a significant role in various data analyses, including determining the distribution of dam base pressure. The relationship between dam base pressure and leakage often hinges on the disparity between reservoir and downstream water levels [12].

Various methods are used to measure water levels. Simple water level gauges include fixed water level gauges on permanent reservoir structures or painted water level markings. These gauges provide a visual indication of water levels and are particularly useful for quick reference and visual assessment. Modern water level sensors provide accurate and automated measurements of water levels. These sensors use various technologies, such as pressure sensors, ultrasonic sensors, or radar sensors, to detect water levels with precision [13].

As shown in Figs. 3.1 and 3.2, manually observed reservoir level gauges are a simple yet effective method to visually determine water levels. This visual assessment serves as an integral component of dam safety inspections and contributes to a comprehensive understanding of the dam's hydraulic conditions.

3.2.2 *Seepage and Leakage*

Under the influence of hydrostatic pressure, water from the reservoir can infiltrate the dam's base and the surrounding area. This water flows through the pores of the dam fill and any cracks or crevices in the rock. This infiltration generates pressure that is distributed uniformly in all directions, known as pore water pressure. The upward component of this pressure is termed buoyancy pressure. Buoyancy pressure can counteract the effective pressure on the dam's foundation, potentially reducing the slip resistance of the foundation and impacting dam stability. The infiltration pressure within the dam body or foundation has a direct bearing on the dam's overall stability [14]. Figure 3.3 shows a schematic diagram of the arrangement of seepage monitoring

Fig. 3.2 Water level

instruments for concrete dams and earth dams, and the specific arrangement should be considered according to the actual situation.

Pressure monitoring instruments, including various types of seepage gauges, are employed to measure infiltration pressure at specific locations within the dam body, foundation, and embankment. Seepage gauges are used to monitor the total hydrostatic pressure on various surfaces, such as embankment bodies, pipes, spillway structures, dam foundations, and retaining walls. These measurements help understand the pressure distribution across the dam's components. Seepage gauges can assess the effectiveness of seepage curtains by positioning them ahead of and behind the curtain. These measurements reveal how well the curtain is managing seepage and pressure distribution.

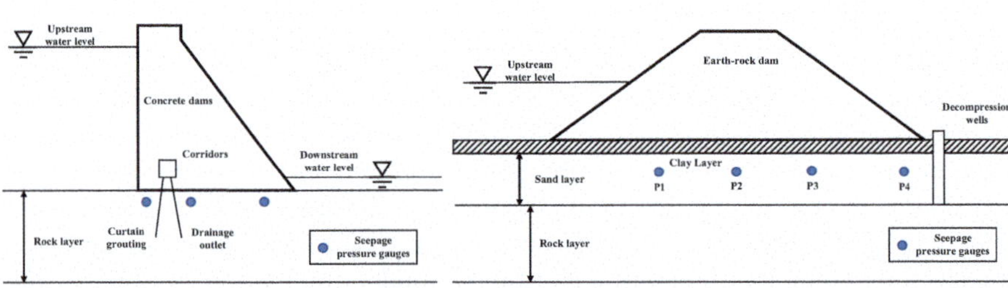

Fig. 3.3 Typical cross-sections for placing seepage pressure gauges in concrete dams and earth-rock dams

Fig. 3.4 Water measuring weir

The amount of seepage is usually related to the reservoir level and is an important factor in assessing the seepage behavior of a dam. Sudden changes in the amount of seepage collected for no apparent reason, such as corresponding changes in reservoir levels or recent heavy rainfall, indicate problems with the dam. Similarly, serious seepage problems in dams may be indicated if the seepage water is turbid or discolored, contains increasing amounts of sediment, or if there is a dramatic change in chemical composition. The presence of seepage on the embankment, shoulder or downstream of the dam base may also indicate a problem. The most common methods of measuring seepage include the weir method, the volumetric method and the flow rate method [15] (Fig. 3.4).

3.2.3 Movement

All dams are subject to a variety of displacements and deformations. Horizontal displacements usually occur upstream and downstream of the embankment, but may also occur along the axis of the dam and may involve displacement of the entire dam relative to its shoulder or foundation, or may involve displacement of one part of the dam relative to another. Vertical downward displacement caused by consolidation of the dam body or foundation is known as settlement. Vertical displacement can also be upward, usually caused by buoyancy or structural stress concentrations at the dam face, and is called bulging, which tends to occur at the toe of the dam. The water level in the reservoir has a significant effect on the displacement. Dam settlement displacements are shown in Fig. 3.5 and horizontal displacements upstream and downstream of the dam are shown in Fig. 3.6. Careful observation of dam displacements is essential when assessing the performance and safety of dams, particularly concrete dams. Concrete dams allow little deformation and even small displacements can be extremely damaging to a concrete dam [16].

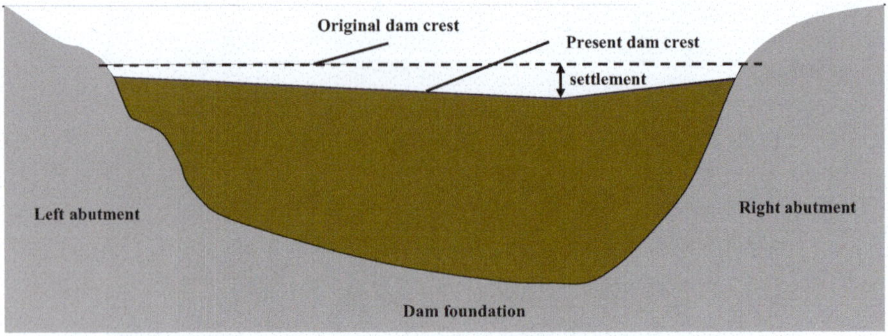

Fig. 3.5 Schematic diagram of settlement and displacement

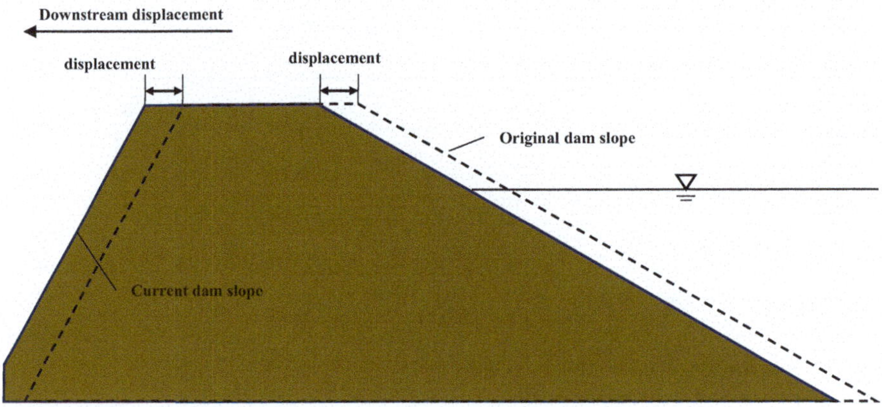

Fig. 3.6 Diagram of downstream horizontal displacement

The dams also deform due to external loads produced by reservoir water, rapid drawdown, earthquakes, undermining, swelling clays, and piping. Concrete structures deform due to internal loads such as pore pressure, cooling, and alkali aggregate reaction of concrete; and external loads caused by air and reservoir temperature, solar radiation, reservoir levels, uplift pressure, wind, earthquakes, undermining, ice, overflowing water, swelling clay, and foundation settlement.

Movement can be divided into three types: surface movement, internal movement, and crack or joint movement. Since it can occur in any direction, measurements in three mutually perpendicular directions are necessary to accurately determine vector movement. Measurements are typically made in vertical, transverse horizontal, and longitudinal horizontal directions. Figure 3.7 shows the changes in vertical and horizontal displacement measurement points over time as the construction period changes.

(1) **Surface movement** is defined as horizontal or vertical movement of a point on the surface of a structure relative to a fixed point off of the structure. It is

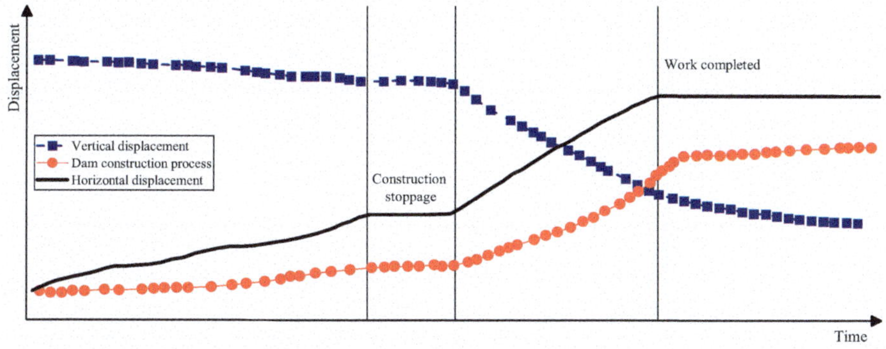

Fig. 3.7 Measurement point displacement versus time

usually determined by some type of surveying. Modern surveying equipment has increased the number and type of surveys that are available.

(2) **Internal movement** is defined as horizontal or vertical movement within the structure. It is usually determined relative to some point on the structure or in the foundation.

(3) **Joint or crack movement** is defined as horizontal or vertical movement of one part of a structure relative to another part of a structure. It is usually measured across block joints or cracks in concrete structures or cracks in earth structures.

Movements in response to such loads are normal and acceptable, provided they are within tolerable ranges and do not cause structural distress. Embankments are less brittle than concrete structures and can undergo larger movements without distress. As a result, measurements of surface movements of embankment dams are typically less precise than those for concrete structures. Sudden or unexpected direction, magnitude, or trend of surface movement could indicate developing problems. Internal movement measurements of both concrete and embankment dams and their foundations should be detailed and precise. Measuring points for all movement surveys should be installed so that they are not subject to movement from freeze–thaw action or traffic.

Displacement measurements of concrete dams can be carried out using a number of different instruments. Surface deformations can be measured using total stations, tachymeters, GNSS systems, etc. Internal deformations can be measured using plumb coordinators, lead tensioners, inclinometers, displacement meters, seamometers, bedrock deformation meters, static levels, vacuum laser collimation systems, etc. Internal deformation measurements monitor the relative displacement between the various parts of the dam or dam foundation blocks, while absolute displacement of surface deformation can be measured by deformation monitoring networks [17, 18].

The main displacements to be monitored in earth and rock dams include horizontal displacement of the dam body, settlement and deformation of the dam foundation. The main instruments used for measurement are total stations, levels, GNSS systems, settlement gauges, rock deformers, inclinometers, displacement gauges, surface displacement measuring points and various other measuring instruments and devices [19].

3.2.4 Stress and Strain

The internal forces within a dam during operation differ from the design. Special stress measuring instruments are therefore used to monitor and determine the actual stresses at specific locations, for example between the dam and its shoulder or foundation, or between specific components of the dam [20]. The purpose of stress monitoring is to measure the stresses (loads) at the contact surface of the embankment or within the overall embankment. Different types of stress monitoring sensors are used to monitor dam stresses. The main stresses in the embankment can be determined from the monitoring data from specially positioned directional sensors. The main value of the stress data is to verify design assumptions and to provide data for future design improvements.

There are many different types of strain gauges, including differential resistance strain gauges, vibrating wire strain gauges and fiber optic strain gauges. These strain gauges are subdivided into earth pressure gauges, rebar strain gauges, anchor strain gauges, etc. for different measurement targets [21].

Earth pressures within fill and against concrete structures are commonly measured with earth pressure cells. These are also known as total pressure cells. They consist of two flexible diaphragms sealed around the periphery, with a fluid in the annular space between the diaphragms (Fig. 3.8). Pressure is determined by measuring the increase in fluid pressure behind the diaphragm with pneumatic or vibrating-wire sensors. Earth pressure cells should have similar stiffness as the surrounding soil to avoid inaccurate measurements of in-situ stress caused by arching. Soil pressures against structures can also be measured with a Carlson-type cell. It consists of a chamber with a diaphragm on the end. Deflection of the diaphragm is measured by a Carlson-type transducer and is converted to stress. Stress in concrete structures can be measured with total pressure cells or Carlson-type cells designed to have a stiffness similar to concrete.

The modulus of elasticity, creep coefficient, and the Poisson's ratio for concrete can be determined from the laboratory testing of concrete field cylinders. These values are required for converting strain measurements to stress. A variety of mechanical and electrical strain gages are used to measure strain in concrete structures. Some of the instruments are designed to be embedded in the dam during construction and others are surface mounted following construction. Strain gages are often installed in groups so that the three-dimensional state of strain can be evaluated (Fig. 3.9).

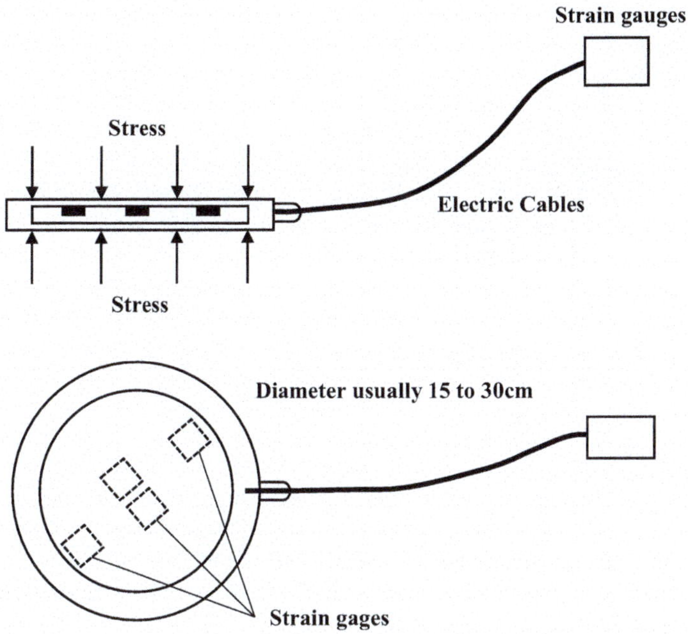

Fig. 3.8 Diagram of the earth pressure gauge

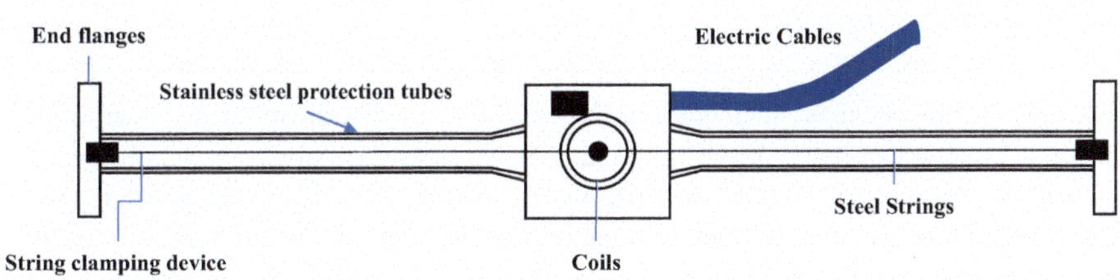

Fig. 3.9 Schematic diagram of a rebar strain gauge

3.2.5 Environmental Quantities

Environmental quantities, including upstream and downstream water levels of the reservoir and meteorological conditions (e.g. temperature and precipitation), are the dependent variables affecting dam deformation and seepage. Environmental variables must be monitored to provide a continuous historical record of the influencing factors for dam deformation and seepage analysis. Therefore, these environmental variables must be recorded simultaneously with each instrumental observation.

Temperature measurements of a dam, foundation, or instruments, are often required to reduce data from instruments, increase precision, or to interpret results. For example, movements of concrete dams and changes in leakage at concrete dams

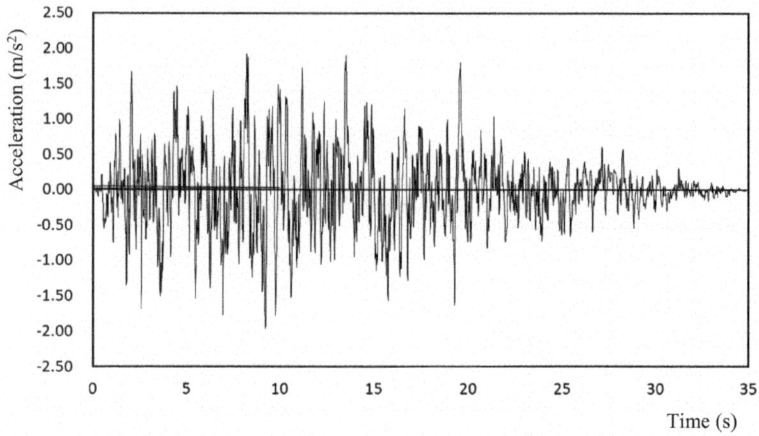

Fig. 3.10 The spectrum of seismic waves

are commonly related to changes in temperature. Temperature is also commonly measured in concrete dams under construction to evaluate mix design and placement rates, to time grouting, and to assess thermal loads.

3.2.6 Seismic Loads

Seismic strong motion instrumentation records acceleration from earthquake shaking. The data are used to evaluate the dynamic response of dams. Seismic acceleration and velocity are usually recorded with strong-motion accelerographs. These devices typically consist of three mutually-perpendicular accelerometers, a recording system, and triggering mechanism. To prevent accumulation of unwanted data, the instruments are usually set to be triggered at accelerations generated by nearby small earthquakes or more distant, larger earthquakes. They are expensive, especially considering that multiple instruments are necessary to record dynamic response at several locations on a structure, a foundation, or abutments. The devices must be properly maintained, so that they operate if an earthquake occurs [22]. Figure 3.10 shows the frequency spectrum of seismic signals, which can provide a basis for subsequent numerical analysis and calculation.

3.2.7 Automated Monitoring Systems

The automated system is an important part of dam safety monitoring and its evaluation includes data acquisition devices, computer and communication facilities, information acquisition and management software, operating conditions, and operation and maintenance. The evaluation of the data acquisition equipment includes the function, the average time without failure, the rate of missing data acquisition and the accuracy of the star measurement [3]. The focus is on the functions that are frequently used during the operation of the data acquisition device and ensure the long-term stable operation of the data acquisition device, such as inspection, selection, time measurement, communication, data storage, power failure protection, lightning protection, anti-interference and other essential functions, and the secondary functions such as system self-test, self-diagnosis, manual star measurement interface and corrosion prevention. Evaluation of computer and communication equipment includes operating status, power failure protection, average uptime, etc. Evaluation of information collection and management software includes functional completeness, correctness and operability. Evaluation of operating conditions includes temperature and humidity, operating power supply, power supply lightning protection and grounding network, etc. Operation and maintenance evaluation includes data backup, clock calibration, comparison test, spare parts, equipment inspection and maintenance, etc. [23].

Automated monitoring systems have the advantages of low human intervention, high efficiency and independence from weather and environmental conditions. In the twenty-first century, automated monitoring systems, including remote data acquisition and processing systems, have developed rapidly. Automated monitoring systems often require trained personnel to maintain and repair them [7, 24]. However, no automated monitoring system, no matter how advanced and comprehensive, can replace human on-site monitoring and inspection.

3.2.8 Instrument Installation Suggestions

Minimum instrumentation recommendations for all dams are established in this section. The recommended minimums are considered to be generally applicable; however, since each dam is unique, the recommendations should be applied using engineering judgement and common sense. See Table 3.1 for main monitoring items and parts of dam safety monitoring. Table 3.2 summarizes the routine monitoring items and related monitoring instruments for dams.

- Those with $\star$ are mandatory monitoring items, while those with $*$ are optional monitoring items, which can be selected as required, and the blank is not required.
- For high concrete dams or concrete dams with soft rock strata on the foundation bank, it is recommended to conduct deep deformation monitoring.
- In case of cracks in Grade 1–3 dams, it is necessary to set up crack opening and closing monitoring items.
- For Grade 1 dam with a dam height of more than 70 m, the stress and strain are mandatory items.

Table 3.1 The main monitoring items and parts of the dam safety monitoring

Dam type	Monitoring category	Monitoring items	Dam grade				
			Grade 1	Grade 2	Grade 3	Grade 4	Grade 5
Concrete dam	Deformation monitoring	Dam surface deformation	★	★	★	★	★
		Deformation inside dam body	★	★	★	∗	∗
		Deformation of dam foundation	★	★	★	∗	∗
		Tilt	★	∗	∗		
		Joint, crack opening	★	★	∗	∗	
	Seepage monitoring	Seepage flow rate	★	★	★	★	∗
		Uplift pressure	★	★	★	★	∗
		Seepage pressure of dam body	∗	∗	∗	∗	
		Seepage around dam	★	★	∗	∗	∗
		Water quality analysis	★	★	∗	∗	
	Stress–strain and temperature monitoring	Stress	★	∗			
		Strain	★	★	∗		
		Temperature of concrete	★	★	∗		
		Temperature of dam foundation	★	★	∗		
Earth-rock dam	Deformation monitoring	Horizontal deformation of dam body surface	★	★	★	∗	∗
		Vertical deformation of dam surface	★	★	★	★	★
		Deformation inside dam body	★	★	∗	∗	∗
		Deformation of dam foundation	★	★	∗		
		Interface displacement	★	★	∗		
		Joint, crack opening	★	★	∗		
	Seepage monitoring	Seepage flow rate	★	★	★	★	★
		Seepage pressure of dam foundation	★	★	∗	∗	∗

(continued)

Table 3.1 (continued)

Dam type	Monitoring category	Monitoring items	Dam grade				
			Grade 1	Grade 2	Grade 3	Grade 4	Grade 5
		Seepage pressure of dam body	★	★	*	*	*
		Seepage around dam	★	★	*	*	*
		Water quality analysis	★	★	*		
	Stress–strain and temperature monitoring	Pore water pressure	★	*	*	*	*
		Soil pressure	★	*	*		
		Stress–strain and temperature	★	*	*		

3.3 Completeness of Monitoring System and Reliability of Monitoring Data

The Completeness evaluation of the monitoring system shall include the following points:

(1) Whether the monitoring project meets the regulatory requirements and whether the layout of measurement points is reasonable.
(2) Whether a monitoring data information management system has been established and whether the system functions are complete.
(3) Whether the observation frequency meets the regulatory requirements, and whether the monitoring data is timely compiled and analyzed according to the regulatory requirements [25].

The reliability of monitoring data is actually the authenticity of the monitoring data, whether it reflects the actual characteristics of the dam. In the process of evaluating the reliability of monitoring data, various abnormal measurement values caused by instrument failures, manual measurements, and input errors should be identified and deleted or corrected to ensure the reliability of the analysis results [26]. The reliability evaluation of monitoring data should include the following key points:

(1) Is the selection of monitoring instruments appropriate?
(2) Check whether the performance of the monitoring instrument is stable or intact, and whether the observation accuracy of the instrument meets the design requirements.
(3) Whether the automatic monitoring system operates stably, whether the average time between failures and the missing rate of collected data meet the requirements?

Table 3.2 Summary of measurement conditions, causes and instruments used

Monitoring quantity	Monitoring purpose	Using instruments
Hydraulic pressure	Pore water pressure in the dam body or foundation directly affects the stability of the dam. Abnormal water pressure data may indicate unexpected displacements or leakage	Osmometer
Pressure (stress)	Design stresses may not always occur as expected within a completed dam. Therefore, use pressure data to validate design assumptions and provide data for future design improvements	Pressure gauge Stress gauge
Leakage	Measuring flow over a period of time helps determine whether the flow is increasing or decreasing. Increased flow may indicate that the leakage problem is becoming more serious and a decrease in flow may indicate that the anti-seepage measures are losing their effectiveness. Seepage turbidity indicates that soil particles are suspended in the water. Internal erosion caused by uncontrolled seepage is the main cause of dam break	Seepage flow meter Weir for measuring water Water tank Flowmeter
Water level of reservoir and downstream (tail water level)	Uplift pressure and permeability rate are usually related to elevation difference between reservoir water level and tailwater level	Manual water gauge Water level sensor
Weather conditions • Temperature • Rainfall	These data are useful in assessing day-to-day problems and tracking problem formation Temperature affects the joint and crack width of concrete structures, resulting in permeability and internal stress. Rainfall affects seepage and pressure level	Temperature: thermometer, preferably a self-recording thermometer Rainfall: rainfall gauge, preferably self-recording
Seismic activity	Large vibrations can cause liquefaction of foundations or embankments, leading to serious stability problems. Excessive deformation of earth-rock dam and cracking of concrete structure will be caused without liquefaction. These data can be used to improve the design of future dams, to assess damage following major earthquakes at existing dams, and to assess potential damage in the event of stronger earthquakes	Seismology, including strong vibration accelerometer

(continued)

Table 3.2 (continued)

Monitoring quantity	Monitoring purpose	Using instruments
Displacement • Horizontal displacement • Vertical displacement	These data sets are collected in specific areas where the dam has moved, alerting inspectors to look for possible problems during the inspection process. Movements may cause damage to concrete dams or structural damage, resulting in cracking and sliding of earth-rock dams	Measuring device Bedrock metamorphometer Crossarm settler Settlement meter Inclinometer Displacement meter Inclinometer Vertical line device Crack and joint measuring equipment Stress, strain and buried thermometers

(4) Check whether the physical meaning of the monitoring data is reasonable, whether it exceeds the physical limits of the instrument range and material, and whether the inspection results are within the limits.
(5) Whether the monitoring data complies with the principles of continuity, consistency, relevance, etc.?

3.4 Observation Data Analysis

Dam safety monitoring is an important measure to understand the operational status and development trend of dams, and is also an effective means to verify the design results, check the quality of construction and understand the law of change of various parameters of dams. It is an indicator that reflects the operational status of the dam and is the key information for analyzing and assessing the safety of dams in reservoirs. Dam safety monitoring is a highly effective real-time process, capable of detecting and identifying any anomalous conditions that may arise within the dam structure [27, 28].

The analysis of dam monitoring information allows for the rapid evaluation and monitoring of the safe operation status of dams. This analysis also enables the understanding of the working status and changes in hydraulic structures under the action of adverse loads and various factors. This understanding allows for the accurate assessment and evaluation of the safety of engineering structures [29].

Consequently, the effective analysis of monitoring data is conducive to timely research and early warning, as well as the implementation of emergency removal measures. This is achieved by overcoming the uncertain consequences of delaying removal measures through manual intermittent inspection, which is employed to detect abnormal conditions outside the dam. Furthermore, this approach ensures the safe operation of reservoir dams [30].

In the Engineering Fronts 2022, the topic of "***Intelligent monitoring and risk early-warning for dam safety***" was selected. Additionally, the report highlighted the necessity of developing technologies for the identification of dam damage, the construction of scenarios, and the rapid diagnosis of dam safety. It was also emphasized that future research should aim to enhance the thorough perception, risk assessment, and risk warning capabilities to effectively deal with emergencies [31, 32].

3.4.1 Main Contents of Monitoring Data Analysis

Analysis on dam safety monitoring data mainly includes the following tasks:

- All previous walkaround inspection data shall be analyzed. Qualitative judgment is made to the possible connection with the safety of the project based on the abnormality of the dam appearance, and its location, its change pattern and tendency of its development.
- The pattern of effect quantity change with respect to time is analyzed and changing tendency and stability of the data under the same operation conditions is investigated to make judgment on whether there is any abnormality in operation behavior of the dam and whether there exist any time related effect which develops in the direction prejudicial to safety.
- Spatial distribution and the related characteristics of the effect quantity are analyzed to judge if there is any abnormal or unsafe part or zone in the dam.
- The characteristic and abnormal values of effect quantity are analyzed and compared with design values, test values, model forecast values and range of change in previous years. When the effect quantity exceeds the threshold, the reason and the impact to the dam safety shall be analyzed.
- By utilizing correlation diagrams and mathematic models, the main impact factors of effect quantity and the associated quantitative relationship and the change pattern are analyzed to find out the main causes of abnormality of the effect quantity, to investigate the stability of the correlation between the effect quantity and the cause quantity and to predict the varying tendency of the effect quantity and judge whether the abnormality influences the safe operation of the dam. When the monitoring data sequence is long, statistic modal can be used. Deterministic model or hybrid model can be used if the situation allows.

3.4.2 Monitoring Data Analysis Method

All monitoring data must be collected, processed and evaluated in a timely manner. Untimely, incorrect or improperly evaluated data can be misleading and lead to erroneous conclusions about dam performance and safety. Typical data evaluation methods include plotting monitoring data as a series of continuous historical series, process line plots (e.g. reservoir level vs. seepage level process line plots), correlation line plots, etc. Data plots can also be used to correlate different monitoring quantities to indicate possible trends or anomalies (e.g. seepage flow, seepage pressure level and displacement data can be correlated on the same data plot) [4].

Dam safety patrol inspectors should be aware of thresholds (dam safety monitoring should be designed with a range of monitoring data to indicate acceptable levels of dam structural performance) that may be recommended in some cases where these values do not exist. Management units need to be aware that thresholds should be exceeded and that investigation and response measures should be taken to find and correct the problem to avoid possible structural damage [33].

3.4.2.1 Methods by Statistics of the Characteristic Values

Characteristic values include the maximum and minimum values, amplitude of change, cycle of change, nanual mean value and anual change trend, etc. of various physical variables in previous years (including time of occurance). It can be seen from statistical analysis of characteristic values whether the changes in quantity among the monitored physical variables are consistent and reasonable. The statistical analysis of the above characteristics is used to understand the range of variation in the measured values and their relationship to the main environmental factors, and to determine whether there is consistency and reasonableness in the quantitative changes between the monitored impact quantities.

1. **Environmental volume characteristic value statistics**: for the upstream and downstream water level usually statistically annual, monthly, and daily average; the highest and lowest value and its occurrence date during the year, the annual water level change range; the number of days and dates that exceed a certain high water level, the number of days and dates that are lower than a certain water level; according to some monitoring project information analysis needs, sometimes can also count a certain period of time water level average, variation, the highest and lowest value, the rate of change, etc. Temperature characteristic value statistics mainly statistics for one year of daily, monthly, annual average temperature, maximum temperature, minimum temperature, and the corresponding range of change; multi-year daily, monthly, annual average temperature, maximum temperature, minimum temperature, and the corresponding range of change; above (or below) a certain temperature of the number of days and dates.
2. **Physical quantity characteristic value statistics**: for dam deformation, seepage, stress strain and other monitoring physical quantities, usually statistics of the maximum and minimum values at different stages and the corresponding occurrence time, the average value, etc., in order to understand the range of change of the corresponding physical quantity, the sensitivity to changes in the corresponding environmental quantity, and so on.

3.4.2.2 Methods by Graphing

By graphing method, it is to draw process diagram, corelation diagram, distribution diagram and integrated process diagram (for instance, the upstream water level, precipitation, seepage pressure and seepage flow are put into one same diagram). From these diagrams, the magnitude and pattern of change of monitored values, load factors that influnce monitored values and the degree of influence to the monitored values and whether there is abnormal monitored value can be intuitively understood and analysed.

Plots facilitate screening of data and comparison with expected data. Plots are also useful to summarize data. All reduced data should be summarized in graphical form. All plots should include sufficient previous data to identify any long-term trends. Furthermore, the plots should be self-explanatory. They should show the project name, type of instrument, and what is being measured. If more than one set of data are on one plot, different symbols or line types should be used to distinguish the data and a legend should be provided. Scales should be consistent to allow comparison of data between plots and they should be labeled. Threshold limits, scatter, and magnitude of significant changes should be considered when selecting scales. Plan and section drawings showing the number, location, and details of each instrument should be included with the plots. Where practical, threshold limits should be shown on the plots. Plotting software is available to facilitate this step (Table 3.3).

1. **Process line analysis**

The process line analysis is based on the change process and spatial distribution of the monitored dam effect quantities, combined with the change process of the corresponding environmental quantities and the structural conditions of the dam base and body, to analyse whether the change process of the effect quantities follows the normal rules, whether the values are within the normal range of change, and whether the distribution rules correspond to the structural conditions of the dam body (Fig. 3.11).

2. **Correlation plot analysis**

If there are relevant factors for each monitored physical quantity, then correlation diagrams for correlation diagram analysis shall be drawn, such as seepage pressure water level and upstream water level correlation line, infiltration line, vertical displacement contour line, etc., as follows

(a) According to the correlation diagram to know the process of change of the effect measurement value, the influence of the change of environmental factors and the extent to which the effect measurement value lags behind the change of environmental factors, etc.
(b) If another environmental factor is marked next to a point on the correlation diagram between the effect measurement and the environmental factor (e.g. temperature on the water level–displacement relationship diagram), the influence of the environmental factor on the change in the measurement can be seen (Fig. 3.12).

Table 3.3 The relationship between the various factors affecting dam safety

Interlinked factors	How the relationship between these factors affects data analysis
Infiltration pressure/ reservoir level and downstream water level/ rainfall	Infiltration pressure at the base of earth and rock dams/infiltration pressure at the base of concrete dams (lift pressure) in relation to reservoir level and downstream water level
	The infiltration pressure at the shoulder of the dam is related to the reservoir level and the downstream water level
	Infiltration pressure on the slope of the dam is directly related to the reservoir water level
	Rainfall and groundwater may affect infiltration pressures in the base and body of earth-rock dams and in the base and shoulders of concrete dams
Penetration pressure/ displacement	The effective shear strength decreases as the infiltration pressure within the embankment, at the base or in the shoulder of the embankment increases. If the effective shear strength is less than the horizontal water pressure, displacement will occur
Infiltration pressure/ seepage volume/ reservoir level/ rainfall	Leakage is usually directly related to the level of the reservoir
	Rainfall often leads to a temporary increase in seepage
	Reduced seepage and increased infiltration pressure in the embankment or shoulder indicate that the drainage system or outlet control system is failing
	Reduced seepage and infiltration pressure at the base or shoulder of the dam indicates that the infiltration of reservoir water into the base or shoulder of the dam is decreasing (possibly due to siltation of the reservoir area)
Displacement/ reservoir level	Under normal conditions, the concrete dam will deflect downstream as the reservoir level rises
	Under normal conditions, the rise in reservoir level will not cause significant displacement of the earth and rock dam
Displacement/ temperature	In summer, as the temperature of the downstream face of the dam rises, the concrete near the downstream face expands. At the same time, the concrete near the upstream face hardly expands because the reservoir maintains a constant temperature at the upstream face. This local expansion of the concrete can cause the top of the dam to move upstream or deflect slightly
	Where there is a large temperature difference between summer and winter, the concrete monoliths in the dam can move, causing joints and cracks to open in winter and close in summer
Permeate flow/ permeate pressure/ turbidity/ dissolved matter	An increase in turbidity of seepage water may indicate changes in seepage volume and infiltration pressure
	After a sudden change in permeate pressure, a seepage flow check should be carried out
	A steady increase in percolation without an increase in turbidity should be checked for the presence of dissolved material in the percolation

(continued)

Table 3.3 (continued)

Interlinked factors	How the relationship between these factors affects data analysis
Seismic activity/ seepage volume/ displacement/ infiltration pressure	Earthquakes can cause dams to move or deform, resulting in damage or structural instability and destruction. This movement can also cause damage to the impermeability function
	An increase in seepage, an increase in turbidity or a significant increase in infiltration pressure all indicate that the structure has been damaged
Construction process	Sudden changes in infiltration pressure or seepage volume or unexpected deformation of the structure during construction or reconstruction work indicate a potentially unsafe situation

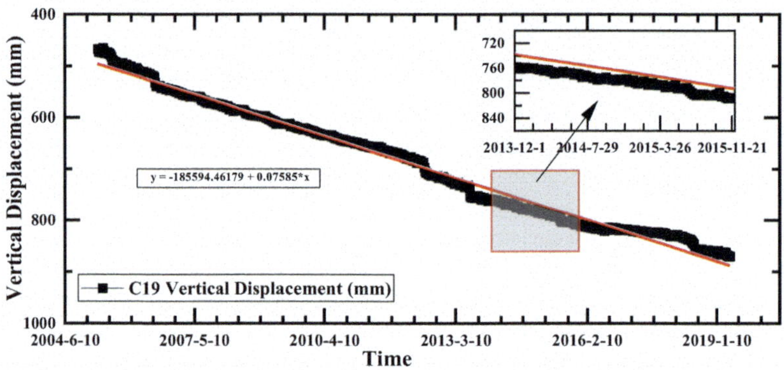

Fig. 3.11 Trend analysis of monitoring data

(c) If the effect is obvious, the factor can also be plotted as a contour (i.e. a complex correlation plot) expressing the relationship between the two environmental factors and the measured value of the effect measure.

(d) The presence of 'islands' on the correlation plot should be identified and eliminated, often due to measurement error, data recording error or changes in operating conditions.

3. **Analysis of the distribution map**

The distribution maps show the regularity of the distribution, the location of the maximum and minimum values, the size of the differences between the measurement points, especially between adjacent measurement points, and whether there are sudden changes, etc. For the distribution maps with embankment height, modulus of elasticity and geological parameters, it is important to know whether the distribution of the measured values corresponds to them and what the relationship is. For maps with multiple distribution lines for different periods of the same project, it is important to understand the evolution of the measured values. For families of distribution

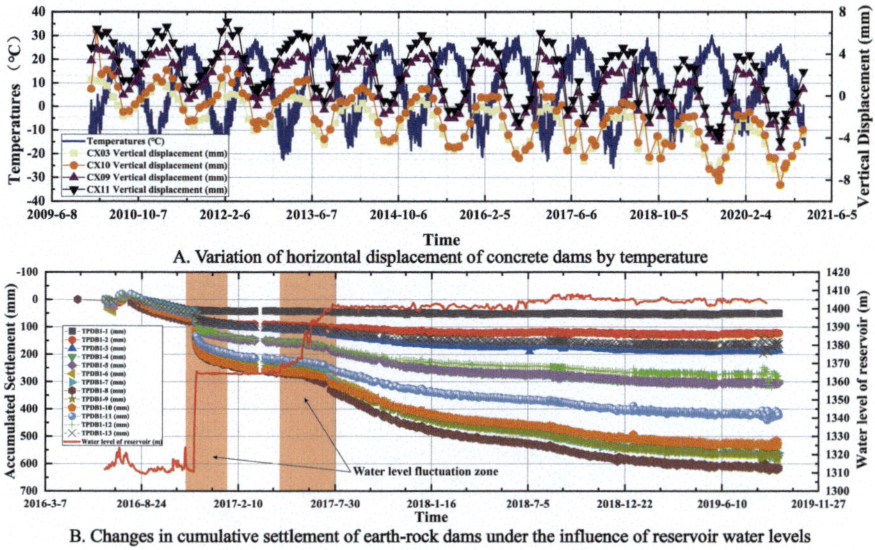

Fig. 3.12 Comparative analysis of multiple monitoring effect quantities

lines with multiple effects at the same time, the differences are compared to determine whether the effects are closely related and whether the changes are synchronous (Figs. 3.13 and 3.14).

3.4.2.3 Methods by Comparison

The comparative method compares monitored physical quantities, monitoring results with theoretical or experimental results (or curves), monitored values with designed and calculated values, monitoring indicators with predicted values, and changes and trends in the occurrence of various anomalies.

1. **Comparison with historical values**: Compare the measured values with their historical values, first with the last measured value to see if there is a continuous, gradual or abrupt change; compare with the historical maximum and minimum values to see if there is a breakthrough; compare with historical measured values under the same conditions (with similar environmental quantities acting) to see the degree of difference and the direction of the deviation (positive or negative). If there is an abrupt change in the measured value from the previous measurement, or if it breaks through the historical extremes, or if it deviates significantly from the measured value under the same conditions in the past, analyse the reasons for this.
2. **Compare with relevant information**: The information relevant to the measured value is mainly the information on the measured value of similar effects in the vicinity, the information on the measured value of other effects in the vicinity

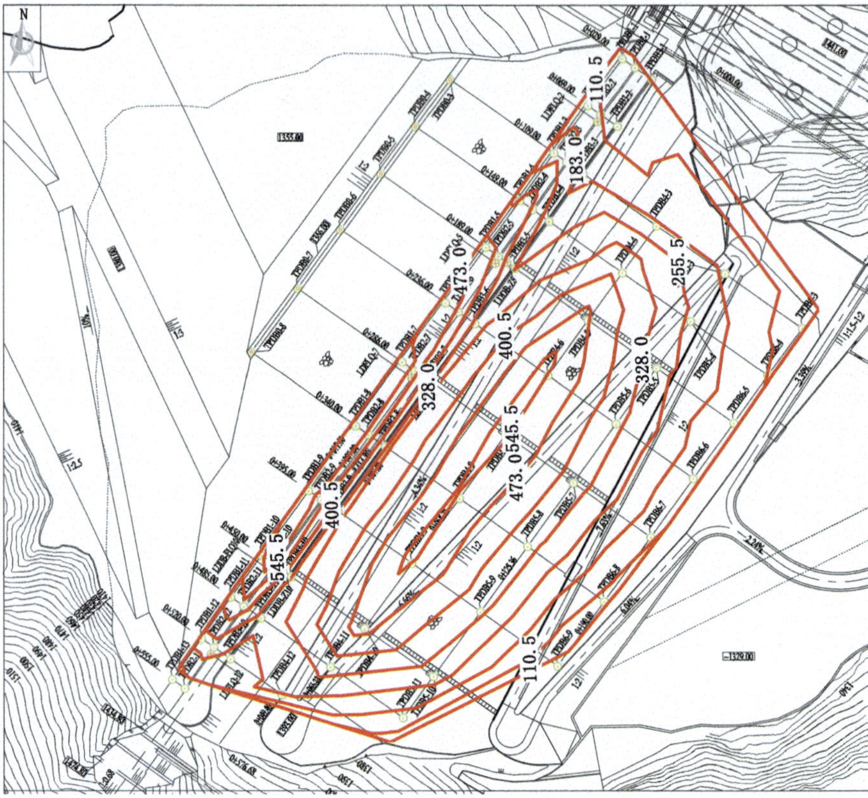

Fig. 3.13 The Contour map of dam settlement

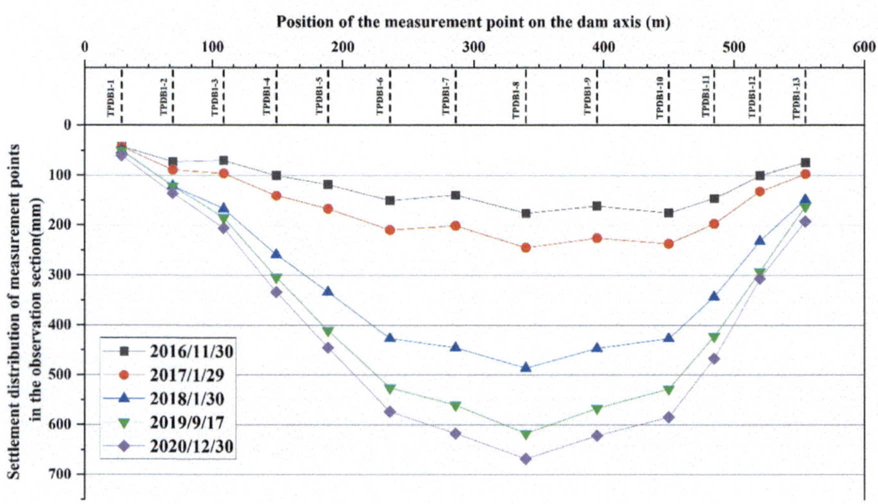

Fig. 3.14 Settlement distribution of measurement pointsin the observation section

that can corroborate each other, and the information on the measured value of the corresponding environmental variables. Comparison with similar effect measurements at neighbouring points shows whether the difference is within the normal range and whether the distribution is consistent with historical patterns. Comparison with other effect measures that can corroborate each other to see if there are any inconsistent anomalies between them. Comparison with corresponding environmental measures is mainly based on whether the change in the effect measure is consistent with the change in the environmental measure.

3. **Comparison with design and model test values**: The design and model test values are quantitatively different from the dam safety monitoring values. When comparing the values, attention should be paid to the effect of the respective operating conditions in the calculation of the design and model test baseline values and the physical quantities for dam safety monitoring, and the corresponding transformation process to make the two comparable.
4. **Comparison with specified safety monitoring values and predicted values**: Compare the monitoring values with the corresponding safety monitoring values to see if the measured values exceed the limits; compare with the predicted values to see if the difference is on the safe or dangerous side, etc.

3.4.2.4 Mathematic Model

Mathmatical model method includes statistical model, deterministic model and mixed model [34]. By using mathmatical models, quantitative analysis is carried out on monitoring data throuhg establishing relation between effect quantities like displacement, seepage pressure, seepage flow, etc. and cause quantities as reservoir water levl, temperature, ageing, etc. By establishing a mathematical relationship between effect and cause quantities, the effects of different stresses are separated, the degree and pattern of influence of each cause quantity on the effect quantity is understood, and potential unsafe factors and trends are identified.

Statistical models are commonly used when monitoring data are available for a long period of time. Mixed model or deterministic model can also be used when the deterministic relationship between effect quantity and cause quantity is obtained conditionally (usually by finite element calculation) [35].

The mathematical model of operation period includes three parts: water pressure component, temperature component and time component. The change form of time-dependent component is an important basis for evaluating whether the effect quantity is normal or not, and the cause of abnormal change needs to be found out as soon as possible [36].

3.4.3 Mathematical Model Analysis Methods for Dam Safety Monitoring Information

The monitoring data is quantified by establishing a functional relationship between the monitoring effect quantities (e.g. displacement, seepage, head pressure, pressure tube level, etc.) and the cause quantities (e.g. reservoir level, air temperature, time duration, etc.). The mathematical model for the operating period contains the components of the cause quantity (e.g. water pressure component, temperature component and time effect component), of which the variation pattern of the time effect component is an important basis for assessing whether the effect quantity is normal or not. For abnormal changes, the causes must be identified [32, 33].

3.4.3.1 Statistical Modelling of Deformation Monitoring Volumes

Deformation refers to displacements and the opening and closing of surface cracks etc. Displacement is divided into horizontal displacement in the downstream direction δx (or radial displacement δr), horizontal displacement in the lateral direction δy (or tangential displacement $\delta\theta$) and vertical displacement δz, and openness in the three-way direction *Kx*, *Ky*, *Kz*. By dam type, concrete dams (including gravity dams, arch dams, abutment dams or continuous arch dams), earth and rock dams, etc. are classified [37].

1. **Statistical model for concrete dam deformation measurement points**

Under loads such as water pressure, lift pressure, sediment pressure and temperature, any point of the dam produces a displacement vector δ, which can be decomposed into horizontal displacement δx, lateral horizontal displacement δz and vertical displacement δy. According to its causes, the displacement can be divided into three components: the water pressure component (δ_H), the temperature component (δ_T) and the ageing component (δ_θ).

$$\delta\left(\delta_x \text{ and } \delta_y \text{ and } \delta_z\right) = \delta_H + \delta_T + \delta_\theta \tag{3.4.1}$$

Some dams produce a large range of horizontal cracks in the downstream face, which also have an effect on the displacement, and how the effect of the cracks is considered requires an additional crack displacement component (δ_J).

$$\delta\left(\delta_x \text{ and } \delta_y \text{ and } \delta_z\right) = \delta_H + \delta_T + \delta_\theta + \delta_J \tag{3.4.2}$$

The mathematical expression for the water pressure component δ_H can be summarised in Table 3.4. The mathematical expressions for the temperature displacement components are summarised in Table 3.5.

Table 3.4 Mathematical expressions for the water pressure component

The dam type	Reservoir water pressure	Dam base lifting pressure	Dam lifting pressure
Gravity dams	$\sum_{i=1}^{3} a_i H^i$	$a_f \Delta \overline{H}_j$	$a_b \left(\Delta \overline{H}_j\right)^2$
Arch dams	$\sum_{i=1}^{4(5)} a_{1i} H^i + \sum_{i=1}^{3} a_{2i} H_1^i$	$a_f \Delta \overline{H}_j$	$a_b \left(\Delta \overline{H}_j\right)^2$

H, H_1, ΔH_j are the upstream water depth at the time of monitoring, the downstream water depth and the difference between the reservoir level at the time of monitoring and the average reservoir level on day j before monitoring, respectively. a_i, a_{1i}, a_{2i}, a_f, a_b are the fitting coefficients

Table 3.5 Mathematical expressions for the temperature displacement component

The situation	Cyclical items	Depth reflects water temperature	Air temperature reflects water temperature
Only air temperature information	$\sum_{i=1}^{m} \left(b_{1i} \sin \frac{2\pi it}{365} + b_{2i} \cos \frac{2\pi it}{365}\right)$	$\sum_{i=1}^{3} b_i H^i$	$\sum_{i=1}^{m_1} b_i T_i$
Concrete temperature information available	$\sum_{i=1}^{m_2} b_i T_i$ and $\sum_{i=1}^{m_3} b_{1i} \overline{T}_i + \sum_{i=1}^{m_3} b_{2i} \beta_i$		

m, m_1, m_2 and m_3 are the number of periods, the number of temperature factors, the number of thermometer branches and the number of equivalent temperature layer layers respectively; b_{1i}, b_{2i}, b_i are the fitting coefficients

According to a large number of monitoring data analysis shows that the dam body and dam foundation under long-term load produces a time-dependent displacement, its influencing factors are extremely complex, comprehensively reflect the dam body concrete creep and bedrock creep, plastic deformation, dam body cracks and dam foundation fracture caused by irreversible displacement and its own volume deformation and so on. It generally varies greatly during storage and early operation and stabilises later. As δ_θ reflects the change in deformation and stability of the dam and its foundation, it is an important basis for analyzing and evaluating the safety of dams. Based on a large amount of monitoring data analysis results, its simulation when the utility of exponential function, hyperbolic function, polynomial, logarithmic function and linear function, etc., commonly used linear term and logarithmic term expressed.

$$\delta_\theta = c_1 \theta + c_2 \ln \theta \tag{3.4.3}$$

where

c_1 and c_2 the regression coefficients for the linear and logarithmic terms, respectively;

θ the number of days from the observed displacement to the starting date, and θ is usually divided by 100 to reduce the numerical volume.

Some cracks (e.g. transverse, longitudinal and horizontal cracks) vary regularly with external loads and may also affect the displacement of the dam, for which the measured value of the opening and closing of these crack gauges is used as a factor.

$$\delta_J = \sum_{i=1}^{m_4} d_i J_i \tag{3.4.4}$$

where

d_i the regression coefficient;
J_i the openness of the seamometer, where δx, δy and δz are the openness measurements in the x, y and z directions respectively.

2. **Statistical modeling of earth and rock dam deformation measurement points**

The deformation of earth and rock dams is divided into three components: vertical displacement (i.e. settlement), horizontal displacement in the upstream and downstream directions, and displacement along the dam axis. As earth and rock dams are loose granular materials, the vertical displacements are much larger than those of concrete dams under load, and the magnitude and time effect of the displacements are important influences on the safety of the structure and the prevention of cracking.

Settlement in earth and rock dams is mainly caused by consolidation, but is also influenced by factors such as reservoir level and temperature, with consolidation-induced settlement also reflecting the characteristics of time effects. Settlement during operation is

$$\delta_z = \delta_H + \delta_T + \delta_\theta \tag{3.4.5}$$

Water pressure component (δ_H): After water storage, the dam body is under the action of water pressure, resulting in a seepage field whose vertical infiltration force causes vertical displacement and horizontal infiltration force causes horizontal displacement; the capacitance of the soil below the infiltration line becomes a buoyant capacitance, reducing the vertical displacement and causing wetting deformation. Thus, the vertical displacement caused by the above factors is a complex function related to the upstream water level H.

$$\delta_{H_1} = f(H) \tag{3.4.6}$$

Applying a Taylor series expansion to the formula and ignoring the higher order terms gives the following results.

$$\delta_{H_1} = \sum_{i=1}^{3} a_{1i} H^i \tag{3.4.7}$$

At the same time, as the seepage from the dam lags behind the reservoir level, the average reservoir level of the previous i days is used as a factor to account for the effect of the lag.

$$\delta_{H_z} = \sum_{i=1}^{m_1} a_{2i}\overline{H}_i \tag{3.4.8}$$

The water pressure component (δ_H) is

$$\delta_H = \sum_{i=1}^{3} a_{1i}H^i + \sum_{i=1}^{m_1} a_{2i}\overline{H}_i \tag{3.4.9}$$

Temperature component ($\boldsymbol{\delta_T}$): The variation in linear soil swelling caused by temperature changes is small, but the effect on subsidence caused by frost swelling at low temperatures in alpine regions is more significant. In general, the freezing period varies mainly in an annual cycle as well as in relation to negative temperature values. For this reason, the temperature component is expressed as the average temperature and the period of *i* days when subsidence is observed.

$$\delta_T = \sum_{i=1}^{m_2} b_i T_i + \sum_{i=1}^{m_3} \left(b_{1i} \cos \frac{2\pi it}{365} + b_{2i} \sin \frac{2\pi it}{365} \right) \tag{3.4.10}$$

where

b_i, b_{1i}, b_{2i} regression coefficients;
T_i the temperature of the observation day and the average temperature of the previous *i* days;
m_2, m_3 the number of factors taken, respectively, generally taken as 9–10.

Ageing component ($\boldsymbol{\delta_\theta}$): The ageing component is the main component of earth and rock dams, and the ageing during the operating period is mainly caused by subconsolidation, so using consolidation theory, we deduce that δ_θ is:

$$\delta_\theta = \begin{cases} c_1\theta + c_2 \ln\theta \\ \theta/(c_1\theta + c) \\ \sum_{i=0}^{n} c_i\theta^i \\ c_1 e^{c_2/\theta} \end{cases} \tag{3.4.11}$$

where

c_1, c_2, c_i regression coefficients;
θ the number of days from the observed displacement to the starting date, and θ is usually divided by 100 to reduce the numerical volume.

In addition to subsidence, horizontal displacement will also occur, such as horizontal displacement due to water pressure or the horizontal component of the infiltration force; horizontal displacement due to lateral expansion in the process of soil consolidation and sub-consolidation; soil wetting; horizontal displacement of the upstream slope that faces upstream, turns downstream when the reservoir level is high and persists for a period of time, and turns upstream again when the reservoir level drops. As earth and rock dams are loose grained, irreversible deformation accounts for most of the horizontal displacement. In summary, the horizontal displacement of earth and rock dams is influenced by time, water level and temperature, and the expressions of the statistical model are the same as those of the statistical model of subsidence.

3.4.3.2 Statistical Modeling of Seepage Monitoring Volumes

Percolation observations include pore water pressure, dam base uplift pressure, percolation pressure coefficient and percolation volume for concrete dams, and percolation line and percolation volume for earth and rock dams.

1. **Statistical model for seepage pressure in concrete dams**

The seepage pressure of the dam P is related to water pressure P_H, temperature P_T and ageing P_θ.

$$P = P_H + P_T + P_\theta \tag{3.4.12}$$

Factor expressions for each component were obtained after analysis.

$$P = a_0 + \sum_{i=1}^{4} a_i H^i + b_1 T + c_1\theta + c_2 \ln\theta \tag{3.4.13}$$

2. **Statistical model for lifting pressure in concrete dam foundation gauges**

Based on the results of the analysis of a large number of head pressure observations, the head pressure at the base of the dam Y (i.e. the water level in the head pressure measurement hole) is mainly related to the influence of upstream Y_{Hu} and downstream Y_{Hd} water levels, rainfall Y_p, temperature Y_T and timing Y_θ.

$$Y = Y_{Hu} + Y_{Hd} + Y_p + Y_T + Y_\theta \tag{3.4.14}$$

Upstream water level component Y_{Hu}: According to seepage theory and analysis of measured data, the lift pressure lags behind the upstream water level. For this reason, there are two treatment methods.

(1) the average reservoir level (H_{ui}) on the i days prior to the measured value is used as a factor.

$$Y_{Hu} = \sum_{i=1}^{m_1} a_{ui}\overline{H}_{ui} \quad (i = 1, 2, 5, 10, 15, \ldots, m_1) \tag{3.4.15}$$

(2) Using the effective water level as a factor, the effective water level is derived from the curve of the effect of the upstream water level on the head pressure.

$$Y_{Hu} = a_1 \int_{-\infty}^{0} \frac{1}{\sqrt{2\pi}x_2} e^{\frac{-(t-x_1)^2}{2x_2^2}} H_i(t)dt = a_{1u}H_{ium} \tag{3.4.16}$$

where

- $H_i(t)$ the upstream water level at time t;
- H_{um} the effective upstream water level;
- x_1 the number of days lagged upstream water level;
- x_2 the standard deviation of the normal distribution of upstream water levels, i.e. the number of days of influence;
- a_{1u} the regression coefficient.

Downstream water level component Y_{Hd}: The downstream water level generally varies little and the number of measurements is small. For this reason the current day's measurement (H_d) is taken as a factor.

$$Y_{Hd} = a_d H_d \tag{3.4.17}$$

The rainfall component Y_p: The buoyancy pressure at the base of the dam section on both banks is influenced by the groundwater on both banks, which in turn is mainly influenced by rainfall. As infiltration is a time-dependent process, the buoyancy pressure lags behind the rainfall. For this reason, two treatment methods are used:

(1) The average rainfall of the previous period before the measurement ($\overline{p}_i$, $i = 1$, 2, 5, 10, …, m_2) was used as a factor.

$$Y_p = \sum_{i=1}^{m_2} d_i\overline{p}_i \tag{3.4.18}$$

(2) Effective rainfall (p') is used as a factor.

$$Y_p = d_1 p' \tag{3.4.19}$$

$$p' = \int_{-\infty}^{\infty} \frac{1}{\sqrt{2\pi}x_4} e^{\frac{-(t-x_3)^2}{2x_4^2}} (p(t))^{\frac{2}{5}} dt \tag{3.4.20}$$

where

$p(t)$ the rainfall per unit time period at moment t;
x_3 the number of days that rainfall lags behind the lifting pressure;
x_4 the standard deviation of the normal distribution of rainfall, i.e. the number of days of influence.

Temperature component Y_T: The bedrock temperature variation is small, generally in annual cycles, and no measured temperature information is available. For this reason it is expressed as a periodic function.

$$Y_T = \sum_{i=1}^{m_3} \left(b_{1i} \sin \frac{2\pi it}{365} + b_{2i} \cos \frac{2\pi it}{365} \right) \quad i = 1, 2 \tag{3.4.21}$$

The time component Y_θ: Based on the analysis of the measured data, the expression of Y_θ is:

$$Y_\theta = c_1\theta + c_2 \ln \theta \tag{3.4.22}$$

3. **Statistical model of the total lift pressure at the base of a concrete dam**

In order to grasp the total lift pressure at the base of the dam, more than three pressure measurement holes are arranged at the base of the dam, so that the total lift pressure is force is

$$U_i = \sum_{i=1}^{m+1} \frac{(H_i + H_{i+1})}{2} \Delta b_i \tag{3.4.23}$$

where

H_i the water level (lift pressure) at the ith borehole, upstream and downstream respectively at the heel and toe of the dam;
Δb_i the distance between the two boreholes;
m the number of measurement holes.

4. **Statistical modeling of seepage pressure coefficients at the base of concrete dams**

The seepage pressure coefficients α_1 and α_2 after curtain and drainage are the basis for the seepage control effect of the curtain and the pressure reduction effect of the drainage pipe, respectively. The seepage pressure coefficient is calculated as the formula for calculating the seepage pressure coefficient is

$$\alpha = \frac{H_i - H_d}{H_u - H_d} \tag{3.4.24}$$

where

H_i the measured value of the lift pressure behind the curtain or drain;
H_u the upstream water level;
H_d the downstream water level, or if the downstream water level is lower than the built-up surface elevation, the built-up surface elevation is used.

5. **Statistical model for seepage from concrete dams**

According to the analysis of a large amount of monitoring data and seepage theory, the amount of seepage Q from the concrete dam and the dam base is mainly influenced by the upstream and downstream water levels (Q_{Hu}, Q_{Hd}), temperature Q_T and timing Q_θ.

$$Q = Q_{Hu} + Q_{Hd} + Q_T + Q_\theta \tag{3.4.25}$$

According to the analysis of the rock seepage field, the seepage at the base of the dam is related to the primary and quadratic side of the water depth upstream and to the downstream the primary side of the water depth.

$$Q_H = Q_{Hu} + Q_{Hd} = \sum_{i=1}^{2} a_{ui} H_u^i + \sum_{i=3}^{m_1} a_{ui} \overline{H}_{ui} + a_d H_d \tag{3.4.26}$$

where

H_u the upstream water depth on the observation day;
H_{ui} the average upstream water depth on day i of the observation day;
m_1 the number of days lagged;
H_d the depth downstream.

$$Q_T = \sum_{i=1}^{2} \left(b_{1i} \sin \frac{2\pi it}{365} + b_{2i} \cos \frac{2\pi it}{365} \right) \text{ and } Q_\theta = c_1\theta + c_2 \ln \theta \tag{3.4.27}$$

6. **Statistical model of water levels in earth and rock dam gauges**

According to the analysis of measured data and seepage theory, the water level in the gauging holes of earth-rock dams h is mainly influenced by the upstream and downstream water level (h_u, h_d), rainfall h_p and timing h_θ.

$$h = h_u + h_d + h_p + h_\theta \tag{3.4.28}$$

The extent of the influence of the upstream water level is obtained from the analysis of the seepage theory of earth and rock dams.

$$h_u = \sum_{i=1}^{m_1} a_{ui}\bar{h}_{ui} \tag{3.4.29}$$

where

m_1 the number of days before, generally taken as the first 1, 2, 3, 4 ~ 7 days.

The amount of influence of the downstream water level is taken as the downstream water level h_d on the day of observation.

$$h_d = a_d h_d \tag{3.4.30}$$

The rainfall component is taken as a factor from the previous rainfall. The average of the rainfall on the day of the observation, the previous 1 day and the previous 2 days is generally taken.

$$h_p = \sum_{i=1}^{m_2} d_i \bar{p}_i \tag{3.4.31}$$

$$h_\theta = c_1\theta + c_2 \ln\theta \tag{3.4.32}$$

7. **Statistical modeling of seepage volumes in earth and rock dams**

The factors influencing the seepage volume of earth and rock dams are the same as the water level in the infiltration line gauge hole, and the statistical model is

$$\begin{aligned} Q &= Q_{Hu} + Q_{Hd} + Q_p + Q_\theta \\ &= \sum_{i=1}^{2} a_{ui} H_u^i + \sum_{i=1}^{m_1} a_{ui}\overline{H}_u + a_d H_d + \sum_{i=1}^{m_2} d_i \bar{p}_i + c_1\theta + c_2 \ln\theta \end{aligned} \tag{3.4.33}$$

3.4.4 *Methodology for the Development of Early Warning Indicators for the Safe Operation of Reservoirs and Dams*

The purpose of determining the early warning indicators for the safe operation of reservoir dams is to select suitable safety warning items for the key parts of reservoir dams, select the appropriate research and evaluation threshold formulation method, and determine the early warning indicators and criteria for the safe operation of each reservoir dam, in order to identify the potential risks of dams and achieve rapid early warning for the safe operation of reservoir dams [38].

The determination of early warning indicators for the safe operation of reservoir dams is based on the structure, geological conditions and operating environment of each dam, the results of comprehensive analysis of dam design, construction and operation data and safety monitoring data, the identification of key areas of concern, and the selection of appropriate safety warning items such as deformation and seepage [39, 40].

3.4.4.1 Comprehensive Comparison Method

A comprehensive analysis and comparison of the monitoring effect is a basic method of determining whether the structural condition is normal or not. It is important to monitor whether the magnitude, range and rate of change of the effector are consistent with historical measurements and the pattern of change; whether the trend, magnitude and rate of change of the temporal part of the effector indicate abnormal changes in the structure or foundation.

3.4.4.2 Confidence Interval Method

The confidence interval method is based on a sequence of effect quantity monitoring by establishing a mathematical model of "effect quantity ~ load" to monitor whether the difference between the measured value and the fitted value falls within the proposed confidence band to determine whether the service condition of the dam has changed [41]. The confidence interval method is a fault-free monitoring indicator for effect quantity monitoring, which is a method using statistical theory to handle small probability events to obtain the boundary between normal and abnormal monitoring values by establishing a mathematical model for effect quantity monitoring to fit the measured values, and when the fitting accuracy reaches a certain standard, the monitoring indicator Y_m, i.e.:

$$Y_m = Y \pm \Delta \tag{3.4.34}$$

where

Y the monitoring model for monitoring the effect size;
Δ the confidence interval half-band width.

As can be seen from Eq. (3.4.24), the confidence interval method transforms the problem of formulating monitoring indicators into the problem of determining a monitoring model for monitoring effect sizes; the problem of evaluating the safety status of a dam using monitoring indicators is transformed into the problem of evaluating the safety status of a dam using the predicted values of effect sizes. Therefore, to ensure that the monitoring indicators are correct and valid, it is necessary to establish a good quality mathematical model [42]. This requires sufficient accuracy of the monitoring data and sufficient number of measurements to avoid large residual standard deviations caused by poor monitoring accuracy or too few measurements, which affect the accuracy of the model; at the same time, the correct model factors should be selected to fully extract the contribution of the factors to the monitoring quantity, otherwise the unextracted part will be attributed to the residuals, resulting in excessive residuals.

Also, the selection of the confidence interval half bandwidth is a key issue in the confidence interval method. Δ is too small and easy to false positives and too large is easy to miss. Δ value is

$$\Delta = i\sigma \tag{3.4.35}$$

where

σ the standard deviation of the monitoring model;
I a natural number, determined by the significance level *a*.

a indicates the probability of treating a normal value as an abnormal value as such an error, when $a = 1\%$, $i = 2.567$; when $a = 5\%$, $i = 1.96$. $1 - a$ then indicates the probability that a forecast value will occur, called the guarantee rate, when $i = 1$, $1 - a = 68.3\%$; when $i = 2$, $1 - a = 95.4\%$; when $i = 3$, $1 - a = 99.7\%$. It can be seen that the guarantee rate corresponding to $i = 1$ is only 68.3%, which is far from the actual application requirements of the project. Therefore, the value of *i* should not be too small and should be combined with the actual situation of the project when choosing. Usually if the monitoring items are more reliable and important, such as concrete dam displacement, deflection, foundation subsidence, etc., take $i = 2$, corresponding to $1 - a = 95.4\%$; if the monitoring items have strong ability to indicate the danger, more sensitive and reliable observation, such as the sliding of the dam shoulder, dam base, etc. along the soft sandwich, dam crack opening, etc., take $i = 1 \sim 1.5$; if the monitoring items are discrete and strong, low reliability, if the monitoring items are discrete and less reliable, take $i = 2.5$.

In addition, the confidence interval method is based on the assumption that the dam is in normal and stable operation. If the trend of the monitoring values varies significantly and is not yet stable, it is not appropriate to establish a mathematical monitoring model of the effector quantities, nor is it appropriate to use the confidence interval method to develop monitoring indicators [43]. Therefore, to correctly and reasonably model the time-dependent deformation of the dam, when the time-dependent component is stable, the monitoring indicators developed by the confidence interval method can be used for the safety monitoring of the dam.

The confidence interval method is based on an interval $[Y - \Delta, Y + \Delta]$. If the measured displacement value is within the monitoring interval $[Y - \Delta, Y + \Delta]$ and there is no significant trend change in the measured value, the dam is considered to be operating normally or safely. If the measured value exceeds the monitoring interval or if there is a trend, the dam should be analyzed immediately to find the cause and take measures to solve the problem.

3.4.4.3 Typical Small Probability Method

The small probability method takes the measured sequence of the effective volume of the dam under the action of the adverse load combination as the sample space, selects a reasonable probability distribution model (normal and lognormal distribution, gamma distribution, Weibull distribution, etc.) to represent the sample space, and obtains the effective volume safety warning value by formulating the probability of failure of the service state of the dam [44, 45].

Depending on the type of dam and its specific situation, the monitoring effect quantity for adverse load combinations or the individual load components of its mathematical model (e.g. water pressure component, temperature component, time-dependent component, etc.) is selected from previous monitoring data as the typical monitoring effect quantity E_{mi}. Obviously, E_{mi} is a random variable and a sub-sample is selected each year to obtain a small sub-sample space, i.e.:

$$E = \{E_{m1}, E_{m2}, \ldots, E_{nm}\} \tag{3.4.36}$$

$$\overline{E} = \frac{1}{n}\sum_{i=1}^{n} E_{mi} \tag{3.4.37}$$

$$\sigma_E = \sqrt{\frac{1}{n-1}\left(\sum_{i=1}^{n} E_{mi}^2 - n\overline{E}^2\right)} \tag{3.4.38}$$

where

$\overline{E}$ the first-order moment of origin in the sample space;
σ_E is an unbiased estimate of the standard deviation of the sample space;
n the number of samples in the sample space.

Calculate the numerical eigenvalues of the small sample space E. Based on the statistical properties of the random variable, assume its probability density function $f(E)$ and distribution type, and determine the distribution function $F(E)$ by testing with the small sample statistical test method.

Let E_m be the extreme value of the monitoring effect or a load distribution, i.e. the safety monitoring indicator, and determine the probability of failure $P_a(a)$ based on the importance of the dam. If $E > E_m$, the dam is about to experience an abnormal or dangerous condition, and the probability of failure based on $f(E)$ and $F(E)$ of the determined typical monitoring effect is

$$P_\alpha = P(E > E_m) = \int_{E_m}^{\infty} f(E)dE \tag{3.4.39}$$

The distribution function of Emi is used to find the monitoring indicator E_m under the corresponding crash probability a. If E_{mi} is the individual components of the monitoring effect quantity, then the individual components are iterated to become the safety monitoring indicator.

$$E_m = F^{-1}(E, \alpha) \tag{3.4.40}$$

Depending on the individual dam, the monitoring effect quantity E_{mi} or its individual components are selected according to the load combinations that are unfavourable for strength and stability [46].

3.4.4.4 Structural Analysis Method

The structural properties of a dam can be divided into 3 stages: elastic, elasto-plastic and destabilisation damage [47–49]. Therefore, the monitoring indicators can also be divided into one, two and three stages.

(1) **The control criteria for Tier 1 indicators**

$$\text{Strength}\, \sigma_t \leq [\sigma_t], \sigma_s \leq [\sigma_s], \text{ stabilization}\, K \geq [K] \tag{3.4.41}$$

(2) **The control criteria for Tier 2 indicators**

$$\begin{aligned}&\text{Strength}\, \sigma_t \leq \sigma_{tR}, \sigma_z \leq \sigma_{zR}, \text{ stabilization}\, F \leq R_{m2},\\&\text{crack resistance}\, Kc \leq [Kc]\end{aligned}$$

(3) **The control criteria for Tier 3 indicators**

$$\text{Strength}\, \sigma_t \leq \sigma_{\text{im}}, \sigma_s \leq \sigma_{\text{sm}}, \text{ stabilization}\, F \leq R_{m3} \tag{3.4.42}$$

where

σ_t and σ_s — the calculated tensile and compressive stresses, respectively, at the monitoring point under adverse load conditions;

$[\sigma_t]$, σ_{tR} and σ_{tm} — the allowable tensile stress, the tensile yield strength and the yield strength limit of the material, respectively;

σ_z, σ_{zR} and σ_{zm} — the allowable compressive strength, the compressive yield strength and the yield strength limit of the material, respectively;

F — the slip force on the slip cracking surface corresponding to the adverse load combination;

R_{m2} and R_{m3}	the sliding resistance when the shear strength (f, c) of the material on the corresponding sliding surface is at its yield or strength limit, respectively;
K and $[K]$	the sliding stability factor of safety and the allowable sliding stability factor of safety for the corresponding adverse load combination, respectively;
K_c and $[K_c]$	the stress intensity factor and the fracture toughness of the crack tip, respectively.

3.5 Evaluation on Dam Safety Performance

Based on the results of the analysis of the deformation, seepage and pressure (stress) monitoring data, combined with the inspection data, a qualitative assessment of the overall safety status of the reservoir dam is made to identify the key parameters and key areas of concern for dam safety.

In the analysis on dam safety monitoring data, clear conclusions shall be drawn on:

- Whether the deformation of the dam accords with the general pattern and becomes stable; whether the seepage field of the dam is stable, e.g. whether the seepage line (surface) of the earth-rock fill dam or the uplift pressure of the concrete dam foundation is normal; whether the stress (pressure), strain of the dam is smaller than the allowed value of the design and the specification. Based on that, the dam safety performance is comprehensively assessed.
- Walkaround inspection or the monitoring data shall reflect the location, attribution and characteristic of the abnormality of dam safety performance, and the associated occurrence time and operation condition, as well as the process and effect of the disposal of the abnormality.
- Suggestions for improvement of monitoring instruments, methods, observation frequency, etc. are proposed based on the problems existing during the monitoring process.
- Hidden troubles which may undermine the dam safety and the causes thereof are pointed out based on the results of analysis of monitoring data and suggestions to improve the dam operation management, maintenance and reinforcement are proposed.

Classification of the dam safety performance according to the results of analysis of monitoring data shall accord with the following principles:

- When the pattern of change of all the monitoring data is normal, the measured values fall within empirical value and the allowed range of specification, design and test and there is no abnormality during the operation, the dam safety performance can be considered as **normal**.

- When partial monitoring data shows a varying tendency, but the values still fall within empirical value and the allowed range of specification, design and test, the dam safety performance can be considered as **basically normal**.
- When the monitoring data shows an obvious tendency of variation in the direction prejudicial to safety of the dam, or the measured values show sudden changes, or exceed the threshold or empirical value and the allowed range of specification, design and testing, the dam safety performance can be considered as **abnormal**.

A qualitative evaluation on whether the dam safety performance is normal can be carried out in a whole through analysis of monitoring data. Whether the dam is safe or not shall be determined also on check calculation of seepage safety, structure safety and seismic safety.

References

1. Jiang C, Xiao C (2021) Detailed analysis and countermeasure research on the present situation of reservoir dam safety monitoring in China. Hydro-Sci Eng 6:97–102 (in Chinese)
2. Zheng S, Shao C, Gu C et al (2022) An automatic data process line identification method for dam safety monitoring data outlier detection. Struct Contr Health Monit 29(7):e2948. https://doi.org/10.1002/stc.2948
3. Wang S, Gu Y, Ge C (2019) Technical framework for dam safety monitoring system evaluation. Hydro-Sci Eng 176(4):63–67. https://doi.org/10.16198/j.cnki.1009-640x.2019.04.009
4. Li B, Yang J, Hu D (2019) Dam monitoring data analysis methods: a literature review. Struct Control Health Monit 27. https://doi.org/10.1002/stc.2501
5. Adamo N, Al-Ansari N, Sissakian V et al (2020) Dam safety: use of instrumentation in dams. J Earth Sci Geotech Eng 145–202. https://doi.org/10.47260/jesge/1115
6. On C (1983) Guidelines for dam instrumentation and monitoring systems, May 1983. Australian National Committee on Large Dams, Brisbane
7. Han Z, Li Y, Zhao Z, Zhang B (2022) An online safety monitoring system of hydropower station based on expert system. Energy Rep 8:1552–1567. https://doi.org/10.1016/j.egyr.2022.02.040
8. Avci O, Abdeljaber O, Kiranyaz S et al (2021) A review of vibration-based damage detection in civil structures: from traditional methods to Machine Learning and Deep Learning applications. Mech Syst Signal Process 147:107077. https://doi.org/10.1016/j.ymssp.2020.107077
9. Lixia Ji LJ, Lixia Ji XZ, Xiao Zhang YZ, Yao Zhao ZL (2022) Anomaly detection of dam monitoring data based on improved spectral clustering. J Internet Technol 23:749–759. https://doi.org/10.53106/160792642022072304010
10. He Q, Gu C, Valente S et al (2022) Multi-arch dam safety evaluation based on statistical analysis and numerical simulation. Sci Rep 12. https://doi.org/10.1038/s41598-022-13073-9
11. Gantz SD, Philpott DR (2013) Continuous monitoring. In: FISMA and the risk management framework, pp 367–401. https://doi.org/10.1016/b978-1-59-749641-4.00014-x
12. Dou G, Chen R, Han C et al (2022) Research on water-level recognition method based on image processing and convolutional neural networks. Water 14:1890. https://doi.org/10.3390/w14121890
13. Younos TM, Heyer CJ (2015) Advances in water sensor technologies and real-time water monitoring. In: The handbook of environmental chemistry, pp 171–203. https://doi.org/10.1007/978-3-319-14212-8_7
14. Encardio Rite (2022) Seepage monitoring: needs, methods & solutions. In: Encardio Rite. https://www.encardio.com/blog/seepage-monitoring. Accessed 16 Sept 2023

15. Alam MM, Bhutta MN (2004) Comparative evaluation of canal seepage investigation techniques. Agric Water Manag 66:65–76. https://doi.org/10.1016/j.agwat.2003.08.002
16. Jia D, Zhou Y, He X et al (2021) Vertical and horizontal displacements of a reservoir slope due to slope aging effect, rainfall, and reservoir water. Geodesy Geodyn 12:266–278. https://doi.org/10.1016/j.geog.2021.06.004
17. Lienhart W (2017) Geotechnical monitoring using total stations and laser scanners: critical aspects and solutions. J Civ Struct Health Monit 7:315–324. https://doi.org/10.1007/s13349-017-0228-5
18. Zhou J, Shi B, Liu G, Ju S (2021) Accuracy analysis of dam deformation monitoring and correction of refraction with robotic total station. PLoS ONE 16:e0251281. https://doi.org/10.1371/journal.pone.0251281
19. Jänichen J, Schmullius C, Baade J et al (2022) Monitoring of radial deformations of a gravity dam using sentinel-1 persistent scatterer interferometry. Remote Sens 14:1112. https://doi.org/10.3390/rs14051112
20. Huang P, Guan J-C (2022) Stress and strain behavior monitoring of concrete through electromechanical impedance using piezoelectric cement sensor and PZT sensor. Constr Build Mater 324:126685. https://doi.org/10.1016/j.conbuildmat.2022.126685
21. Wu C, Xiang H, Jiang S-F, Ma S (2022) Stress monitoring of concrete via uniaxial piezoelectric sensor. Sensors 22:4041. https://doi.org/10.3390/s22114041
22. Wang F-M, Tulamaiti Y, Fang H et al (2023) Seismic response characteristics of polymer anti-seepage wall in earth dam based on earthquake wave motion input method. Structures 47:358–373. https://doi.org/10.1016/j.istruc.2022.11.060
23. Ma C, Xu X, Yang J, Cheng L (2023) Safety monitoring and management of reservoir and dams. Water 15:1078. https://doi.org/10.3390/w15061078
24. Ren Q, Li M, Kong T, Ma J (2022) Multi-sensor real-time monitoring of dam behavior using self-adaptive online sequential learning. Autom Constr 140:104365. https://doi.org/10.1016/j.autcon.2022.104365
25. Nelson WB (2021) Statistical methods for reliability data, second edition. Technometrics 63:437–440. https://doi.org/10.1080/00401706.2021.1945328
26. Laiba S (2022) Seepage monitoring: needs, methods & solutions. In: Encardio Rite. https://www.encardio.com/blog/seepage-monitoring. Accessed 16 Sept 2023
27. Prakash G, Dugalam R, Barbosh M, Sadhu A (2022) Recent advancement of concrete dam health monitoring technology: a systematic literature review. Structures 44:766–784. https://doi.org/10.1016/j.istruc.2022.08.021
28. Bukenya P, Moyo P, Beushausen H, Oosthuizen C (2014) Health monitoring of concrete dams: a literature review. J Civ Struct Health Monit 4:235–244. https://doi.org/10.1007/s13349-014-0079-2
29. Hu J, Wu SH (2019) Statistical modeling for deformation analysis of concrete arch dams with influential horizontal cracks. Struct Health Monit 18:546–562. https://doi.org/10.1177/1475921718760309
30. Su H, Wen Z, Sun X, Yan X (2017) Multisource information fusion-based approach diagnosing structural behavior of dam engineering. Struct Control Health Monit 25:e2073. https://doi.org/10.1002/stc.2073
31. Civil, Hydraulic, and Architectural Engineering Project Team (2022) Civil, Hydraulic, and Architectural Engineering Front 2022. Eng Front 6(1):124–149
32. Yang X, Xiang Y, Shen G, Sun M (2022) A combination model for displacement interval prediction of concrete dams based on residual estimation. Sustainability 14:16025. https://doi.org/10.3390/su142316025
33. Salazar F, Morán RA, Angel M, Oñate E (2017) Data-based models for the prediction of dam behaviour: a review and some methodological considerations. Arch Comput Methods Eng 24:1–21. https://doi.org/10.1007/s11831-015-9157-9
34. Song J, Liu Y, Yang J (2023) Dam safety evaluation method after extreme load condition based on health monitoring and deep learning. Sensors 23. https://doi.org/10.3390/s23094480

35. Zhao E and Wu C (2020) Unified egg ellipse critical threshold estimation for the deformation behavior of ultrahigh arch dams. Eng Struct 214:110598. https://doi.org/10.1016/j.engstruct.2020.110598
36. Li B, Yang J, Hu D (2020) Dam monitoring data analysis methods: a literature review. Struct Control Health Monit 27(3):e2501. https://doi.org/10.1002/stc.2501
37. Chen S, Gu C, Lin C et al (2020) Prediction, monitoring, and interpretation of dam leakage flow via adaptative kernel extreme learning machine. Measurement 166:108161. https://doi.org/10.1016/j.measurement.2020.108161
38. Liu X, Wu Z, Yang Y et al (2012) Information fusion diagnosis and early-warning method for monitoring the long-term service safety of high dams. J Zhejiang Univ Sci A 13:687–699. https://doi.org/10.1631/jzus.a1200122
39. Qiu J, Zheng D, Xu P et al (2022) Online structural damage state identification of concrete arch dams under dynamic loads using a recursive TVARX approach. Struct Health Monit 147592172211197. https://doi.org/10.1177/14759217221119709
40. Sevieri G, De Falco A (2020) Dynamic structural health monitoring for concrete gravity dams based on the Bayesian inference. J Civ Struct Health Monit 10:235–250. https://doi.org/10.1007/s13349-020-00380-w
41. Li X, Li Y, Lu X et al (2019) An online anomaly recognition and early warning model for dam safety monitoring data. Struct Health Monit 19:796–809. https://doi.org/10.1177/1475921719864265
42. Huaizhi S, Jiang H, Zhongru W (2011) A study of safety evaluation and early-warning method for dam global behavior. Struct Health Monit Int J 11:269–279. https://doi.org/10.1177/1475921711419993
43. Zhao E, Gu C (2021) Review on health diagnosis of long-term service behaviors for concrete dams. J Hydroelectr Eng 40(5):22–34
44. Lei P, Chang X, Xiao F et al (2011) Study on early warning index of spatial deformation for high concrete dam. Sci China Technol Sci 54:1607–1614. https://doi.org/10.1007/s11431-011-4373-5
45. Pagano L, Sica S (2012) Earthquake early warning for earth dams: concepts and objectives. Nat Hazards 66:303–318. https://doi.org/10.1007/s11069-012-0486-9
46. Xiang Y, Sheng J, Liu C, Fang Z, Zhang K, Cheng Z (2018) Research progress in long-term service and risk assessment of earth-rockfill dams. Adv Sci Technol Water Resour 38(5):86–94
47. Chen W, Wang X, Tong D et al (2021) Dynamic early-warning model of dam deformation based on deep learning and fusion of spatiotemporal features. Knowl-Based Syst 233:107537. https://doi.org/10.1016/j.knosys.2021.107537
48. Sang L, Wang JC, Sui J, Dziedzic M (2022) A new approach for dam safety assessment using the extended cloud model. Water Resour Manage. https://doi.org/10.1007/s11269-022-03124-1
49. Su H, Wen Z, Ren J (2020) A kernel principal component analysis-based approach for determining the spatial warning domain of dam safety. Soft Comput 24:14921–14931. https://doi.org/10.1007/s00500-020-04845-4

Chapter 4
Evaluation on Construction Quality

4.1 Introduction

Due to the project's age and early construction, it is possible that the original design and current state may not meet contemporary design and construction standards [1]. The impact of this on the structures' safety must be evaluated, and a thorough analysis of specific items such as seepage safety, structural integrity, and seismic safety should be conducted. If the evaluation indicates that the structure of the dam fails to meet the necessary safety standards for the engineering project or experiences significant quality deficiencies, an evaluation of the impact on the dam's safety shall occur to determine the necessity of addressing them.

The purpose of evaluating the construction quality is to confirm that the dam foundation treatment is reliable, seepage control treatment is effective, and the dam structure is complete, durable, and safe enough to meet the requirements for the project's safe operation. For dams with available data and information on geological surveys, design, construction, inspection and acceptance, and operation, the evaluation should concentrate on the impact of addressing construction quality defects, construction quality defects that persist after inspection, and construction quality defects discovered during dam operation. This should be based on relevant data analysis and information. Abbreviations of technical terms should be explained when first used. If the necessary data for assessing the construction quality of dams is inadequate or anomalies arise during operation, the safety inspection and boring test must be performed. Moreover, the quality evaluation of dam construction should be carried out in conjunction with operational performance.

Y. Xiang et al., *On-site Inspection and Dam Safety Evaluation*, Hydroscience and Engineering, https://doi.org/10.1007/978-981-97-3987-5_4

4.1.1 The Main Contents of Construction Quality Evaluation

During the reservoir dam safety assessment process, the conclusions of project quality evaluations serve as a reference for determining whether the dam is hazardous. The evaluation of quality can also provide original parameters in line with the project's actual situation for reviewing or evaluating the structure's safety in terms of seepage, seismic activity, and metal structures during the reservoir dam safety evaluation stage. Therefore, the evaluation of reservoir dam quality should focus on the following issues:

- Before the project quality evaluation, the original construction data of the project should be collected as much as possible, and the construction methods, raw materials, quality control and other aspects of each project should be evaluated. For the medium and small projects that have been operating for several decades and lack original construction data, geological exploration, and concrete testing should be arranged as needed.
- Geological Conditions and Foundation Treatment Assessment: Assess whether the geological conditions and foundation treatment of the dam meet the requirements of the safe operation of the engineering project. This involves evaluating whether the geological conditions and the foundation treatment processes implemented during construction adhere to the requirements for the safe operation of the engineering project. It ensures that the foundation is stable and capable of supporting the dam structure effectively [2].
- Dam Quality Evaluation: Evaluate whether the quality condition of the dam meets the requirements of the safe operation of the engineering project. The quality of the dam's construction is thoroughly assessed to verify whether it meets the necessary standards for safe operation. This includes examining the integrity of materials used, construction methods employed, and overall structural soundness [3].
- Analysis of Quality Changes: Analysis of the change in the quality of the dam according to the operating behaviour of the dam in order to find out whether there is a quality defect and to assess its impact on the safety of the dam. The evaluation analyzes any changes in dam quality based on the behavior of the dam during operation. This analysis aims to identify potential quality defects and assess their impact on dam safety. Any deterioration or unexpected shifts in quality can be addressed promptly.
- Providing Actual Parameters for Safety Evaluation: Provide actual parameters in accordance with the actual situation of the dam to evaluate the safety of the dam. Utilizing real-world data and parameters derived from the actual condition of the dam, an assessment of the dam's safety is performed. This includes considering aspects like structural integrity, seepage behavior, and overall stability.

4.1.2 The Basic Methods of Construction Quality Evaluation

1. **On-site inspection**: This section aligns with the matters addressed in Chap. 3. An on-site inspection is conducted to evaluate the physical dimensions, appearance quality, and operational status of the dam. Simple measurements, tests, and analysis of safety monitoring data support this evaluation, enabling an appraisal of the dam's construction quality [4].
2. **Analysis of historical data**: The construction quality of the dam is assessed based on current specifications by reviewing and analyzing data pertaining to construction quality control, quality inspection, acceptance, safety assessment, operation, and safety monitoring. Historical data from both the initial construction phase and subsequent project phases are also examined to address any safety concerns. Safety assessment encompasses storage safety assessment, final acceptance safety assessment, as well as all previous dam safety assessments. More information should be obtained through consulting relevant data, and the engineering quality should be evaluated in combination with the problems in the actual operation process [5].
3. **Drilling test and safety test**: If an evaluation of the construction quality of the dam cannot be executed using the aforementioned methods, a drilling test and a safety test are conducted to obtain in-situ parameters that enable the evaluation of the construction quality of the dam [6]. The procedure employed in this step is expounded in Chap. 3.

4.2 Evaluation on Engineering Geological Conditions

4.2.1 Objects, Tasks and Requirements of Engineering Geological Survey

Engineering geology is the foundation of dam construction. Dam engineering and geologic conditions interact and constrain each other, so the engineering geologic condition is very important for dam safety. In dam safety assessment, engineering geology investigation and review is one of the most important links. During the investigation and review, the geological data of existing engineering areas should be collected and analyzed as much as possible, and the conditions since the operation of the reservoir should be comprehensively analyzed from the point of view of geological investigation for foundations and buildings, to understand the hazards and hidden hazards related to geological conditions, and to analyze the causes of hazardous sections, hazardous situations and hidden hazardous areas from the point of view of geology, so as to obtain basic geological data for the safety assessment of reservoir dams [7]. Therefore, the main investigation requirements should be closely related to the actual geological conditions of the reservoir and the potential hazards of the reservoir.

The main reconnaissance requirements of safety assessment [8] ***are:***

- Understand the general situation of the regional geology in the project area.
- Collect and analyze the engineering geological survey data, construction records, hazards or hidden hazards at an early stage.
- Basically find out the engineering geology, hydrogeological conditions of the reservoir dam foundation and other building foundations, and the changes after water storage operation.
- Check the design quality of the dam body.
- Determine the nature and distribution of potential hazards to the dam body and foundation.

4.2.2 Main Contents of Engineering Geological Survey and Review

The geomorphology and landform, stratigraphy, geologic structure, seismicity, and hydrogeology of the nodal region shall be evaluated to ensure that there are no geologic deficiencies or problems affecting the safety of the project and that an earthquake hazard assessment has been performed for major projects. (Geological problems and deficiencies include faults, structures, fractures, permeable strata, weak interbeds and intrusions, corrosive groundwater, etc.) For those dams that have experienced earthquakes or significant changes in geological conditions during operation, the change in engineering geological conditions and its impact on the safety of the project shall be evaluated [9].

The main tasks of engineering geological investigation for safety assessment are:

- Complete review of the hydrogeology and engineering geology of the reservoir project site area, review of project operation and changes in geological conditions; preliminary evaluation of the engineering geology and hydrogeology of dam foundations, drainage and water conveyance structures [10].
- Examine the current situation of each hydraulic structure and check whether the quality of the dam body meets the requirements of the safe operation of the engineering project; evaluate the quality of the dam foundation and embankment treatment.
- Preliminary analysis of the distribution and geological causes of various diseases affecting reservoir safety; preliminary assessment of the degree of damage to the project.
- Provide geological data and evaluation calculation parameters for engineering quality evaluation, structural safety evaluation, seepage safety evaluation and seismic safety evaluation in reservoir dam safety evaluation.
- Review the seismic parameters of the project area.

4.3 Evaluation on Construction Quality of Earth-Rock Fill Dam

4.3.1 Common Quality Problems with Earth-Rock Fill Dam

(1) The seepage protection of the dam foundation may not meet the requirements of the safe operation of the engineering project. The main manifestations are incomplete foundation clearance of the dam, insufficient depth of vertical seepage prevention facilities, insufficient length, thickness and permeability coefficient of horizontal blanket, and insufficient water catchment tank or water catchment tank for dam foundation.
(2) The seepage protection of the dam body may not meet the requirements of the current regulations. The main manifestations are that the permeability coefficient of the dam body or core wall does not meet the requirements of the safe operation of dams, the seepage resistance label of the concrete seepage-proof plate or core wall does not meet the requirements of the safe operation of dams, or the concrete seepage-proof plate or core wall is cracked due to large deformation, resulting in water leakage, etc.
(3) The compactness or relative density or porosity of the dam body cannot meet the requirements of the specifications.
(4) The gradation of filter material and transition layer material in the dam body cannot meet the requirements of the specifications, and the drainage body is ineffective [11].

4.3.2 Evaluation Content

When assessing the construction quality of earth and rock fill dams, the treatment of the dam foundation, the selection of materials for the embankment dam and its fill, the dam structure, the construction of the anti-seepage body, the connections between the dam body and the foundation and the connections between the embankment slope and other infrastructure shall be checked to ensure that they comply with the requirements of the design and construction requirement. Treatment of the connections between the dam body and the foundation and between the embankment and other structures shall be in accordance with the design and construction requirement.

The excavation of the dam foundation and embankment, seepage control for sand-gravel fill dam foundations, treatment of rock dam foundations, treatment of liquefiable soil, soft clay and collapsible loess dam foundations shall be inspected to verify the quality of the dam foundation treatment. Soil properties, particle content, permeability and degree of compaction, relative density and porosity of fill shall be determined when selecting embankment dam materials and verifying the quality of dam fill.

When inspecting the dam structures, the dam zoning, impermeable body, filter layer and transition layer, dam body drainage, slope protection, etc. shall be carefully examined. The construction quality of the impermeable body shall meet the requirements of the safe operation of the engineering project. In addition, the curtain grouting, the filling of the earthen impermeable body, the construction of the concrete cut-off wall, and the construction of the concrete slabs and toe plates of the concrete rock fill dam shall comply with the requirements of the safe operation of the engineering project.

For rock fill dams where uneven settlement, collapse, cracking, sliding, concentrated seepage, leaching, etc. occur during operation, a geological survey and safety inspection shall be carried out as necessary to identify and analyse the quality defects and to assess their impact on the stability of the dam structure and the seepage stability [12].

4.4 Evaluation on Construction Quality of Concrete Dam

4.4.1 Common Quality Problems

(1) The dam's anti-seepage curtain does not meet the requirements of the safe operation of the engineering project. The main performance is that the dam foundation is not thoroughly cleaned, the curtain grouting depth does not meet the current requirements of the safe operation of the engineering project, the anti-seepage curtain is not sufficiently extended to both banks, and the water stop is damaged [13].
(2) The strength of the dam concrete does not meet the design requirements, the impermeability and freeze–thaw resistance index are not present, and the dam drainage is not functioning.
(3) The treatment of the arch dam abutment cannot meet the design requirements, resulting in large deformation of the dam, which causes cracks.

4.4.2 Evaluation Content

In evaluating the construction quality of concrete dams, measures for treating the dam foundation, dam structure, concrete pouring, temperature control and crack prevention shall be reviewed to ensure that the requirements of the relevant design and construction are met. For concrete dams where cracking, erosion, carbonisation, slanting and leakage occur during operation, investigations and tests shall be carried out to find and analyse the quality defects, and their effects on the stability, durability and overall safety of the dam shall be assessed.

To check the quality of the dam foundation treatment, the excavation of the dam foundation, the consolidation grouting, the seepage control and drainage of the dam foundation, the treatment of the fault rupture zone and the weak structural plane, and the treatment of the karst seepage control, etc., shall be inspected and determined. For embankment structures, check whether the embankment crest, gallery and passage inside the embankment, embankment joints, water retention and drainage for the embankment body, embankment concrete material and embankment zoning meet the requirements of the safe operation of the engineering project. For concrete grouting quality, concrete strength, permeation resistance index, frost resistance index (grade), erosion resistance, abrasion resistance, corrosion resistance, deformation modulus, etc. should be checked to ensure that they meet the requirements of the safe operation of the engineering project. Temperature control and crack prevention measures for the dam body shall meet the requirements of the safe operation of the engineering project [14].

4.5 Evaluation on Construction Quality of Masonry Dam

4.5.1 Common Quality Problems

(1) The anti-seepage curtain of the dam foundation cannot meet the requirements of the safe operation of the engineering project. The main performance is that the dam foundation is not completely cleaned, the curtain grouting depth cannot meet the requirements of the safe operation of the engineering project, and the length of the anti-seepage curtain extending to both banks is not sufficient.
(2) The anti-seepage curtain of the dam body cannot meet the requirements of the current specifications. The main performance is that the concrete anti-seepage plate or the core wall or the concrete anti-seepage plate or the anti-seepage label of the core wall cannot meet the requirements of the specifications, and the concrete anti-seepage plate or the core wall has cracks, water stop damage, etc.
(3) The masonry materials and cementitious materials of the dam body cannot meet the requirements of the safe operation of the engineering project, or they are severely weathered after years of operation [11].

4.5.2 Evaluation Content

When evaluating the construction quality of a masonry dam, the treatment of the dam foundation, dam embankment materials, dam body seepage control, dam structure, dam body design, temperature control, etc. shall be checked to ensure that the requirements of the safe operation of the engineering project are met. For masonry dams where cracks, leakages occur during operation, investigations and tests shall be carried out to find and analyse the quality defects, and their impact on the stability, durability and overall safety of the dam shall be assessed.

To check the quality of the treatment of the dam foundation, the requirements for checking concrete gravity dams can be applied to masonry gravity dams, and those for concrete arch dams can be applied to masonry arch dams. For embankment dam materials, the stone and cementing material should be checked to see if they meet the requirements. For dam structure, mainly the arrangement of dam crest and traffic, gallery and openings inside the dam, dam body joints, water drainage and foundation bedding shall be checked to make sure that they meet the requirements. For the quality of the dam body construction, the strength, permeation resistance index, frost resistance index (grade), corrosion resistance property of the cementing material and the strength, bulk density, porosity and density of the masonry shall be checked to ensure that they meet the requirements.

4.6 Hydraulic Structures Rehabilitation Project Post Evaluation

The whole life cycle of a dam refers to the entire process from the planning and decision making of a dam construction project to its removal, rather than to a specific stage in the dam construction process [15]. It is generally divided into stages such as design and construction, water storage and operation, operation and management, de-risking and strengthening, downgrading and scrapping. The aim of whole-life management is to add value to dam projects and to build low-risk, high-quality and cost-effective dam projects (Fig. 4.1).

The engineering and non-engineering measures to repair and reinforce the major engineering defects and hidden dangers affecting the safety of the reservoir and maintain the safe operation of the reservoir are called rehabilitation, in response to the problems of poor operation, low flood protection standards and various safety hazards of water conservancy facilities [17].

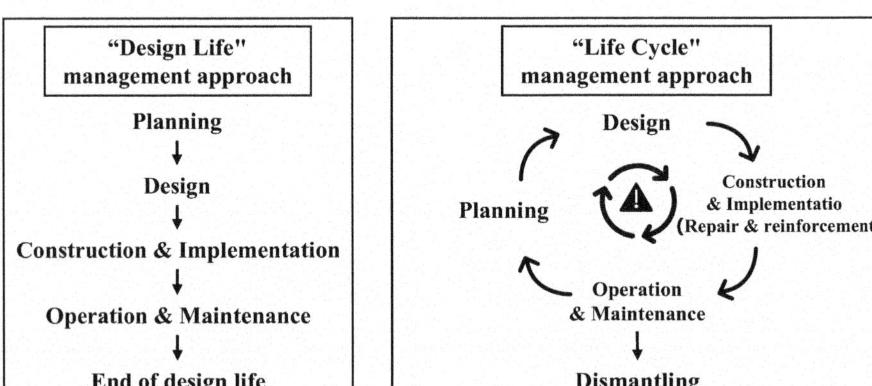

Fig. 4.1 The management method based on "design life" and "life cycle" [16]

After the completion acceptance of the rehabilitation project, through the comprehensive evaluation of the rehabilitation process, economic benefits, social and environmental impacts, soil and water conservation, objectives and sustainability, the results are compared with the objectives set in the project decision, the experience and lessons learned are further summarized, and the corresponding countermeasures and suggestions are made. Finally, the degree of rehabilitation of the project is evaluated according to indicators such as reservoir function and operation management.

In the actual rehabilitation process, the links involved are complicated and the contents are numerous. If the links are not implemented according to the design requirements, or if problems occur in any of the links during the construction process, they will have adverse effects. Therefore, the quality evaluation of the rehabilitation project is a multi-element, multi-objective and comprehensive system engineering [18].

4.6.1 Relevant Data Collection

Documents related to the rehabilitation project, such as rehabilitation information, self-evaluation by the construction management unit, and initial operation information, must be gathered. The technical information collected must be accurate and comprehensive in order to satisfy the project's post-evaluation requirements [19].

Preliminary understanding of the results of rehabilitation through status quo survey. The investigation of the current situation is being conducted through expert surveys, discussions, and on-site inspections to grasp the project's particularities. The inspection of the status quo centers on verifying the primary rehabilitation contents, and any issues, risks, or uncertainties discovered during the survey should be initially analyzed for their effect on the rehabilitation objectives.

The status quo investigation covers project quality, operational status, operation and maintenance expenses, project financial status, social and economic benefits, and environmental and ecological impacts. According to the rehabilitation plan approved for the project, in conjunction with the preliminary design, field research conducted on the buildings, metal structures, and electromechanical equipment related to rehabilitation, and the operational status of management facilities, among other factors.

4.6.2 The Evaluation Index System

Comprehensive evaluation of rehabilitation efforts should be based on the actual situation of the reservoir project, and consider screening indicators, establishing an evaluation index system, and employing a comprehensive evaluation method. This method should cover process evaluation, economic benefits, social and environmental benefits, and sustainability. Comprehensive evaluation indicators should be screened according to the different types of reservoir dams and their own characteristics to ensure the objectivity of the evaluation results. The evaluation should strive to reach a conclusion by applying a comprehensive approach [20] (Fig. 4.2).

Adopt a comprehensive evaluation methodology based on process evaluation, economic benefit evaluation, social and environmental benefit evaluation, and sustainability evaluation, and formulate an evaluation level and conclusion.

Establish a comprehensive evaluation index system for the effectiveness of rehabilitation from five perspectives: rehabilitation program, degree of functional rehabilitation, engineering construction, operation and management, and the benefits of rehabilitation. According to the current state of reservoir engineering and rehabilitation, it is necessary to use methods such as data analysis, site surveys, test analysis, and numerical simulation to determine indicators at all levels (Table 4.1).

4.6.3 The Evaluation Method

The factors affecting safe and stable operation of water conservancy projects after danger removal and reinforcement are very complex, and there are many kinds of influencing factors, which interact and correlate with each other. On this basis, according to different levels and dynamic characteristics of reservoir reinforcement effect, the evaluation index system can be established by using AHP, fuzzy evaluation method, principal component analysis method, situation analysis method and other methods. According to the different levels and dynamic characteristics of reservoir reinforcement effect, qualitative and quantitative indexes can be used to describe multiple influencing factors, so as to carry out stability evaluation [21].

- Analytical Hierarchy Process [22, 23]: Analytical Hierarchy Process (AHP) was born in 1970s. It is a mathematical calculation method combining qualitative and quantitative analysis. It can quantitatively calculate, analyze and judge the decision-making problems of multiple target factors, multiple criteria and multi-layer complex structure, and provide replication support for decision-makers. It is systematic, concise, structured, and Logic and small amount of data required can greatly reduce the difficulty of decision analysis and calculation and is widely used in safety science and environmental science. The main calculation steps of AHP are: (1) Establishing the hierarchical structure model; (2) Construct the judgment (contrast) matrix; (3) Hierarchical single ranking and its consistency check; (4) Total hierarchical ranking and its consistency check [24]. Figure 4.3 illustrates a comprehensive evaluation index system for concrete dams.
- Situation Analysis [25]: Situation Analysis includes S (strengths), W (weaknesses), O (opportunities), T (threats), referred to as SWOT Analysis (See Fig. 4.4). SWOT analysis method is based on situation analysis in internal and external environment and competitive environment, arranged in matrix form, and then analyzed according to system matching thought, resulting in corresponding decision conclusions, which have structural and systematic characteristics. The

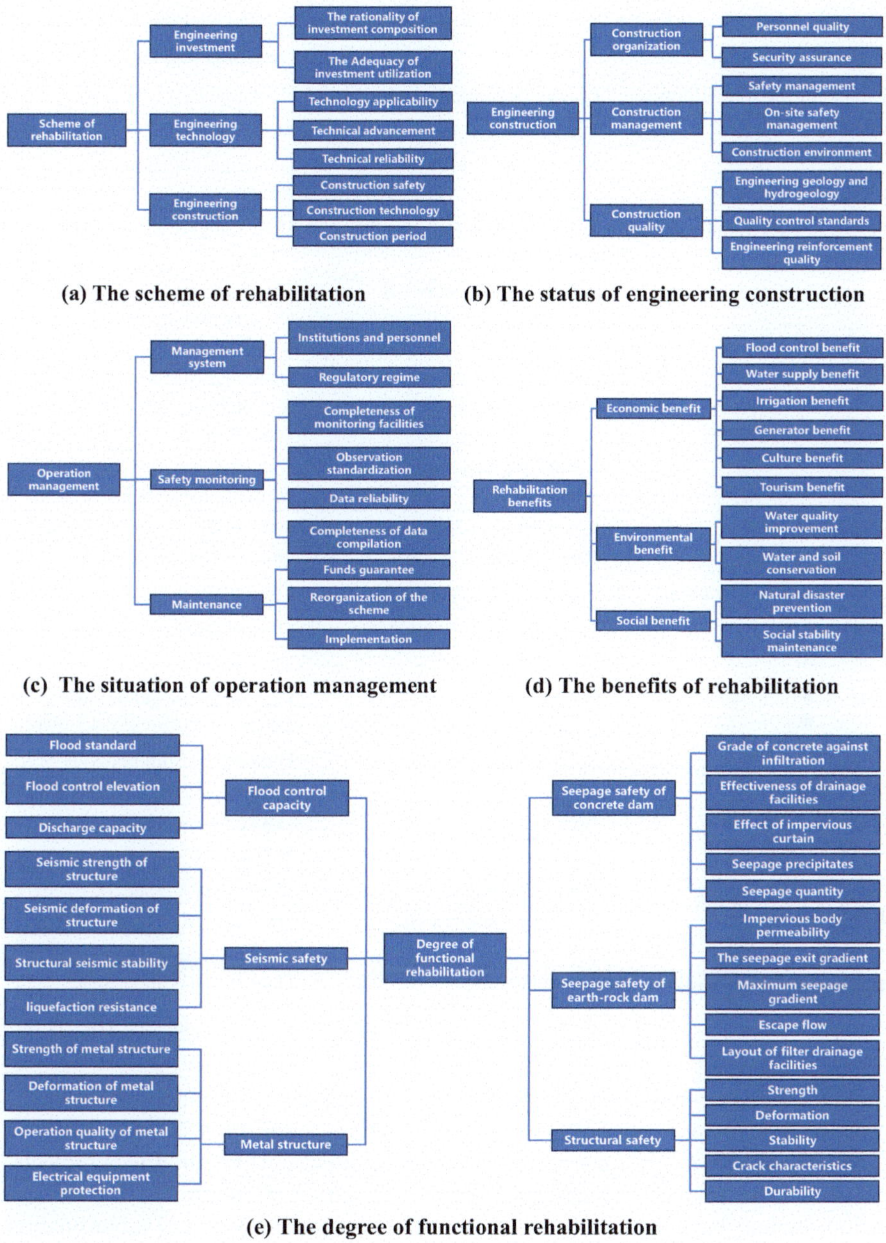

Fig. 4.2 Comprehensive evaluation index system for rehabilitation effect

Table 4.1 Methodology for determining indicators for evaluating the effectiveness of rehabilitation

Evaluating indicator		Determination method
Scheme of the rehabilitation project	Project investment	Data analysis
	Engineering technology	Data analysis
	Engineering construction	Data analysis and on-site survey
Degree of functional recovery	Rehabilitation degree of flood control capacity	Data analysis and on-site survey
	Rehabilitation degree of seepage safety	Data analysis, on-site survey, test analysis, numerical simulation
	Rehabilitation degree of structural safety	Data analysis, on-site survey, test analysis, numerical simulation
	Rehabilitation degree of seismic safety	Data analysis, test analysis, numerical simulation
	Rehabilitation degree of metal structure safety	Data analysis, on-site survey, numerical simulation
Engineering construction	Construction organization	Data analysis
	Construction management	Data analysis
	Construction quality	Data analysis
Operation management	Management system	Data analysis and on-site survey
	Safety monitoring	Data analysis and on-site survey
	Maintenance	Data analysis and on-site survey
Benefits of the rehabilitation project	Economic benefits	Data analysis
	Environmental benefit	Data analysis and on-site survey
	Social benefit	Data analysis and on-site survey

main analysis steps of SWOT analysis method include: (1) Environmental factor analysis; (2) Construct SWOT matrix; (3) Make action plan. Table 4.2 lists a SWOT analysis model for studying the impact assessment of hydropower projects.

- Fuzzy Comprehensive Evaluation (FCE) [27, 28] is a comprehensive evaluation method based on fuzzy mathematics. This method uses the subordination degree theory of fuzzy mathematics to convert qualitative evaluation problems into quantitative evaluation problems (See Fig. 4.5). It is characterized by strong systematicness and clear structure. It can quantify complex, fuzzy and difficult to quantify uncertain problems well. The analysis steps of the fuzzy comprehensive evaluation method can be summarized as follows: (1) Construct the fuzzy evaluation index system; (2) Construct weight vector; (3) Construct evaluation

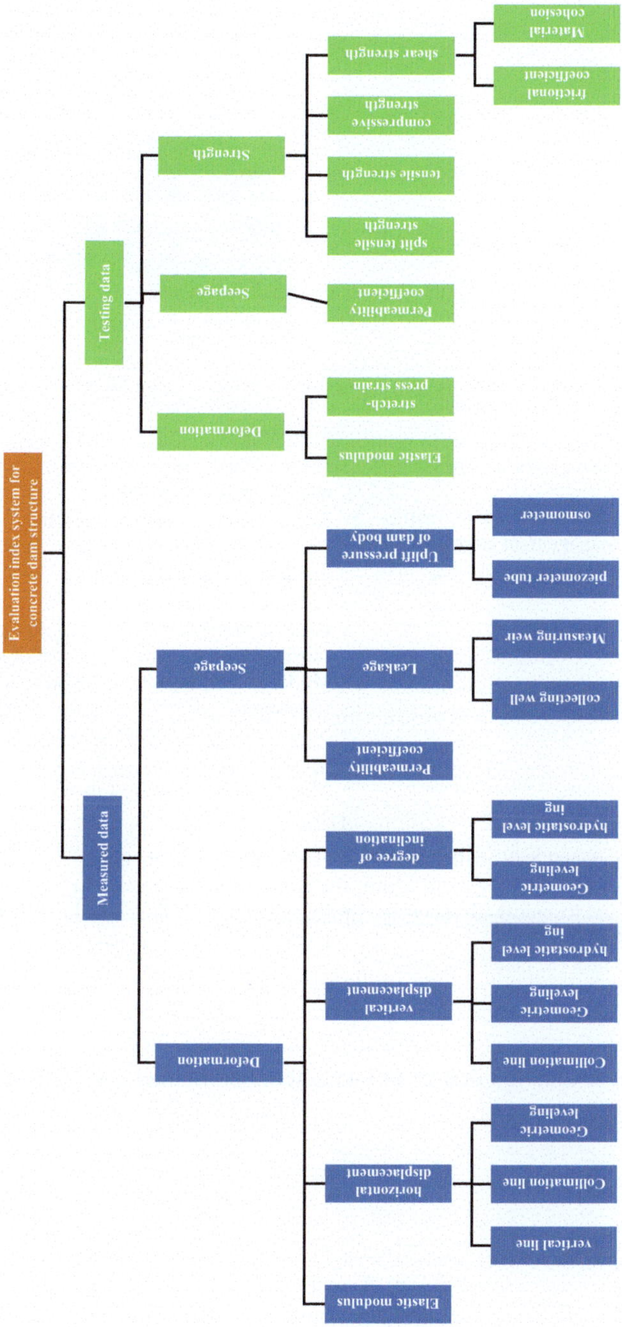

Fig. 4.3 Comprehensive diagnosis and evaluation index system for concrete dam structure

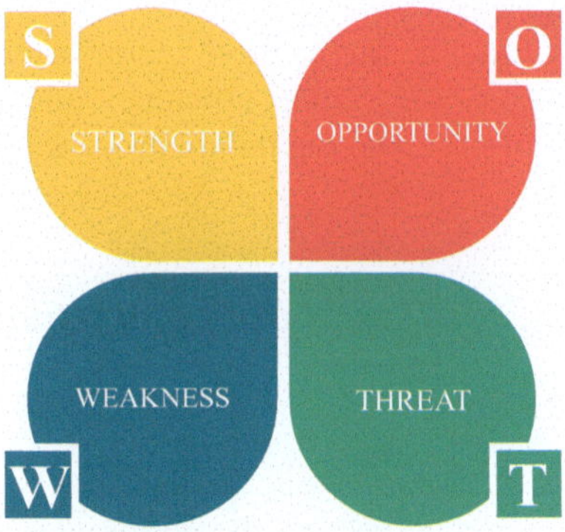

Fig. 4.4 Graphical representation of SWOT analysis

Table 4.2 SWOT analysis of models used in studies on hydropower dam impact assessment [26]

Strengths	Weaknesses
• Suitability for exploring various types of dam impacts • Explore water quality and habitat suitability for different dam management strategies • Enabled syntheses of expert knowledge and empirical data • Applicable at different scales	• Does not take into account the relevant biotic and abiotic interactions • Non-transparent variable selection • Lack of clear statistical criteria to assess the models and related thresholds • Lack of model validation • Predicts biotic responses (e.g., abundance, survival) using a limited number of variables • Model assumptions may not be valid in some situations • Some models are either too complex or overly simplified • Models a low number of data
Opportunities	**Threats**
• Increasing environmental data quality and availability • Growing interest and technical advances in ecological modelling • Integrated models become better and more reliable	• Over- or under-prediction • Expenses for data collection • Extrapolation of the model to other regions, biotic element and spatial scale

matrix; (4) The evaluation matrix and weight are combined and the result vector is explained.

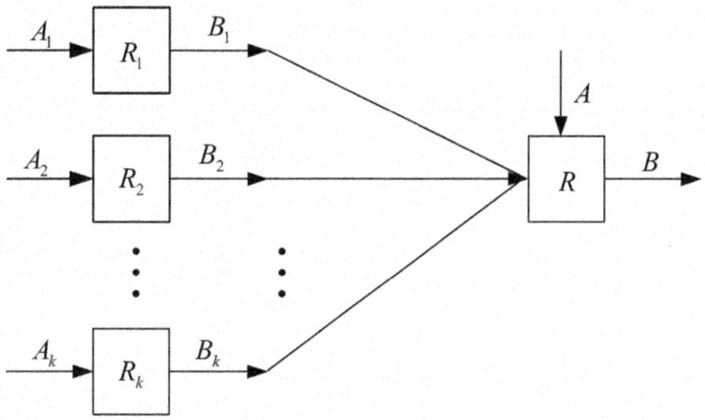

Fig. 4.5 Graphical representation of FCE analysis

4.6.4 Comprehensive Evaluation Recommendations and Countermeasures

According to the importance of the issues of concern to the decision-making and management departments, the experience is summarized mainly in terms of preliminary decision-making, implementation process, functional rehabilitation, benefits and sustainability.

Suggestions should be made on the main problems of rehabilitation projects, including macro suggestions to the state, industry and local government, and micro suggestions to the project legal person and the reinforcement project (Table 4.3).

Table 4.3 Classification of effectiveness of rehabilitation

Effect level	Evaluation value	Meaning
(A) Complete success	[0.9 ~ 1.0]	The goals of rehabilitation have been fully achieved or exceeded, and the project has achieved great benefits
(B) Basic success	[0.8 ~ 0.9]	After rehabilitation, the main diseases have been eliminated, the main objectives have been achieved, and the expected benefits have been basically achieved
(C) Partial success	[0.7 ~ 0.8]	After rehabilitation, the main diseases have been basically eliminated, only some goals have been achieved, and only certain benefits have been achieved
(D) Unsuccessful	[0.6 ~ 0.7]	The goal of rehabilitation is very limited, the seepage and structural hazards have not been effectively eliminated, and the project has basically not produced benefits
(E) Fail	[0.0 ~ 0.6]	The goal of rehabilitation has not been achieved, and there are still major hidden dangers in flood control, seepage and structural safety, so the project has no benefits at all

4.7 Conclusions on Construction Quality Evaluation

The project quality is divided into **being qualified**, **being basically qualified** and **being unqualified** according to the quality problems exposed in the actual construction and operation, and the grading principles are as follows:

- It is rated as being qualified if construction quality meets the requirements of design and the safe operation of the engineering project, and no obvious quality defects have been revealed during operation.
- It is rated as being basically qualified if construction quality basically meets the requirements of design and the safe operation of the engineering project, and local quality defects have been found during operation but caused no significant impact to safety of the project.
- It is rated as being unqualified if construction quality fails to meet the requirements of design and of the safe operation of the engineering project, and serious quality defects have been revealed during operation. Most of the results from safety inspection do not meet the requirements of design and the safe operation of the engineering project. As the result, safe operation of the project is seriously affected.

It should be pointed out here that the quality evaluation of dam safety evaluation is different from the original acceptance or grade evaluation criteria of the project. If the conclusions are inconsistent, the latest review conclusion shall prevail. However, attention shall be paid to the representativeness and typicality of the sample taken during the test or inspection.

References

1. Zhang J, Sheng J, Cai Y et al (2015) Key technologies for safety guarantee of reservoir dam. Water Resour Hydropower Eng 46(1):1–10
2. Jain P, Naithani AK, Singh LG et al (2021) Engineering geological and geotechnical assessment of the foundation of Yaragol gravity dam—a case study form India. J Geol Soc India 97(5):491–500
3. Cui B, Liu N, Zhao C. Comprehensive evaluation of the coupling construction progress and quality of high core rockfill dam. Adv Civ Eng II, PTS 1–4; Zhang XD, Li HN, Feng XT et al (2013) 2nd international conference on civil engineering and transportation (ICCET 2012), pp 256–259, 2439
4. Hu LX, Yang F, Zhu XD (2013) Numerical analysis of time dependent behavior for celadon clay rock. Appl Mech Mater. https://doi.org/10.4028/www.scientific.net/amm.353-356.1067
5. Peyras L, Royet P, Boissier D (2006) Dam ageing diagnosis and risk analysis: development of methods to support expert judgment. Can Geotech J 43:169–186. https://doi.org/10.1139/t05-096
6. Najib N, Pulung A, Santi N et al (2018) Slope stability of Cengklik earthfill dam, Boyolali Regency, Central Java Province, Indonesia. Nucl Atmos Aerosols. https://doi.org/10.1063/1.5047288

7. Lin S, Cheng SG, Chen XH (2014) The engineering geological defect treatment of Yueliangwan reservoir dam site. Appl Mech Mater 638–640:609–613. https://doi.org/10.4028/www.scientific.net/amm.638-640.609
8. Jiang Y, Zhang Q. Research on the grey assessment method of dam failure risk. Adv Mater Comput Sci PTS 1–3; Zhu G (2011) International conference on advanced materials and computer science, pp 474–476, 1690–1695
9. Adamo N, Al-Ansari N, Sissakian V, et al (2020) Dam safety: general considerations. J Earth Sci Geotech Eng 10(6):1–21
10. Sun H (2014) Study on engineering geological exploration methods for risk-elimination and strengthening of rolled earth-rock fill dam of reservoir. Water Resour Hydropower Eng 45(4):127–129
11. Fei T, Anan Z, Jie Y et al (2023) Inversion analysis of rock mass permeability coefficient of dam engineering based on particle swarm optimization and support vector machine: a case study. Measurement 221. https://doi.org/10.1016/j.measurement.2023.113580
12. Du C, Tao H, Yi F (2024) Seepage stability analysis of geogrid reinforced tailings dam. Sci Rep 14:1814 (2024). https://doi.org/10.1038/s41598-024-52472-y
13. Li J, Chen X, Gu C, Huo Z (2019) Seepage comprehensive evaluation of concrete dam based on grey cluster analysis. Water 11:1499. https://doi.org/10.3390/w11071499
14. Ren Q, Xu LY, Wan YH (2007) Research advance in safety analysis methods for high concrete dam. Sci China Technol Sci 50:62–78. https://doi.org/10.1007/s11431-007-6008-4
15. Su H, Gao J, Wen Z (2021) Life cycle cost optimisation model for design and reinforcement of dams based in fuzzy clustering and a backtracking search algorithm. Struct Infrastruct Eng 17(9):1257–1270
16. Annandale GW, Dinar A, Johndrow TB, Kawashima S, Palmieri A, Shah F. Reservoir conservation: economic and engineering evaluation of alternative strategies for managing sedimentation in storage reservoirs: RESCON approach (English). World Bank Group, Washington, DC. http://documents.worldbank.org/curated/en/819541468138875126/RESCON-approach
17. Bettzieche V, Heitefuss C. Monitoring as a basis of cost-effective rehabilitation of an old masonry dam: dams in a European context; Midttomme GH, Honningsvag B, Repp K et al (2001) ICOLD European symposium on dams in a European context, pp 391–399
18. Budweg F. Some particularities of dam rehabilitation projects. In: Symposium—rehabilitation of dams: ICOLD 66th annual meeting, proceedings (PT A); Varma C, Visvanathan N, Rao A (1999) Symposium on rehabilitation of dams, at the 66th ICOLD annual meeting, pp 1–6
19. Choi J-H, Jun C, Liu P et al (2020) Resolving emerging issues with aging dams under climate change projections. J Water Resour Plan Manag 146:04020025. https://doi.org/10.1061/(asce)wr.1943-5452.0001204
20. Choi J-H, Yoon T, Kim J, Moon Y (2018) Dam rehabilitation assessment using the Delphi-AHP method for adapting to climate change. J Water Resour Plan Manag 144. https://doi.org/10.1061/(asce)wr.1943-5452.0000877
21. Wang Y, Han M, Li B, et al (2022) Stability evaluation of earth-rock dam reinforcement with new permeable polymer based on reliability method. Constr Build Mater 320. https://doi.org/10.1016/j.conbuildmat.2021.126294
22. Bai D, Dang W, Ai S (2009) Study on the safety evaluation method of dam system project in Jiuyuangou watershed. J Water Resour Water Eng 20(1):54–57
23. He J, Gao Q, Shi Y (2016) A multi-hierarchical comprehensive evaluation method of dam safety based on cloud model. Syst Eng Theory Pract 36(11):2977–2983
24. Shin D-W (2016) Comparison of risk assessment for a nuclear power plant construction project based on analytic hierarchy process and fuzzy analytic hierarchy process. J Build Constr Plan Res 4(03):157–171
25. Yan Z, Zhang Y, Zhang G (2013) Present situation analysis for reinforcement of dangerous reservoirs in China. Prog Ind Civ Eng II, PTS 1–4; Yang W, Liang J (2013) 2nd international conference on civil, architectural and hydraulic engineering (ICCAHE 2013), pp 405–408, 2452

26. Nguyen THT, Everaert G, Boets P et al (2018) Modelling tools to analyze and assess the ecological impact of hydropower dams. Water 10:259. https://doi.org/10.3390/w10030259
27. Zhang P, Feng G (2018) Application of fuzzy comprehensive evaluation to evaluate the effect of water flooding development. J Petrol Explor Prod Technol 8:1455–1463. https://doi.org/10.1007/s13202-018-0430-y
28. Du YW, Wang SS, Wang YM (2019) Group fuzzy comprehensive evaluation method under ignorance. Expert Syst Appl 126. https://doi.org/10.1016/j.eswa.2019.02.006

Chapter 5
Evaluation on Operation Management

5.1 Introduction

Inadequate or poorly standardized operation and management practices can contribute significantly to dam break accidents and compromise overall dam safety. Effective dam operation management is essential for ensuring the safety, stability, and functionality of dams [1, 2]. Developing clear and comprehensive rules and regulations for dam operation management is a fundamental step. These rules should be aligned with relevant laws, regulations, and technical standards for dam safety. They should outline specific procedures, protocols, and guidelines for different operational tasks, from routine maintenance to emergency response [3].

Dam operation management is not a static process; it requires continuous improvement based on lessons learned, evolving technologies, and changing conditions [4]. Each dam and reservoir has unique characteristics and challenges. As such, the rules and regulations for dam operation management should be tailored to the specific conditions of the reservoir, the dam structure, the surrounding environment, and potential risks. Regular reviews and updates to operational protocols should be conducted to incorporate new knowledge and best practices. Adopting international best practices in dam operation management is essential. Learning from the experiences of other countries and sharing knowledge can contribute to safer and more effective management of dams.

Dedicated and trained personnel should be assigned to oversee and implement the dam operation management. These individuals should have the necessary expertise and experience to ensure that the dam is operated and managed safely and effectively. Responsibilities should be clearly defined to avoid confusion and ensure accountability. Establishing a strong safety culture among dam operators and management personnel is essential. This includes providing ongoing training and professional development to ensure that all individuals involved are well-equipped to handle routine operations, as well as unforeseen emergencies.

Y. Xiang et al., *On-site Inspection and Dam Safety Evaluation*, Hydroscience and Engineering, https://doi.org/10.1007/978-981-97-3987-5_5

Conducting periodic evaluations of dam operations management is critical to identifying any deficiencies or areas for improvement. This evaluation should cover various aspects, including evaluation of reservoir operation management capability, reservoir regulation operation, dam maintenance and repair, and safety monitoring. The purpose of the dam operation management evaluation is to assess the existing management conditions, management work, and level to ensure that its meet the requirements of relevant safety management laws, regulations, and technical standards to ensure the safe operation of the dam, and to provide guidance and suggestions for improving the dam operation management.

5.2 Evaluation of Operation Management Ability

To assess the operational management capability, the institutional mechanism, the organizational structure, and the management procedures of the management shall be reviewed to ensure that the dam's operation management aligns with the requirements of safety regulations and technical standards. Determine if there are mechanisms in place for continuous improvement of dam operation management. Encourage the organization to learn from past experiences, adopt best practices, and adapt to changing conditions [5].

5.2.1 Institutional Mechanism

For the institutional mechanism, it is necessary to check whether the appropriate scope of management and protection has been defined. If the scope of management and protection is not defined for a dam that has already been built, the competent authority of the dam shall submit an application to the government at the national level and above to do so in accordance with the requirements of safety management.

The administrative leadership responsibility system for operation safety shall be put into practice for all reservoirs, and the responsible persons from three levels—government, water resources administrative department and management unit—shall be appointed and identified and publicised through public media. The responsible person shall accept supervision from all walks of life in society. The dam operation safety responsibility system shall be established and implemented by the local township (city) government for those small reservoirs owned by rural collective organizations or water use cooperative organizations. Whether an administrative responsibility system for dam safety has been established; whether responsible persons from the government, competent authorities and administrative units have been appointed [6].

Whether the reform of the reservoir management system is completed as required; whether the management system is put in order; and whether personal basic expenses and expenses for project maintenance and repair are determined. These requirements are designed to ensure that both manpower and money are available to manage the reservoirs.

5.2.2 Management Organization

As for the management organization, it shall be checked whether the management unit suitable for reservoir operation management is organized and established in accordance with the requirements of relevant laws and regulations and normative documents, and whether administrative and technical personnel who have appropriate professional qualifications and meet the requirements for reservoir operation management are provided in sufficient numbers [7].

The main management body (including the management organization and management team) plays a central role in the operational management of the reservoir, is crucial components of effective reservoir operational management, and is the primary guarantor of the safe operation of the reservoir. The main management organization is responsible for making important decisions, implementing operational strategies, and overseeing the day-to-day management of the reservoir.

For important small reservoirs, the reservoir management body should be appointed by the competent authority (or owner). For other small reservoirs, the management personnel should be specially appointed and the reservoir management and protection personnel should be defined.

Management personnel of small reservoirs should participate in professional training organized by the water resources management department. This training helps enhance their knowledge and skills related to reservoir operation, safety protocols, emergency response, and relevant regulations.

The management organization should be open to continuous improvement. Regular assessments and feedback mechanisms can help identify areas for enhancement, address challenges, and adapt to changing conditions.

5.2.3 Management System

The primary purpose of establishing a reservoir management system is to minimize the risk of dam failure caused by human factors. This involves creating a set of rules, regulations, procedures, and protocols that cover various aspects of reservoir operation, safety, maintenance, monitoring, emergency response, and more. Reservoir management systems include rules and regulations set by both the competent authority (government agency overseeing water resources) and the management unit responsible for the reservoir. These systems collectively ensure that the reservoir's operations align with legal requirements and industry best practices. The essential management systems encompass a wide range of areas, including: Reservoir Registration System; Reservoir Operation, Maintenance, and Monitoring

System; Dam Safety Evaluation System: Assessing the Safety of the Dam Structure; Reservoir Emergency Response Management System; Water Project Management Performance Evaluation System; Reservoir Operation Management Supervision System; Responsibility System and Performance Management System; Information Management System [8].

As for the management system, it is necessary to check whether the reservoir management organization is in accordance with the requirements of relevant laws and regulations and standard documents, and whether management systems for regulation operation, safety supervision, maintenance and repair, flood control and emergency rescue, gate operation, and for administrative management, water policy supervision, technical archives, etc. are formulated and strictly implemented in accordance with the actual conditions of the reservoir [9].

The reservoir dam safety management system typically includes the following elements:

- **Government Regulatory Authorities**: Positioned at the top of the diagram, government regulatory authorities, such as the national water resources management agency or relevant ministries, are responsible for issuing regulations, policies, and standards, as well as overseeing and guiding the safety management of reservoir dams.
- **Reservoir Dam Owners**: Reservoir dam owners, which can be government entities, private companies, or other organizations, are responsible for the operation, maintenance, and safety of the dam.
- **Dam Safety Regulatory Agencies**: These specialized agencies are responsible for overseeing and evaluating dam safety. They conduct regular safety inspections and monitoring to ensure the safe operation of the dam.
- **Professional Engineers**: Professional engineers are part of the reservoir dam safety management team. They are responsible for conducting safety assessments, monitoring dam conditions, providing recommendations, and taking necessary actions to ensure dam safety.
- **Technical Standards and Guidelines**: This includes the development of technical standards and guidelines related to dam safety to assist management and engineering teams in executing safety management tasks.
- **Monitoring and Data Collection**: This section involves the dam safety monitoring system, which includes sensors, monitoring equipment, remote sensing technologies, etc., used for real-time monitoring of dam conditions and the collection of crucial data.
- **Emergency Plans and Crisis Management**: Emergency plans are designed for taking action in urgent situations to protect people and the environment surrounding the dam and to mitigate potential risks.

These elements constitute the basic schematic representation of the reservoir dam safety management system architecture (Fig. 5.1). In practice, these elements may vary depending on the specific reservoir and national regulations [10].

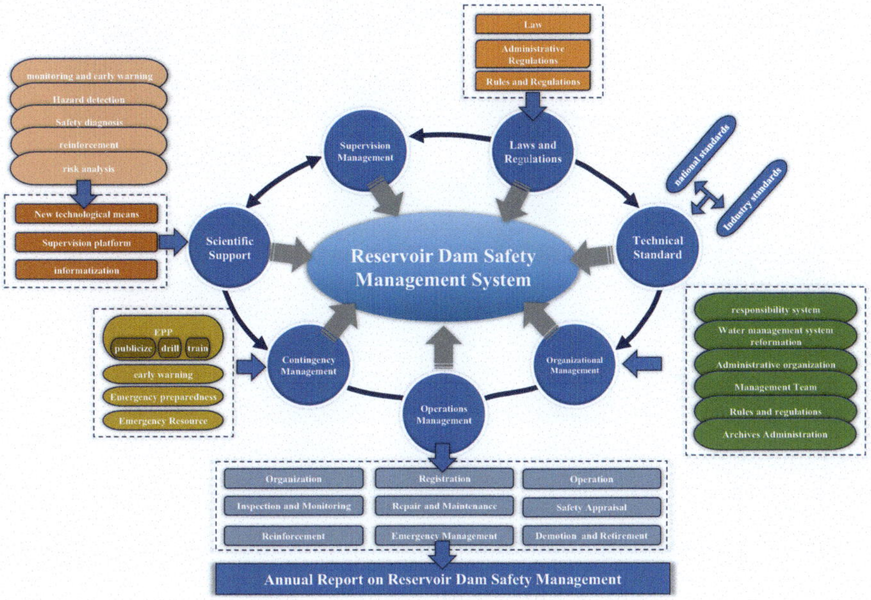

Fig. 5.1 Schematic diagram of reservoir dam safety management system architecture

5.2.4 Management Facility

The management facilities are the necessary conditions and material base to ensure the normal operation management of the reservoir and the safety of the dam, including the network of hydrological observation stations, the project safety monitoring facility, the reservoir regulation automation system, the flood control conveyance and communication facilities, the alarm system, the maintenance and repair equipment, and the flood control equipment, the water supply structures and automatic metering equipment, the water quality monitoring equipment, and the reservoir management office space [11].

The management facilities must be checked to ensure that they are complete, ready and in normal operating condition. The key points for assessing the completeness of the management facilities are described below:

1. **Hydrological observation and forecasting** is the basis for monitoring and warning of emergency flood events, flood control and emergency rescue, and scientific regulation of reservoirs. For large and important medium-size reservoirs, **a network of hydrological observation stations and an automatic monitoring and forecasting system** shall be established in accordance with the requirements and linked to the higher-level system. For general small and medium reservoirs, at least one precipitation observation station shall be established in the reservoir area (Fig. 5.2).

Fig. 5.2 Hydrological observation and forecasting system

2. **Dam safety monitoring facilities** are the "ears and eyes" that reflect the operating condition of the dam, important technical means for inspecting the design and construction quality. Therefore, they have become an essential part of the design, construction and operation management of the reservoir and have received increasing attention. Through monitoring, the quality and structural behavior of the dam can be revealed and potential safety problems can be detected in time to provide early warning of emergencies. Monitoring data can also be used as a basis for inverse calculations and analysis for dam safety assessment. For large and medium-sized reservoirs, a dam safety monitoring system shall be provided as required to meet the needs of reservoir operations management and to reflect the safety performance of the project. For small reservoirs, the necessary safety monitoring equipment shall be provided as required. For reservoirs with a water supply function, a water supply metering facility and a water quality monitoring facility shall be provided (Fig. 5.3).
3. **Flood control transportation roads** shall be constructed to ensure that materials, equipment for flood control and emergency rescue can be sent by vehicles to the dam, dam embankment or under the dam in the event of an emergency (catastrophic flood, earthquake or dangerous situation or even signs of dam failure),

Fig. 5.3 Dam safety monitoring facilities

in addition to meeting the passage requirements for the daily operation management of the reservoir. For large-sized and medium-sized reservoirs, roads to the outside and roads between various structures within the project management area and means of transportation shall be provided in sufficient numbers to meet the requirements of daily operation management and the requirements of flood control and emergency rescue. The standard of roads for large reservoirs shall be Grade III and above, and that for medium reservoirs shall be Grade IV and above. External access roads shall be connected to the formal highway. Turning points, parking areas, road signs and milestones shall be provided at appropriate locations on the roads. In the case of small reservoirs, the necessary communication conditions shall be provided for access to the main structures of the center. Flood control roads shall lead to the dam embankment or under the dam. The standard of the road shall meet the requirements of flood control and emergency rescue (Fig. 5.4).

4. In addition to meeting the requirements of internal and external communication for the daily operation management of the reservoir, **the communication facilities** shall be provided to ensure that flood information, work information, hazard situation and the like can be reported to higher units and made known to related parties, and, in particular, to ensure that alarm signals and signals for evacuation of people can be transmitted in advance to related parties and the public downstream when there is a possibility of dam failure. Large and medium sized reservoirs shall be equipped with reliable internal and external telecommunication facilities to meet the needs of information transfer for daily reservoir management,

Fig. 5.4 Flood control transportation roads

flood information reporting during flood periods, and emergency alarms. For external communication, wired and wireless communication networks shall be established for communication with competent authorities, higher flood control headquarters and main hydrometric stations and related sites downstream and upstream of the reservoir. Small reservoirs shall be provided with the necessary communication facilities to meet the requirements for flood information reporting and emergency alerting. Important small reservoirs shall be equipped with more than two types of effective communication means, while other small reservoirs shall be equipped with more than one type of effective communication means.

5. **The goods and materials used in maintenance and flood control** are divided into two types. One is working equipment, such as construction equipment, cement, reinforcement, sand and gravel, bags, shovels, generators, lighting equipment, and rain gear. The other is rescue equipment, such as life jackets, life rings, rescue boats, portable work lights, ropes and stretchers.

The goods and materials used for maintenance and flood control should be provided for large and medium reservoirs, depending on the size and characteristics of the reservoir. These facilities should include emergency power, lighting equipment, maintenance and flood control supplies and equipment, emergency and rescue equipment, warehouses and storage areas for supplies and equipment. For small reservoirs, materials, tools and equipment for flood control and emergency rescue should be reserved as needed (Fig. 5.5; Table 5.1).

6. For large-sized and medium-sized reservoirs and small reservoirs with operational management organizations, office rooms, production rooms and office facilities and equipment shall be provided. The provision shall be based on the number of management personnel and the requirements of daily management, flood control and emergency rescue. Other small-sized reservoirs shall be provided with the necessary management rooms to meet the needs of the management personnel to supervise during the flood period (Fig. 5.6).

Stone for flood control **Sand for flood control** **Soil bags for flood control**

Fig. 5.5 Flood control transport roads

Table 5.1 Flood control materials stockpile for reservoirs and dams single species base table

Type	Emergency materials						Lifesaving equipment		Small recovery equipment			
	Bags	Geotextiles	Sand-gravel material	Rock block	Lead wire	Pile timber	Life jacket	Rescue lifeboat	Generator	Portable working lamp	Projector	Cable
	piece	m^2	m^3	m^3	kg	m^3	piece	piece	kW	piece	piece	m
I	20,000	8000	2200	2000	2000	4	200	2.5	40	40	1.5	650
II	15,000	6000	1800	1500	1500	3	150	2	30	30	2	500
III	9000	4000	1000	1000	1000	2	100	1.5	20	20	1.5	300
IV	4500	2000	500	500	500	1	50	1	10	10	1	150
V	1500	800	200	150	200	0.5	20		5	5	1	50

Note The reserves of boulders and sand and gravel materials may be adjusted within the total amount depending on the condition of the dam project and emergency needs

Fig. 5.6 The management and office rooms

5.3 Characteristics of Foreign Operations Management

Today, annual report system for dam safety is set up in many developed countries to eliminate such defects as long time interval for regular dam safety evaluation and delayed reflection of change of safety performance of the reservoir.

5.3.1 Operation and Management of Reservoir Dams in Switzerland

The Swiss Federal Government places great importance on dam safety. To date, there has been no record of dam failures in Swiss history, and its dam safety management is at an international leading level. As early as 1916, the federal government enacted laws for the development and utilization of water resources. In 1957, based on the requirements set out in the federal law of 1877 regarding waterbody regulation, the Swiss Federal Government issued the "Swiss Federal Dam Safety Ordinance," establishing the concept and goals of dam safety and specifying the responsibilities of various parties involved in dam safety, including the government (dam safety regulatory agency), owners, engineers, professional associations, and others. Building upon this foundation, in 1998, the Swiss Federal Government enacted new legislation and issued the "Swiss Federal Law on the Comprehensive Monitoring of Reservoirs and

Dam Safety," which applies to the management of reservoir dams of all types (earth-fill dams, concrete dams, roller compacted concrete dams, etc.), various sizes, various purposes (power generation, irrigation, water supply, flood control, etc.), and various types of ownership (public and private) [12].

Switzerland established the framework for dam safety management in the early 1940s, and the "Dam Law" (WRFA) was implemented in 1998. Since then, it has been continuously improved and has reached its current state. Currently, the main laws and regulations governing dam safety monitoring in Switzerland are the "Dam Law" and the "Dam Regulation" (WRFO). To implement the "Dam Regulation," the Federal Office of Energy (SFOE) has issued a series of guidelines on dam safety criteria, structural safety, flood safety assessment, earthquake resistance standards assessment, dam monitoring and maintenance, emergency plans, and more.

Switzerland has a well-structured hierarchical system for dam safety management. Since the 1970s, there has been minimal new construction of dams and hydroelectric plants in Switzerland, with the primary focus shifting to engineering safety and operational management. Through decades of practice, Switzerland has established a comprehensive system for dam safety and hydroelectric plant operation management. As early as 1957, Switzerland developed the "Dam Safety Regulation" based on the "Swiss Federal Hydraulic Engineering Law." The current "Dam Safety Regulation" has been in effect since January 1, 1999, with revisions made to earlier regulations. This regulation outlines the responsibilities of various parties involved in dam safety, such as dam safety regulatory agencies, owners, experienced engineers, and independent experts. Switzerland's four-level dam safety supervisory system (dam owners, experienced engineers, independent experts, and dam safety regulatory agencies) and its comprehensive dam safety management model (structural safety, dam monitoring and maintenance, and emergency plans) provide strong support for dam safety management both at the institutional and technical levels, playing a crucial role in ensuring dam safety [13, 14] (Table 5.2).

Switzerland has adopted a "three-pronged" approach to dam safety (Fig. 5.7). A "safe" dam requires three conditions: structural safety, continuous safety monitoring and emergency planning. Structural safety is ensured through strict and scientific dam safety standards, sound design and good construction quality; continuous safety monitoring and regular safety assessment during operation to understand the safety status of the dam and to eliminate any engineering deficiencies and safety hazards; and emergency planning to prevent possible emergencies and to ensure the safety of the dam and the public. Swiss law explicitly requires all dam owners to prepare and submit annual dam safety reports as the main means for the government supervisory authorities to have a timely and comprehensive understanding of the dam safety performance and as a basis for the dam owner to make plans for maintenance and repair of the reservoirs [15] (Fig. 5.8).

Table 5.2 Swiss dam safety supervision system

No.	Responsibility	Safety supervision jobs	Safety supervision results
1	Owner	Regularly inspect the operational condition of the dam (visual inspection and monitoring); inspect spillways and discharge outlet gates	Monitoring and detection records
2	Senior Engineer	Analyze measured data and observation results; conduct annual safety inspections of the dam	Prepare an annual report on dam safety and operational status
3	Independent Experts	Conduct a safety inspection and assessment of the dam every five years	Compile a report on dam safety and long-term operational status; analysis of specific safety issues
4	Regulator	On-site inspection; review annual reports from senior engineers and assessment reports from independent experts; verify the implementation status of necessary measures	

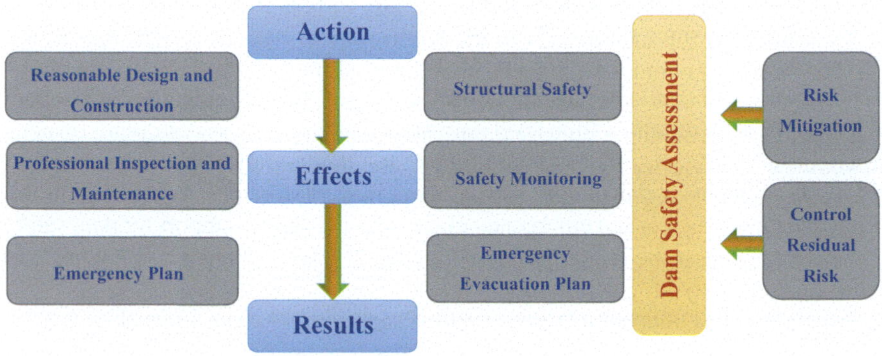

Fig. 5.7 The "THREE-LEGGED" concept of dam safety

5.3.2 Operation and Management of Reservoir Dams in USA

In the United Sates, a dam operation permit system is established on the basis of annual report system for dam safety. The related departments in government will make decision on whether the owner is permitted to operate the reservoir in accordance with the dam safety performance reflected in the annual report [16, 17] (Fig. 5.9; Table 5.3).

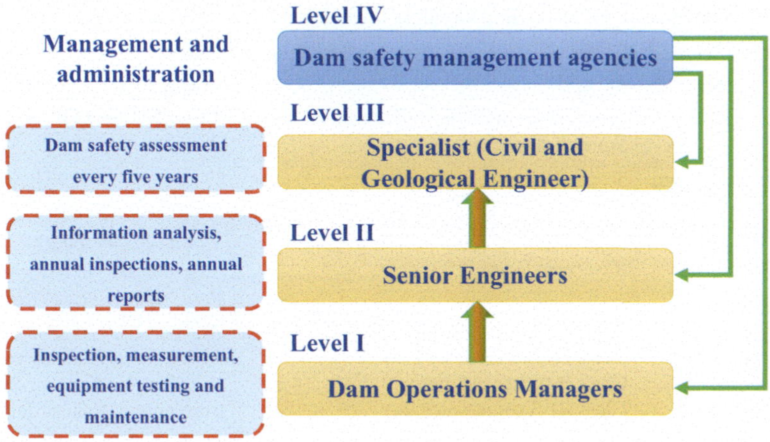

Fig. 5.8 Swiss FOUR-LEVEL system for dam safety supervision

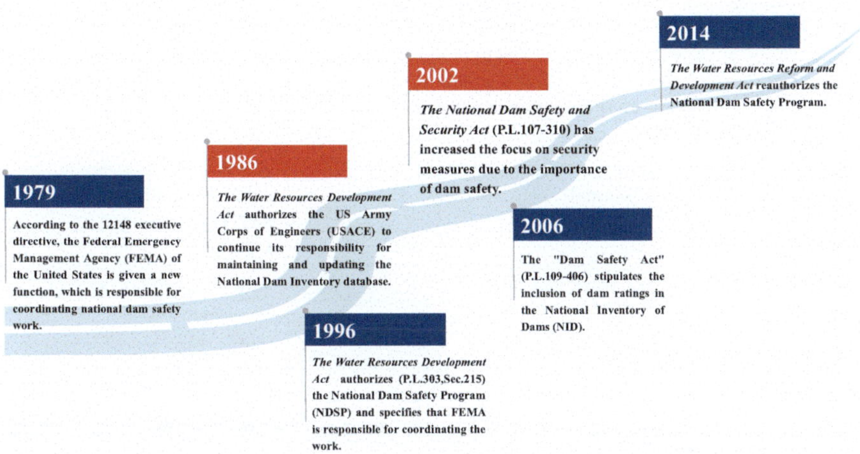

Fig. 5.9 Development of dam safety management in the United States

5.3.3 *Operation and Management of Reservoir Dams in Spain*

Spain has a strong tradition and technical capability in large dams, particularly in dam safety management, due to its over 1200 large dams. Table 5.4 presents the relevant guidelines prepared by the Spanish Management Committee.

In Spain, managers concentrated dams in the basin for combined management and strengthened non-engineering measures such as monitoring, forecasting and early warning. All safety regulations for dams should take into account not only technical aspects but also the available economic resources and technical capabilities

Table 5.3 A comparison of reservoir dam safety management in the US and China

Contents	China	USA
Number of dams	95,296 (From 2023 Statistic Bulletin on China Water Activities)	90,580 (From National Inventory of Dams)
Average age of dam (years)	50	57
Safety legislation	*Regulations on the Safety Management of Reservoir Dams* (1991)	*National Dam Safety Program Act* (1976) *Dam Safety and Security Act* (2002)
Safety classification	Class I (Normal dam) Class II (Diseased dam) Class III (Dangerous dam)	Divided into five classes according to the level of risk, from normal to near damage
Number of dams breached	3558 (1954–2021)	1645 (1848–2017)
Security management	Weak management (before 1978) → Gradual establishment of a sound management system (1979–1999) → Equal emphasis on engineering measures (de-risking and strengthening) and non-engineering measures (institutional reform, EPP) (2000–present)	Initial stage (1929–1970) → Transitional stage (1970–1986) → Period of legalized and scientific management (1986–present)
Technical training	Currently not very systematic and professional	Many years of systematic technical training, with over 9000 people trained by ASDSO
Publicity and education	Water Week of China (22–28 March each year)	National Dam Safety Awareness Day (since May 31, 1889)

of both dam owners and regulators. It is of utmost importance to provide training and capacity building to all personnel involved. A modern Dam Safety Program that is truly effective should be risk-oriented and an ongoing process [18, 19].

Spanish organizations and institutions, such as the General Directorate of Water under the Ministry of Agriculture and Environment and the Spanish Committee on Large Dams SPANCOLD, are delighted to share the knowledge and expertise acquired over many years with other countries through exchange and training programs.

The failure of the Tous Dam [20] in 1982 led to a thorough examination of dam safety regulations in Spain, with a primary focus on the potential risks posed by dams downstream and the implementation of non-structural measures, such as real-time hydrological and meteorological information systems, Emergency Action Plan, The Dam Comprehensive Safety Management Plan. Simultaneously, an extensive examination of the current dam safety conditions will be carried out, along with a re-evaluation of current operation and maintenance procedures, with the aim of

Table 5.4 Guideline collection prepared by the Spanish Committee on Large Dams

SPANCOLD technical guidelines	Year of publication
Dam safety	2005
Design criteria for dams and complementary works	2002
Geologic-geotechnical aspects and material studies	1999
Design flow	1997
Spillways and outlets	1997
Dam construction and quality control	1999
Monitoring of dams and their foundations	2004
Operation and maintenance of dams: vol. I: risk analysis applied to dam safety management	2012

enhancing their efficiency [21]. Safety monitoring activities will be enhanced to cover dams that are currently operational. Detailed tasks and actions including:

- Constitution of Dam File containing all relevant documents
- Potential Risk Classification
- Operation and Maintenance rules
- Emergency Action Plan
- Inspection Reports
- Safety Review
- Annual Reports.

The dam engineering community agrees on the importance of updating and unifying regulations and safety requirements for both state-owned and non-state dams. There is currently a clear trend towards a Risk-Based Dam Safety Decision Process. A new safety law for dams, which will be based on updated technical regulations and the risk analysis methodology, is currently being prepared [21].

5.3.4 Operation and Management of Reservoir Dams in Canada

In Canada, the responsibility for water management is shared by the federal, provincial, and municipal governments, and in some instances, by the territories and by Aboriginal governments under self-government agreements. This shared responsibility necessitates close cooperation and collaboration among all levels of government, Aboriginal peoples, and the public [22].

Canada follows a system where dam safety is the responsibility of the dam owner. Dam owners are accountable for the safety of their dams, and inspections and safety assessments of dam conditions are carried out by independent, licensed professional engineers hired and funded by the dam owner [23]. These engineers are responsible for providing safety assessment conclusions, and in the event of accidents, insurance companies provide compensation (practicing engineers in Canada typically have professional liability insurance). To ensure that dams meet appropriate standards and enhance the level of dam safety management, the Canadian Dam Association (CDA) [24] was established in 1989. The CDA is a voluntary association consisting of dam owners, engineers, consulting firms, and other stakeholders. Subsequently, this industry association developed the *Canadian Dam Safety Guidelines*, which were revised in 1999. Practical experience has shown that this is an effective management system in Canada.

Canada currently primarily follows the provisions outlined in the *Canadian Dam Safety Guidelines*. These guidelines are divided into 11 chapters, covering General, Dam Safety Review, Operation, Maintenance and Monitoring, Emergency Plans, Earthquakes, Floods, Spillway Facilities, Geotechnical Engineering, Concrete Structures, Reservoirs and the Environment, and Documentation. The *Canadian Dam Safety Guidelines* apply to all dams and, while they do not have legal force, they are widely recognized and adopted within the industry. Canada proposes different operational management requirements based on the level of risk, and the corresponding experience is worth learning [25] (Table 5.5).

Table 5.5 Risk-based operational management requirements in Canada

Project	Dam failure consequences			
	Extremely severe	Severe	Low	Extremely low
On-site inspection	Once a week	Once a week	Once a month	Once a quarter
Formal inspection	Once every half a year or a year	Once a year	Once a year	Once a year
Instrument monitoring	Refer to OMS manual	Refer to OMS manual	Refer to OMS manual	Nothing
Metal structure detection	Once a year	Once a year	Once a year	Once a year
EPP	Update address book every six months	Update address book every six months	Update address book once a year	Nothing
Operation, maintenance and monitoring manual	Review once every 7–10 years	Review once every 7–10 years	Review once every 7–10 years	Review once every 7–10 years
Dam safety review	Once every 5 years	Once every 7 years	Once every 10 years	Once every 10 years

5.3.5 *Operation and Management of Reservoir Dams in Japan*

In terms of reservoir management, Japan primarily adheres to the *River Management Facilities and Other Construction Regulations* (ダム定期検査の手引き「河川管理施設のダム版」) issued by the Ministry of Land, Infrastructure, Transport and Tourism, and the *Reservoir Structure Management Standards* (ダム管理用制御処理設備標準設計仕様書) published by the Japan Dam Association. According to relevant legal provisions, each reservoir is required to establish a *Reservoir Facility Management Regulations* based on its design, construction, safety management, scheduling, and other factors. Important reservoirs are also required to develop *Detailed Implementation Rules for Reservoir Facility Management Regulations* based on these regulations.

(1) **A management model characterized by multiple departments working together in collaboration.**

Japan has multiple departments involved in reservoir management, and the responsibilities among these departments are clearly delineated based on relevant laws. The Ministry of Land, Infrastructure, Transport and Tourism is the primary department responsible for water resource management, conservation, and development. It includes regional development bureaus and the Hokkaido Development Bureau, which handle regional river administrative affairs. The regional development bureaus oversee the construction and management of rivers and reservoirs through their offices and branches [26]. The Ministry of Agriculture, Forestry and Fisheries is responsible for agricultural water use and the maintenance of water source conservation forests. The Ministry of Economy, Trade and Industry is in charge of industrial water use and hydropower generation. The Ministry of the Environment is responsible for water quality and environmental protection [27].

(2) **Japan implements a separation of management and maintenance in reservoir operation management.**

The management unit has a small staff, adheres to strict management standards, and utilizes advanced and highly automated facilities that enable remote monitoring (including meteorological and water temperature data) and operations. The responsibility for dam safety monitoring and routine operational maintenance is entrusted to professional companies.

Reservoir operation management not only ensures the safety of reservoir projects through regular inspections and maintenance but also places special emphasis on public safety and ecological safety. To guarantee flood control and public safety downstream of reservoirs, advanced technologies such as radar, remote sensing, CPS (Cyber-Physical Systems), fiber optics, and information systems have been integrated into reservoir operation management. This includes the establishment of real-time monitoring and forecasting systems for engineering safety and rainfall conditions. During periods of heavy rainfall or flooding, continuous data collection on river and rainfall conditions occurs. Predictions are made regarding peak flow rates, disaster preparedness measures, optimal discharge rates, and other factors. Based on these predictions, adjustments are made to the operation of various reservoirs to ensure the safety of downstream areas, prevent significant fluctuations in river water levels, and minimize the impact of flooding disasters.

(3) **Implement strict routine inspections.**

The regular inspection of reservoirs is carried out in accordance with relevant regulations under the jurisdiction of the Ministry of Construction for river management facilities and other construction structures, as well as the Dam Structure Management Standards established by the Japan Dam Association. These inspections are conducted by responsible personnel in charge of management.

Daily inspections of reservoirs are performed by on-site management personnel, following the guidelines specified in each reservoir's manual.

For multi-purpose reservoirs that have been in operation for over 20–30 years, an expert panel consisting of experts from various fields (planning, geology, design, analysis, construction, and operational management) is assembled. This panel includes representatives from the Ministry of Construction, universities, civil engineering research institutes, JDEC (Japan Dam Engineering Center), and other organizations. They conduct safety assessments of reservoir dams, evaluations of operational correctness, and assessments of crisis management measures [28].

5.3.6 Operation and Management of Reservoir Dams in Russia

The Soviet Union initially implemented the *Interim Regulations on Power Plant Safety Supervision* in 1972, which included safety supervision for hydraulic structures. However, following the dissolution of the Soviet Union, social upheaval and economic difficulties in Russia led to insufficient attention to the safety of hydraulic structures. In the 1990s, a series of dam failures in the Urals region drew strong public attention to the safety of Russian hydraulic structures. On July 21, 1997, Russian Federal Law No. 117-FZ, known as the *Hydraulic Structure Safety Law*, officially came into effect. This marked a comprehensive strengthening of safety production and supervision for Russian hydraulic structures [29].

Currently, Russia has established a regulatory framework for hydraulic structure safety based on federal laws such as the *Hydraulic Structure Safety Law* and *the Industrial Safety Law for Hazardous Production Facilities*. This framework includes administrative regulations like the *Federal Regulation on Hydraulic Structure Safety Supervision, Classification and Management Procedures for Hydraulic Structures*, and *Safety Declaration and Registration Procedures for Hydraulic Structures*, among others. Complementary to these, there are departmental regulations, normative documents, and technical standards, forming a comprehensive system of regulations and standards tailored to the Russian context. Additionally, Russia continues to work on revisions and amendments to regulations and standards, including the *Hydraulic Structure Safety Law*.

The water resource management system in Russia consists of three levels: the federal government, local government, and industry regulatory authorities [30].

The federal government in Russia holds the authority to formulate water resource management policies, laws, regulations, and standards. The most significant institution in this regard is the Federal Water Resources Management Agency of the Russian Federation, responsible for the implementation of federal water resource regulations, the development and execution of water resource management plans, and the oversight of water resource development and utilization. Additionally, other federal departments, agencies, and state enterprises play crucial roles in water resource management and utilization, such as the Ministry of Natural Resources and Environment of the Russian Federation, the Ministry of Energy of the Russian Federation, and the Federal Forestry Agency of the Russian Federation, among others.

Local governments are responsible for enforcing federal water resource management policies and regulations, formulating local water resource management plans and policies, and managing water resources within their respective regions. Water resource management authorities at the regional level are supervised by local governments or industry regulatory agencies [31].

Industry regulatory agencies are responsible for the management and oversight of water resource-related industries, including water resource engineering construction, water resource development and utilization, water resource protection, and water pollution control, among other aspects. They operate under the supervision and guidance of both the federal and local governments.

5.4 Evaluation of Reservoir Regulation Operation

In evaluating the operation of the reservoir regulation, the preparation of the reservoir regulation procedures, safety monitoring, formulation of the emergency preparedness plan, chronicle of operating events, technical archives, etc. shall be examined to see whether they comply with the requirements of laws and regulations and technical standards for the safety management of the dam, and whether the regulation for operation is reasonably applied in accordance with the approved regulation procedures.

The overall goal of reservoir operation planning is to optimize the use of available water resources to balance the needs of different stakeholders while ensuring the safety of the dam and downstream communities. Effective planning involves a combination of technical expertise, collaboration with relevant parties, data analysis, and informed decision-making. This process helps maximize the benefits of the reservoir while minimizing risks and potential negative impacts.

5.4.1 The Rules of Reservoir Regulation

The main technical indicators for reservoir operation are critical parameters that guide and govern the management of the reservoir to ensure safety, water supply, flood control, and other operational objectives. These indicators play a crucial role in determining the reservoir's water levels, releases, and responses to various conditions. The main technical indicators for reservoir operation include (See Fig. 5.10): check flood level, design flood level, flood control high level, flood season limit water level, normal pool level, lower limit water level for comprehensive utilization, dead water level, reservoir land acquisition and resettlement height, safety standard of downstream flood control system, urban living and industrial water supply, agricultural and animal husbandry water supply, guaranteed output of hydropower plant, etc. The reservoir regulation includes adjusting dam releases, gate operations, and reservoir levels to balance the requirements of flood control, water supply, hydropower generation, and other uses. The reservoir regulation shall include the operation of flood control and profiting regulation [32].

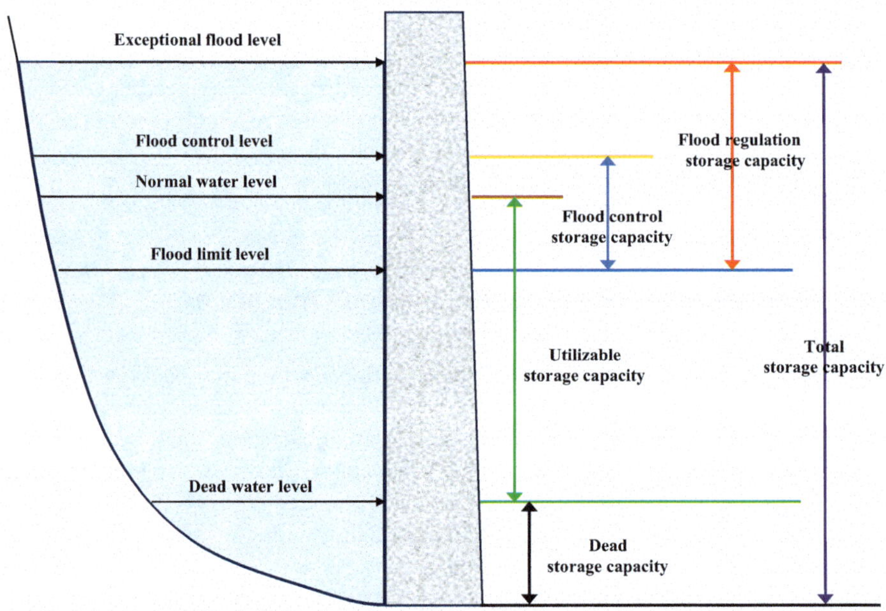

Fig. 5.10 The characteristic water level of reservoir

5.4.1.1 The Operation of Flood Control

The operation of flood control is a crucial aspect of reservoir operation, ensuring the effective management of water levels and releases during flood events, includes the following contents:

1. Confirming the water level at which the reservoir needs to be managed during flood events to ensure safety and prevent overtopping. Developing a plan for reservoirs with seasonal variation in inflow floods. This plan outlines how releases will be managed during different stages of the flood season to ensure effective flood control.
2. Defining the operational strategies and criteria that determine when and how the reservoir releases water during a flood. This involves monitoring hydrological conditions, weather forecasts, and downstream impacts to make informed decisions.
3. Planning for special actions or interventions to manage extreme flood situations that go beyond standard flood control measures. For some critical reservoirs, this may include creating dam break flood inundation maps to predict the potential impact on downstream areas.
4. Planning for special actions or interventions to manage extreme flood situations that go beyond standard flood control measures. For some critical reservoirs, this may include creating dam break flood inundation maps to predict the potential impact on downstream areas.
5. Define the organizational measures and dispatching authority to implement the reservoir flood control dispatching plan. Especially in the flood season, the reservoir should adopt the flood control dispatching mode of staged control flood discharge according to the flood control capacity of the project and the importance of the protection object. For reservoirs with seasonal fluctuation of inflow flood, a staged flood control dispatching plan shall be prepared and implemented after approval by the competent department of the reservoir.
6. Developing and implementing flood forecasting models and systems for larger reservoirs and important medium-sized reservoirs. These schemes provide early warnings and predictions of flood events, allowing for timely preparation and response.

5.4.1.2 The Operation of Profiting Regulation

Efficiently utilizing a reservoir's resources to maximize its benefits requires tailoring strategies to its regulation capabilities and the water demands of different sectors. The ways to promote the benefit of the reservoir have their own emphasis, which should be formulated according to the regulation performance of the reservoir and the water use characteristics of the various departments to promote the benefit. The operation of profiting regulation includes the following contents:

(1) Forecast of Incoming Water: Accurately predicting the expected water inflow into the reservoir for the current year, specific periods, and months. This involves

utilizing hydrological data and forecasting models to estimate the volume of water that will enter the reservoir.

(2) Coordinate Water Supply Requirements: Collaborating with relevant government agencies, industries, and communities that depend on the reservoir for water supply. Understanding their specific water demands and needs is crucial for ensuring equitable distribution and sustainable use of the water resources.
(3) Formulate Operation Indicators: Defining key performance indicators and operational criteria for different periods, which could be daily, monthly, or seasonal. These indicators might include target water levels, reservoir storage, minimum outflows for downstream needs, and maximum allowable water releases.
(4) Develop Water Supply Plans: Based on the forecasted inflow, water supply requirements, and established operation indicators, developing a detailed water supply plan for the year, period, and month. This plan outlines how water will be allocated and released from the reservoir to meet various demands.

5.4.1.3 Rules of Reservoir Regulation

"*The Rules of Reservoir Regulation*" serves as the foundational document guiding the operation of reservoir regulation. This document outlines essential aspects that govern the regulation operation process. It clarifies and defines crucial elements, including the basis for regulation, regulatory tasks and principles, regulatory requirements and conditions, and the methodology for regulation. The formulation and amendment of these rules are subject to specific requirements:

(1) Basis and Analysis for Formulation: The rules of reservoir regulation are developed based on the tasks, principles, parameters, and indexes specified in the preliminary design. In unique cases, regulation conditions and bases may be analyzed and determined according to actual operational and safety conditions. These determinations require thorough examination and approval.
(2) Approval Process: Reservoir regulation rules must receive approval from the administrative department of water resources at or above the county level, according to jurisdiction. If regulation operations cross multiple administrative regions, approval from the higher-level administrative department or the river basin organization is necessary. Regulation plans for the flood season require approval from the flood control and drought relief headquarters, with regulatory power limitations.
(3) Guiding Principles: Reservoir regulation must uphold the principle of "**Safety first and overall planning with full considerations**." This approach leverages comprehensive utilization benefits through the coordination of flood control and regulation, considering societal, economic, and water use sectors. This coordination is done while ensuring project safety and adhering to flood control arrangements. Cascade and multi-reservoir regulations should also be considered.
(4) Amendment and Change Process: Any changes to reservoir regulation methods and regulation rules require submission to the original approving authority

for inspection and approval. When amendments are necessary, the reservoir's competent department coordinates relevant units to review and modify operation indicators, adhering to relevant regulations and specifications. This process involves verifying and revising fundamental data before reporting changes to the superior competent department for approval before implementation.

(5) Monitoring and Modification: In cases where significant shifts occur in regulation tasks, operational conditions, regulation modes, or safety performance, the regulatory procedures must be appropriately modified. These modifications need to be made at suitable intervals and reported to the original examination and approval department for review and approval.

By adhering to these regulations, the process of reservoir regulation is systematic, ensuring safety, efficiency, and proper coordination among various stakeholders involved in the regulation process.

5.4.2 Compilation and Analysis of Dam Safety Monitoring Data

Periodic inspections for dam safety and instrument monitoring shall be performed in accordance with the requirements for earth-rock, concrete, and masonry dams. In addition, monitoring data shall be processed and analyzed in a timely manner to provide guidance for the safe operation of the dam. For reservoirs with a water supply function, water quality shall be monitored. (Detail in Chap. 4.)

5.4.3 Emergency Preparedness Plan for Dam Safety Management

Unexpected events of reservoir dams refer to sudden emergency situations that have the potential to lead to outcomes such as dam break, significant project risks, excessive discharge, reservoir water pollution, endangering public safety, and more. These events require swift and effective emergency responses to mitigate their impact. Depending on the severity of consequences and the extent of their impact, unexpected events are categorized into four classes: depending on the severity of consequence and scope of impact, extraordinarily serious event, serious event, less serious event and common event.

The emergency preparedness plan for dam safety management is a pre-established plan designed to prevent or minimize losses resulting from unexpected events involving reservoir dams. It is a crucial non-engineering measure aimed at enhancing the ability of communities, the public, and reservoir operation management units to effectively respond to unforeseen situations and reduce reservoir-related risks. The key components of an emergency preparedness plan for dam safety management typically encompass: Version No. and objects of issuance, remarks on formulation, analysis of unexpected events and their consequences, emergency response organization system, operation mechanism, emergency security, publicity, training and exercises (drill), appendixes and annexes. The emergency plans usually include:

- Determination and release of emergency situations;
- Possible prevention and rescue measures;
- Notification procedure and notification flow chart;
- Communication and early warning system;
- Composition of rescue team;
- Reserve materials, equipment and materials;
- Traffic, lighting and emergency power supply;
- Response under adverse environment (no lighting, adverse weather, etc.);
- Flood inundation map (Figs. 5.11 and 5.12);
- The organization and route of the evacuated population.

The emergency preparedness plan for dam safety management is a pre-established plan designed to prevent or minimize losses resulting from unexpected events involving reservoir dams. It is a crucial non-engineering measure aimed at enhancing the ability of communities, the public, and reservoir operation management units to effectively respond to unforeseen situations and reduce reservoir-related risks [6]. The emergency preparedness plan shall be publicized, tested and demonstrated (exercised) at regular interval in a proper way so as to enable the people participating in emergency response activity and the public in the area of risk to know the process of emergency response to unexpected events and to fully understand the alarming and evacuation signals.

5.4.4 The Events of Reservoir Operation

The reservoir management unit or the competent authority (the owner) shall prepare a complete and detailed history of the operation of the reservoir, including the following key points: operational characteristics, annual water level and annual discharge, anomalies that occurred during operation and analysis of their causes and solutions, the safety performance of the dam in the event of extreme events such as catastrophic floods, earthquakes, abnormal water levels, the conclusion of the previous dam safety evaluation, and the situation regarding reinforcement and reconstruction.

5.4.5 Establish Technical Archives

The reservoir management unit or the competent authority (the owner) shall improve, establish technical archives concerning the basic information of the reservoir,

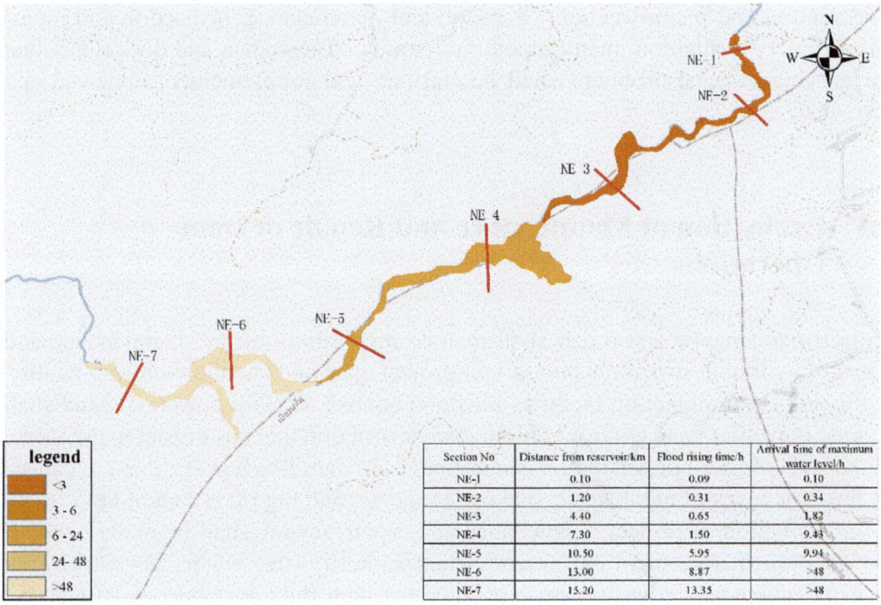

Section No	Distance from reservoir/km	Flood rising time/h	Arrival time of maximum water level/h
NE-1	0.10	0.09	0.10
NE-2	1.20	0.31	0.34
NE-3	4.40	0.65	1.82
NE-4	7.30	1.50	9.43
NE-5	10.50	5.95	9.94
NE-6	13.00	8.87	>48
NE-7	15.20	13.35	>48

Fig. 5.11 Flood arrival time chart

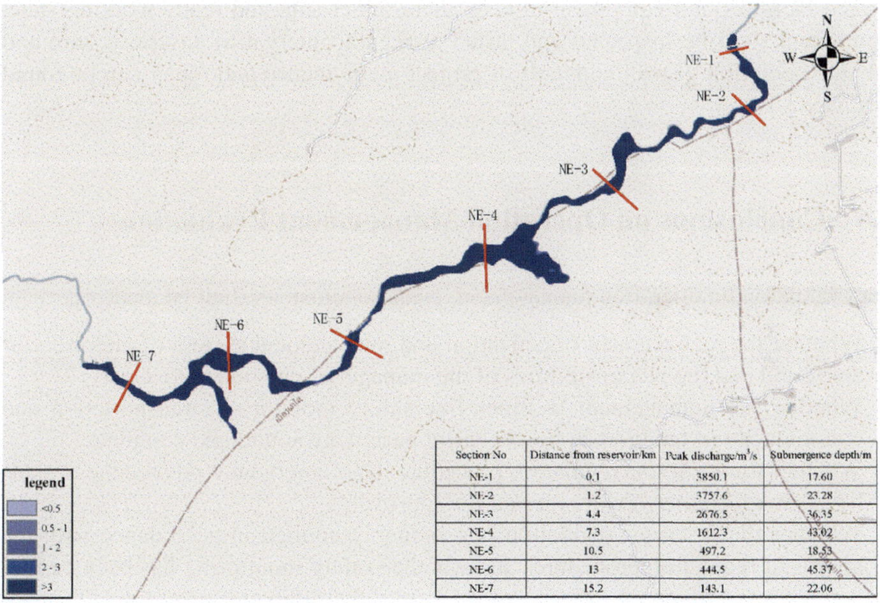

Section No	Distance from reservoir/km	Peak discharge/m^3/s	Submergence depth/m
NE-1	0.1	3850.1	17.60
NE-2	1.2	3757.6	23.28
NE-3	4.4	2676.5	36.35
NE-4	7.3	1612.3	43.02
NE-5	10.5	497.2	18.53
NE-6	13	444.5	45.37
NE-7	15.2	143.1	22.06

Fig. 5.12 Flood arrival depth chart

construction and reconstruction, operation and maintenance, inspection and monitoring, safety evaluation, management system, etc. Those data and documents that are missing or have problems shall be clarified and supplemented, reviewed and corrected.

5.5 Evaluation of Maintenance and Repair of Dam Operations

Project maintenance and repair shall include inspection, testing, maintenance and repair of hydraulic structures in reservoir project, gate and hoist, monitoring facility, traffic and communication facilities for flood control, backup power, etc. and shall include prevention and control of biological destruction that has impact to the safety of dam. (Biological destruction means damages to dam like leakage that are caused by holes or caves of ant, badger, snake, rat, eel etc. and big plant root holes.)

In evaluation of project maintenance and repair, review shall be made to see if the reservoir management unit or competent authority (the owner) has worked out plan for maintenance and repair in accordance with the laws and regulations and technical standards for safety management of the dam, has allocated the expenditure for maintenance and repair, carried out regular maintenance and repair to dam related facilities and equipment to keep them under safe and ready working state. Equipment shall be inspected and tested at regular interval to ensure its safe and reliable operation. (Specific details of project maintenance and repair can be found in Chap. 3.)

5.6 Conclusions on Operation Management Evaluation

For evaluation on operation management, clear conclusions shall be made on:

- Whether the management organization and management system of the reservoir are sound and the responsibilities of the managerial personnel are clear;
- Whether the management facilities like safety monitoring, transportation and communication facilities in flood control period are sound and complete;
- Whether the reservoir regulation procedures and emergency preparedness plans have been formulated and reported for approval;
- Whether the reservoir regulation is reasonably applied in accordance with the approved regulation procedures and whether safety monitoring has been carried out as per specifications to have knowledge of the safety performance of the dam in time;
- Whether the dam has been well maintained and repaired and whether they are under safe and ready working state.

When the five aspects are well done and the reservoir can operate safely in accordance with the designed conditions and functions, the operation management of the dam can be assessed as **Standardized**.

When the five aspects are partly well down, and the reservoir can basically operate safely according to the designed conditions and functions, the operation management of the dam can be assessed as **Fairly Standardized**.

When most of the aspects are failed and the reservoir cannot operate safely according to the designed conditions and functions, the operation management of the dam shall be assessed as **Non-standardized**.

References

1. Silva LC, de Oliveira Mattos UA, da Silva JC (2023) Dams safety: an analysis of safety management and the difficulties of its supervision. Ciênc Nat 45:e6. https://doi.org/10.5902/2179460x72372
2. Ribas JR, Severo JCR, Felix LCG. et al (2021) Prioritization of hydroelectric power plant earth dam safety procedures: a multi-criteria approach. Environ Earth Sci 80:815. https://doi.org/10.1007/s12665-021-10096-9
3. World Bank good practice note on dam safety: new guidance on managing risks associated with dams. World Bank. https://www.worldbank.org/en/topic/watersupply/publication/good-practice-note-on-dam-safety-new-guidance-on-managing-risks-associated-with-dams
4. Zhu X, Bao T, Yeoh JKW et al (2021) Enhancing dam safety evaluation using dam digital twins. Struct Infrastruct Eng 1–17. https://doi.org/10.1080/15732479.2021.1991387
5. Tiller SR (2012) Organizational structure and management systems. Leadersh Manage Eng 12:20–23. https://doi.org/10.1061/(asce)lm.1943-5630.0000160
6. Dam Safety Management Center of the Ministry of water Resources (2017) Reservoir dam safety evaluation guidelines
7. Lai V, Huang YF, Koo CH et al (2022) A review of reservoir operation optimisations: from traditional models to metaheuristic algorithms. Arch Comput Methods Eng 29:3435–3457. https://doi.org/10.1007/s11831-021-09701-8
8. Mustajoki J, Marttunen M (2019) Improving resilience of reservoir operation in the context of watercourse regulation in Finland. EURO J Decis Process 7:359–386. https://doi.org/10.1007/s40070-019-00099-0
9. Giuliani M, Lamontagne JR, Reed PM, Castelletti A (2021) A state-of-the-art review of optimal reservoir control for managing conflicting demands in a changing world. Water Resour Res 57. https://doi.org/10.1029/2021wr029927
10. Jeon J, Lee J, Shin D, Park H (2009) Development of dam safety management system. Adv Eng Softw 40:554–563. https://doi.org/10.1016/j.advengsoft.2008.10.009
11. Sheng J, Xiang Y, Yang D, Yan J, Dong K (2022) Key technologies and applications of safety diagnosis and smart management of reservoir dams. Chin J Geotech Eng 44(7):1351–1366. https://doi.org/10.11779/CJGE202207013
12. SFOE SFO of E dam safety. In: www.bfe.admin.ch. Accessed 17 Sept 2023. https://www.bfe.admin.ch/bfe/en/home/research-and-cleantech/research-programmes/dam-safety.html
13. Metz F, Glaus A (2019) Integrated water resources management and policy integration: lessons from 169 years of flood policies in Switzerland. Water 11:1173. https://doi.org/10.3390/w11061173
14. Miranda GM, Reynard E (2020) Integrated water resources management in federations: the examples of Brazil and Switzerland. Water 12:1914. https://doi.org/10.3390/w12071914

15. Schwager M, Askarinejad A, Friedli B, Oberender P, Pachoud A, Pfister L (2023) Swiss dam safety regulation: framework, recent changes and future perspectives. https://doi.org/10.1201/9781003440420-24
16. U.S. Department of Homeland Security, Federal Emergency Management Agency (2013) Federal guidelines for dam safety. Create Space Independent Pub
17. U.S. Department of Homeland Security, Federal Emergency Management Agency, National Dam (2013) The national dam safety program final report on coordination and cooperation with the European Union on embankment failure analysis (FEMA 602/August 2007). Create Space Independent Pub
18. Morán-Tejeda E, Lorenzo-Lacruz J, López-Moreno JI et al (2012) Reservoir management in the Duero Basin (Spain): impact on river regimes and the response to environmental change. Water Resour Manage 26:2125–2146. https://doi.org/10.1007/s11269-012-0004-6
19. Larraz B, San-Martin E (2021) A tale of two dams: the impact of reservoir management on rural depopulation in Central Spain. Water Resour Manage 35:4769–4787. https://doi.org/10.1007/s11269-021-02938-9
20. Serra-Llobet A, Tàbara JD, Saurí D (2013) The Tous dam disaster of 1982 and the origins of integrated flood risk management in Spain. Nat Hazards 65:1981–1998
21. Alonso Franco M, Yague Córdova J (1988) The management of dam safety in Spain
22. Environment and Climate Change Canada (2016) Water governance and legislation: shared responsibility—Canada.ca. In: Canada.ca. https://www.canada.ca/en/environment-climate-change/services/water-overview/governance-legislation/shared-responsibility.html
23. Renewable Energy World Content Team (2014) How Canada deals with dam safety issues. In: Renewable Energy World. Accessed 17 Sept 2023. https://www.renewableenergyworld.com/baseload/hydropower/how-canada-deals-with-dam-safety
24. CDA-ACB. About us—Canadian Dam Association (CDA-ACB). In: cda.ca. Accessed 17 Sept 2023. https://cda.ca/about-cda/about-us
25. CDA-ACD. Dam safety publications—Canadian Dam Association (CDA-ACB). In: cda.ca. https://cda.ca/publications/cda-guidance-documents/dam-safety-publications
26. Takeuchi H, Tanaka H (2020) Water reuse and recycling in Japan—history, current situation, and future perspectives. Water Cycle. https://doi.org/10.1016/j.watcyc.2020.05.001
27. MAFF Ministry of Agriculture, Forestry and Fisheries. In: www.maff.go.jp. https://www.maff.go.jp/e/
28. Japan Commission on Large Dams—日本大ダム会議. In: jcold.or.jp. Accessed 17 Sept 2023. https://jcold.or.jp/e/
29. Vasilevskii AG (1996) Domestic and foreign experience in monitoring dam safety. Hydrotech Constr 30:117–120. https://doi.org/10.1007/BF02447679
30. Andronova LA, Bleikher OV, Rogaeva IE et al (2016) Interdisciplinary model for Russia and China transboundary water management in the Amur-Heilong River region. Adv Sci Lett 22:1601–1605. https://doi.org/10.1166/asl.2016.6698
31. Tatiana A, Вартумян Арушан A, Irina S et al (2017) Current state and problems of development of the water management in the south of Russia. Юг России: экология, развитие. https://doi.org/10.18470/1992-1098-2017-1-62-72
32. Mao Y, Ye A, Liu X, Ma F, Deng X, Zhou Z (2016) High-resolution simulation of the spatial pattern of water use in continental China. Hydrol Sci J 61(14):2626–2638. https://doi.org/10.1080/02626667.2016.1153102

Chapter 6
Evaluation on Flood Control Ability

6.1 Introduction

A flood is any relatively high streamflow overtopping the natural or artificial banks in any reach of a stream [1, 2]. Floods occur for many reasons, such as long-lasting rainfall over a broad area, locally intense thunderstorm-generated rainfall, or rapid melting of a large snow pack with or without accompanying rainfall. Because floods result from many different circumstances, not all floods are equal in magnitude, duration, or effect [3]. Placing floods in context allows society to address such issues as the risk to life and property, and to study and understand the environmental benefits of floods.

The primary function of water resources is to prevent flood and drought disasters and to protect human life and property. Flood control measures for water conservancy projects refer to the use of water conservancy projects to stop and regulate floods, reduce flood peaks, or divert and delay floods in order to alter the natural movement of floods and achieve the purpose of flood control and damage reduction. The level of flood control is directly related to the importance of the water conservancy project itself or the object of flood control, the severity of the flood disaster and its impact, and the level of development of the national economy [4].

After the reservoir is completed and put into operation, the flood control safety is likely to be affected by the following circumstances. (1) The result of the flood calculation originally designed may change due to the expansion of hydrological data. (2) The human activities in the reservoir controlled area change the discharge and confluence conditions, which further affect the originally designed flood calculation. (3) The dam project and its protection objects may be different from those in the original design; therefore, the flood control standard applicable to the reservoir should be reviewed in accordance with the current status. Based on the hydrological data used during the design phase and the expanded hydrological data during the operation phase, and taking into account the impact of human activities in the upstream and downstream areas after the construction of the dam and the current

Y. Xiang et al., *On-site Inspection and Dam Safety Evaluation*, Hydroscience and Engineering, https://doi.org/10.1007/978-981-97-3987-5_6

status of the reservoir project, the design flood review and flood regulation calculations were carried out to assess whether the current flood resistance of the dam meets the requirements.

Therefore, the safety assessment of the reservoir flood protection should be carried out as part of the regular safety assessment to ensure the safety of the dam, which includes the review of the flood protection standard, the calculation of the designed flood check, the calculation of the flood routing and the review of the flood resistance of the dam.

6.2 Evaluation on Flood Control Standard

According to statistical analysis of the data regarding dam breaks of more than 3500 reservoirs in China since 1954, over 50% of the dam breaks were the results of flood overtopping and the causes of these overtopping are the inadequate flood control standards of the reservoirs. In the first ten days of August in 1963, Haihe Basin encountered catastrophic flood, causing dam breaks in 319 reservoirs in total in Hebei province including 5 medium sized reservoirs, 17 small sized (Scale 1) reservoirs and 297 small sized (Scale 2) reservoirs. The majority of the dam breaks were caused by overtopping. From August 2nd to 8th of 1957, Zhumadian area of Henan province was attacked by Typhoon III and heavy rains fell in the area with an average precipitation of 1028.5 mm and therefore big flood rarely seen in history occurred, causing dam breaks of two large reservoirs, Banqiao Reservoir and Shimantan Reservoir as the result of overtopping, which further led to chained dam breaks of 60 downstream small and medium sized reservoirs. These dam breaks caused a death of more than 26,000 people, the most painful event in the history of dam work in the world [5] (Table 6.1).

Flood control standards are of great importance for dam safety. However, inadequate flood control standard is also the major safety problem commonly existing in reservoir dams in our country [6]. Therefore, during the periodic evaluation of dam safety, the flood control standard shall be adjusted for those reservoirs where the flood control standard does not comply with Table 6.2. The adjusted flood protection

Table 6.1 Statistical analysis of dam breach reason

Causes of dam breach	1954–2021	
	Number of dam break	Proportion of total dam breach (%)
Flood overtopping	1816	51.04
Engineering quality defect	1325	37.24
Misadministration	175	4.92
Other	242	6.80
Total	3558	100.00

standard shall be used as the basis for the flood routing calculation and flood protection capability review. If the current flood control standard of the reservoir is equal to or better than the requirements in Table 6.2, the original flood control standard of the reservoir should be used. Otherwise, the reservoir flood control standard must be adjusted and approval of the adjusted standard is required [7, 8].

In the design for reinforcement of the completed water control project, the standard for emergency operation of flood by permanent hydraulic structures shall be improved. For general water control projects, if it is really difficult to enforce the classification and design standard of the water control project due to too large investment or technical reasons, the flood control standard may be improved in stages after professional discussion and approval by the competent authority for those general water control projects, except for those projects located in exceptionally important places. However, the short-term and emergency flood control standard shall not be lower than that shown in Table 6.3.

Table 6.2 Flood control standard

Hydraulic structure level	Flood control standard (return period)				
	Mountainous and hilly areas			Plain and coastal areas	
	Design	Check		Design	Check
		Concrete dam, masonry dam and other hydraulic structures	Earth dam and rockfill dam		
1	1000 ~ 500	5000 ~ 2000	PMF or 10,000 ~ 5000	300 ~ 100	2000 ~ 1000
2	500 ~ 100	2000 ~ 1000	5000 ~ 2000	100 ~ 50	1000 ~ 300
3	100 ~ 50	1000 ~ 500	2000 ~ 1000	50 ~ 20	300 ~ 100
4	50 ~ 30	500 ~ 200	1000 ~ 300	20 ~ 10	100 ~ 50
5	30 ~ 20	200 ~ 100	300 ~ 200	10	50 ~ 20

Table 6.3 Standard for short term and emergency operation of flood for permanent hydraulic structures

Type	Project scale				
	1	2	3	4	5
Earth fill dam, rock fill dam, dry masonry dam	2000	1000	500		
Concrete dam, cemented masonry dam	1000	500	300		

6.3 Evaluation on Design Flood Calculation

The designed flood includes designed peak discharge, designed flood hydrograph, spatial pattern of designed flood and seasonal designed flood, etc. The designed flood can be derived from precipitation and run-off variation (See Fig. 6.1). For reservoirs with greater natural river-channel storage capacity, calculation of designed flood shall be conducted using data of reservoir inflow flood. If dam-site flood data is used at design stage, reservoir inflow flood data should be used instead, or the adverse impact by the reservoir inflow flood shall be estimated.

For those small and medium sized reservoirs where run-off variation is hard to be gathered, designed storm rain in the area can be calculated based on the precipitation data. Then the designed flood of corresponding frequency is derived from run-off yield and confluence flow calculation [9].

For reservoirs where data of storm rain and flood is not available, the designed flood can be calculated through comprehensive regional analysis by using measured or investigated storm rain data from the adjacent areas [10].

6.3.1 *Basic Data for Flood Control Standard Review*

After putting the reservoir project into operation, with the extension of operation period, the hydrological series is also increasing, and the flood results of the original design may change. With the development of social economy, especially the construction of important facilities downstream of reservoir and the development of cities and towns, the tasks and scales of reservoir projects have changed. With the change of reservoir capacity caused by climate, water and soil loss, reclamation and other reasons over the decades since the reservoir was put into operation, as well as the problems of project structure, the flood control standard and its flood resistance

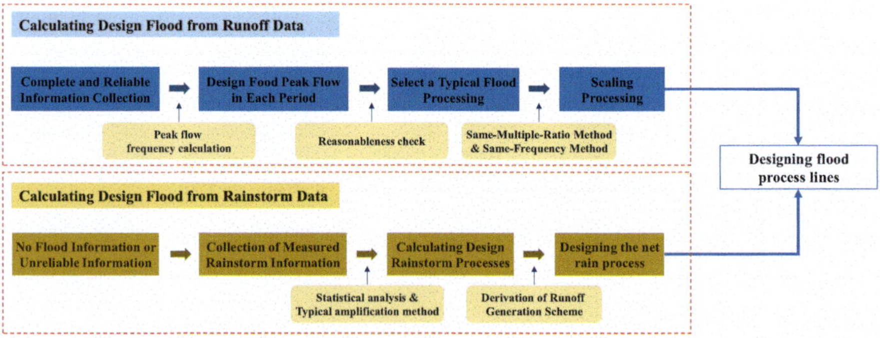

Fig. 6.1 Using runoff and rainstorm information to derive design floods

capacity of the project will be affected [11, 12]. Relevant data should be collected systematically when reviewing and evaluating flood control standards, including:

- Design flood calculation in dam design documents.
- Rainfall data of relevant rainfall stations in the watershed during the application period.
- Flood data measured over the years at relevant hydrological stations and data on the influence of human activities on hydrological parameters in the watershed during the application period.
- Physical geography and river characteristics.
- Water level-reservoir capacity curve.
- Water level-discharge curve.
- Social, economic and demographic data of downstream flood inundated area.
- Relevant data on reservoir catchment area and water diversion works, water and soil conservation works, breach and dam break, etc.
- Engineering operation data.
- Dam acceptance and previous safety appraisal data.
- Relevant data on the structure of the water retaining and discharging structures of the project.

For the above basic data, it is necessary to check before checking the flood control standard, especially the flood control checking is aimed at the current situation of the project, the impact of reservoir discharge on downstream society and economy under current conditions and the flood resistance capacity of the project. The review work mainly includes:

- Review the impact of human activities on hydrological data in the catchment. Generally, the data of human activities in the catchment need to be obtained through various surveys.
- Review the physical geography and river characteristics of the watershed, such as geographical location, catchment area, shape, average slope of the main river, climatic and meteorological characteristics, etc. Field reconnaissance and review are also required if necessary.
- Check the reliability, consistency and representative of flood series calculated by hydrology. The reliability of data series is an important guarantee for the accuracy of hydrological calculation results. The data used should be checked during hydrological calculation to ensure that the data are correct and reliable. Data consistency refers to the fact that the runoff and confluence conditions of watersheds and rivers producing hydrological data for each year have not changed fundamentally during the observation and investigation period, such as the construction of reservoirs in the upstream or the breakdown of embankments, river diversion and other events, which obviously affect the consistency of data, the data should be converted to a uniform basis to ensure consistency; Representation of data series refers to whether the statistical characteristics of existing data series can well reflect the overall statistical characteristics, and the representative of data series should be evaluated.

- Check the calculation conditions and basic data of flood diversion, mainly including water level, reservoir capacity curve and discharge capacity curve.
- Check the structural dimensions of the structures, mainly including the shape measurement of the dam structure (crest elevation, crest elevation of wave-proof wall, slope ratio of upstream and downstream dams, crest length, etc.) and the structural control dimension measurement of discharge structures (crest elevation, elevation of inlet floor, width of discharge orifice, height of side wall, orifice section, etc.).
- Review basic meteorological data, including wind speed and direction near dam site.

6.3.2 Derivation of Designed Flood from Run-Off Information

Where more adequate measured flow information is available at the design reach, the design flood can be derived from the flow information and the procedure for its calculation is broadly as follows:

- Selection of reliable, consistent and representative information through a review of flood information.
- Screening the information and selecting flood flows and volumes from the annual flood that meet the requirements to form various statistical series; Frequency calculation shall be carried out by using measured series data of dam-site flood or reservoir inflow flood at the design stage, historical flood investigation data, adding measured series data of dam-site flood or reservoir inflow flood in the operation period, series data for extension of peak discharge and flood discharge in different stages. If no measured inflow flood data is available in operation period, the inflow flood data can be reversely derived by using records of measured reservoir water level and outflow and water level-storage capacity curve.
- Deriving design flood peaks and design flood volumes through flood frequency calculations; Pearson type III distribution should be used for line type of frequency curve. Other types of lines can be also used under special circumstances, subject to analysis and demonstration. Moment method or other parameter estimation methods can be adopted to estimate preliminarily the statistics parameters of the frequency curve. Following this, the preliminarily estimated statistical parameters can be adjusted by using empirical curve fitting method or optimized curve fitting method. When empirical curve fitting method is used, all the point data should be fitted. When fitting is not well done, more reliable point data of big flood can be taken into account.
- Deriving the design flood process line by selecting a typical flood process line and amplifying it according to the design flood peak and design flood volume, either by the same ratio amplification method or the same frequency amplification method. The designed flood belongs to rare flood. Flood with reliable data and

flood that is representative and is less favorable for flood control shall be chosen as typical flood hydrograph when the designed flood is used for determination of project scale. In the review of designed flood, the typical big flood newly added into the flood series shall be analyzed and the new typical flood hydrograph shall be supplemented if necessary. For enlargement of typical flood hydrograph, the design requirements for flood control and the characteristics of flood in the basin shall be taken into account. If flood peak and flood volume both play a role in the project flood control, a controlled enlargement in same frequency can be adopted in accordance with the designed peak and volume. When flood control is mainly controlled by flood peak or flood volume at a certain period of time, controlled homogeneous multiple enlargement can be adopted as per designed flood peak or flood volume at a certain period of time.

6.3.3 Derivation of Designed Flood from Rainfall Data

In practice, it is often the case that there is no or insufficient flow information at the project site, making it impossible to derive design floods based on flow information. At the same time, in recent decades, the influence of human activities on the hydrological process has become more significant, and many watersheds have been built with a large number of water conservancy projects and soil and water conservation projects, which has affected the consistency of flow information. At the same time, with rapid economic and social development, urbanisation has accelerated, leading to huge changes in the substratum of the basin or region, resulting in significant changes in the original basic hydrological laws [13, 14]. The extrapolation of design flood from storm data is an indirect method that aims to extrapolate the design storm through the local flow production pattern to the design net rainfall process, and the design flood process through the catchment confluence pattern. The following steps shall be taken for derivation of designed flood from rainfall data:

- When there are more and fairly even distributed precipitation stations in the basin area and long-term and reliable storm rain data is available, the average storm rain series for different diachronic areas can be directly selected for calculation of storm rain frequency and derivation of designed storm rain. The designed storm rain shall include the rainfall in different diachronic points or areas in the designed basin area, time distribution and area distribution of storms. When the basin area is small and average storm series for different diachronic areas are not available, the designed area storm can be indirectly calculated by using the corresponding diachronic designed point rainfall and storm point-area relationship. When the basin area is very small, the designed point storm can be used as the designed area average storm in the basin area.
- A number of observation stations shall be selected from the designed basin area or from the adjacent areas to make frequency analysis on the required various diachronic storms. In addition, regional integration shall be carried out to reasonably determine the designed point rainfall in the basin area. Also, designed

diachronic point rainfall required by the project can be calculated with reference to isopleths map of storm statistical parameters.

- As the actual precipitation process changes and is complicated, different rainfall pattern will greatly influence the calculation of flood peak flow. Time distribution of storm shall be designed using integrated rainfall pattern of multiple storms in the region or typical rainfall pattern of heavy storm that is representative. At present, the designed rainfall in varied time durations still adopts same frequency control. However, time durations under control should not be too many in number, generally, two to three are suitable.
- The area distribution of the designed storm rain shall be carried out by homogeneous multiple enlargement in accordance with the integral or typical area distribution that conforms to the characteristic of big storm area distribution, the designed area rainfall in the basin area as control. It can be also calculated by homogeneous frequency enlargement, using the designed area rainfall in sub areas.
- Run-off yield calculation shall be conducted based on the storm calculation result by using storm-run-off correlation method, loss deduction method to get the designed net rainfall duration curve. Then, the designed flood hydrograph is derived based on the designed net rainfall duration curve, by using unit hydrograph method and method of river network confluence curve. If the basin area is small, the designed flood hydrograph can be calculated using rational formula. In respect of runoff yield calculation in humid and semi-humid areas, the influenced rainfall at initial stage or the storage capacity at the beginning of precipitation can be used as parameters when rainfall-runoff correlation method is adopted for run-off yield calculation. For the influenced rainfall at initial stage and the storage capacity at the beginning of precipitation under design conditions, the analysis values based on measured data of a number of heavy storm rains can be adopted. When regional integration method is used for analysis of initial loss in the loss deduction method, the maximum initial loss value can be used to establish relationship with runoff area. The later stage loss can establish relationship with rainfall intensity when conditions are ready. In the basins where the underlying surface conditions and storm rain distribution is uneven, partitioned loss deduction method can be used. The runoff yield parameter is closely related to runoff yield area. If the runoff yield parameter analyzed and calculated without deduction of runoff yield area is used, it would probably lead to a smaller runoff. For confluence flow calculation in the hilly areas where the basin area is within 1000 km^2, unit hydrograph method can be used. Rational formula and unit hydrograph method can be adopted when the basin area is smaller than 300 km^2. When the basin area is above 1000 km^2 and precipitation distribution is very uneven, confluence flow curve of river network can be used.
- When the basin area is smaller than 1000 km^2 and measured storm data is not available, the designed flood can be calculated by using approved storm and run-off chart. When necessary, the parameters may be modified as appropriate.

- For reservoirs where the possible maximum flood is taken as the standard for emergency operation of flood, the calculation results of the possible maximum precipitation and possible maximum flood shall be reviewed.

6.4 Flood Routing Calculation

The mode of operation of the reservoir is reviewed in accordance with the tasks performed by the reservoir and the changes in the operating environment and functions. On this basis, flood routing is carried out. The storage capacity of the reservoir for flood control, flood retention and flood routing and the corresponding characteristic water level for flood control shall be determined in accordance with the standard for flood control confirmed after the review and the standard for short-term and emergency operation of flood.

For reservoirs that do not provide downstream flood control, the flood routing calculation shall be performed in accordance with the flood routing principles determined in the review and based on the discharge capacity of the spillway structures to determine the reservoir's flood retention and flood routing storage, the corresponding flood control characteristic water level, and the maximum discharge volume. For reservoirs providing downstream flood control, in addition to those described above, the reservoir's flood control storage and the corresponding characteristic water level for flood control shall be determined in accordance with the size and type of downstream protected objects, the flood control standard and the permissible discharge volume for downstream river channels, and taking into account other flood control measures.

6.4.1 Flood Routing Calculation Preparation

Before flood routing calculation, the calculation conditions shall be ascertained and related data and information shall be reviewed.

1. To verify the starting water level, the following provisions shall be observed:
    - If the dam design is not modified, the limit water level in flood season determined in the original design shall be used.
    - If the dam has been reinforced or expanded and reconstructed, or the human activities in reservoir control basin area have changed greatly the designed flood, the approved limit water level in flood season that is reconfirmed shall be used.
    - For some reservoirs that have operated for a number of years, where the original designed flood standard is low and does not meet the requirements of current specifications or where exist potential safety problems in dam structure, the flood routing calculation shall be carried out in accordance with the

water level limit for flood season specified in the flood standard as required in the original design or in specifications, subject to approval from competent authority, for operation with lowered water level limit for flood season.

2. In evaluation on flood control safety, flood routing calculation shall be carried out as per the regulation mode specified in the original design. If it is considered that the mode under the original design does not make sense, more reasonable operation modes can be used for flood routing calculation provided that other targets of development of the reservoir are not changed or influenced. The changed operation mode shall be submitted to the original approver for examination and shall be implemented upon approval.
3. Water level-storage capacity curve shall be reviewed. For reservoirs where sediments are heavier in those silt-carrying rivers, the relation curve of reservoir water level and storage would change greatly due to accumulation of sediments after many years' operation. The newest water level-storage curve shall be used for flood routing calculation. And furthermore, the cycle of review shall be correspondingly shortened.

6.4.2 Spillway and Flood Outlet Selection and Design

Flood discharge ability curve of the discharging structures shall be reviewed. For water conveyance structure that has function of flood discharge, flood routing calculation can be carried out by adding its discharge capacity into discharge ability curve.

Spillways and flood outlets should be designed to safely convey major floods to the watercourse downstream from the dam and to prevent overtopping of the dam. They are selected for a specific dam and reservoir on the basis of release requirements, topography, geology, dam safety, and project economics.

1. **Gated or ungated spillways**

An ungated spillway releases water whenever the reservoir elevation exceeds the spillway crest level. A gated spillway can regulate releases over a broad range of water levels. Ungated spillways are more reliable than gated spillways. Gated spillways provide greater operational flexibility and large discharge capacity per unit length. Operation of gated spillways and/or their regulating procedures should generally ensure that the peak flood outflow does not exceed the natural downstream flow that would occur without the dam [15, 16]. The selection of a gated or ungated type of spillway for a specific dam depends upon site conditions, project purposes, economic factors, costs of operation and maintenance, and other considerations. The following paragraphs focus on considerations that influence the choice between gated and ungated spillways:

- Discharge capacity—For a given spillway crest length and maximum allowable water surface elevation, a gated spillway can be designed to release higher discharges than an ungated spillway because the crest elevation may be lower than the normal reservoir storage level. This is a consideration when there are limitations on spillway crest length or maximum water surface elevation.

- Project objectives and flexibility—Gated spillways permit a wide range of releases and have capability for pre-flood drawdown.
- Operation and maintenance—Gated spillways may experience more operational problems and are more expensive to construct and maintain than ungated spillways. Constant attendance or several inspections per day by an operator during high water levels is highly desirable for reservoirs with gated spillways, even when automatic or remote controls are provided. During periods of major flood inflows where automatic or remote controls are not provided, the spillway should be constantly manned. Gated spillways are more subject to clogging from debris and jamming from ice, whereas, properly designed ungated spillways are basically free from these problems. Gated spillways require regular maintenance, and, as a minimum, an annual operation test for safety purposes. However, ungated spillways can have flashboards, trip gates, stop log sections, etc. which can have operational problems during floods and may require constant attendance or several inspections per day during high water levels.
- Reliability—The nature of ungated spillways reduces dam breach potential associated with improper operation and maintenance. Where forecasting capability is unreliable, or where time from the beginning of runoff to peak inflow is only a few hours, ungated spillways are more reliable, particularly for high hazard structures. Consequences of failure of operation equipment or errors in operation are more severe for gated spillways.
- Data and control requirements—Gated spillways require reliable real time hydrologic and meteorological data to make proper regulation possible.
- Emergency evacuation—Unless ungated spillways have removable sections such as flashboards, trip gates, or stop log sections, they cannot be used to evacuate a reservoir during emergencies. The capability of gated spillways to draw down pools from the top of the gates to the spillway crest can be an advantage when emergency evacuation to reduce head on the dam is a concern.
- Economics and selection—Designs to be evaluated should be technically adequate alternatives. Economic considerations often indicate whether gated or ungated spillways are selected. The possibility of selecting a combination of more than one type of spillway is also a consideration. Final selection of the type of crest control should be based on a comprehensive analysis of all pertinent factors, including advantages, disadvantages, limitations, and feasibility of options.

2. **Several common formulas for calculating discharge capacity of spillway**

The purpose of discharge capacity review is to determine the current overflow capacity of spillway and to compare the current overflow capacity with the original design overflow capacity for flood diversion calculation, review of water surface line of discharge tank and evaluation of flood discharge safety. Commonly used flood discharge structures include spillways, tunnels, etc. The calculation formulas for discharge capacity of practical weir, wide-crested weir and hump weir of spillway are introduced below [17].

(1) Formula for calculating discharge capacity of open WES type practical weir (Fig. 6.2):

$$Q = cm\,\varepsilon\sigma_s B\sqrt{2g}H_0^{3/2} \tag{6.4.1}$$

$$\varepsilon = 1 - 0.2[\zeta_K + (n-1)\zeta_0]\frac{H_0}{nb} \tag{6.4.2}$$

where

- Q flow, m^3/s;
- B total net width of overflow weir, m;
- H_0 total head on weir with input flow rate, m;
- m flow coefficient of WES practical weir;
- c correction factor of upstream weir slope influence, lead on upstream weir surface is taken as 1.0;
- ε shrinkage coefficient of pier side;
- ζ_0 shape factor of middle pier;
- ζ_K shape factor of side pier;
- σ_s submergence coefficient.

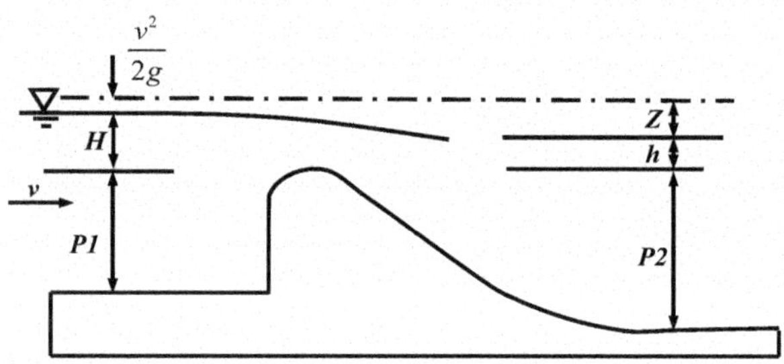

Fig. 6.2 Sketch for calculation of practical weir

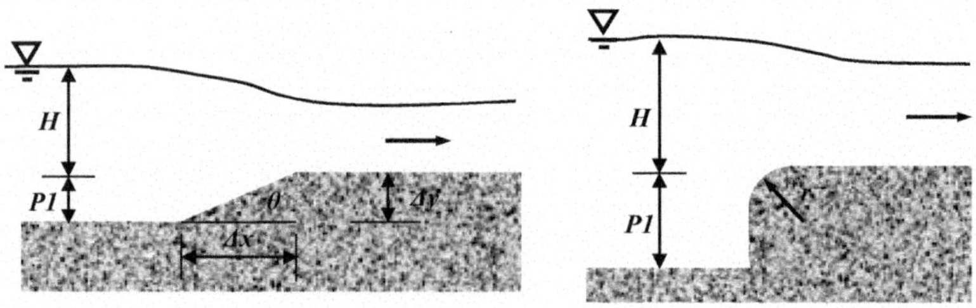

Fig. 6.3 Calculation sketch of wide-crested weir

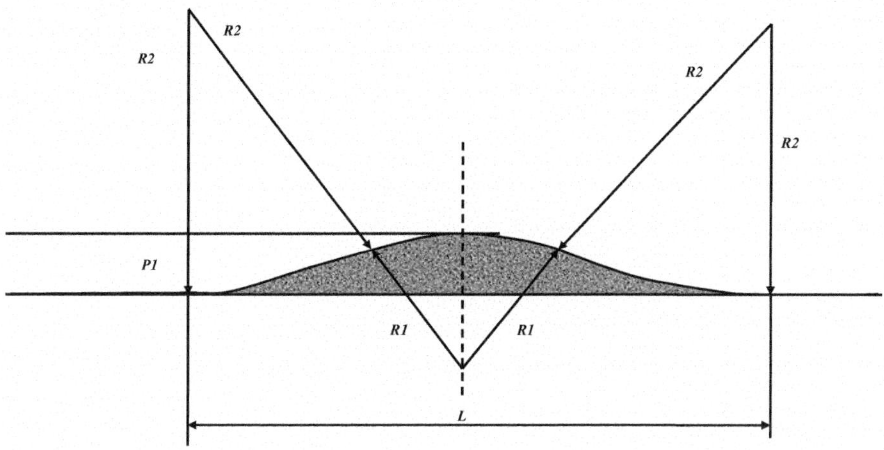

Fig. 6.4 Calculation sketch of hump weir

(2) Formula for calculating discharge capacity of wide-crested weir (Fig. 6.3):

$$Q = m\varepsilon B\sqrt{2g}H_0^{\frac{3}{2}} \tag{6.4.3}$$

$$\varepsilon = 1 - 0.2[\zeta_K + (n-1)\zeta_0]\frac{H_0}{nb} \tag{6.4.4}$$

(3) Formula for calculating discharge capacity of hump weir (Fig. 6.4):

$$Q = m\varepsilon B\sqrt{2g}H_0^{\frac{3}{2}} \tag{6.4.5}$$

$$\varepsilon = 1 - 0.2[\zeta_K + (n-1)\zeta_0]\frac{H_0}{nb} \tag{6.4.6}$$

6.4.3 Flood Regulation Calculation Method

Static capacity method should be adopted for flood routing calculation. For the general large, medium and small sized reservoirs, lake type reservoirs in particular, the result of flood routing calculation is reliable if actually observed data proves that flood routing calculation by only refereeing to static reservoir storage basically reflects actual situation. For large sized reservoirs with dynamic storage accounting for a larger proportion, flood routing calculation can be carried out by reservoir designed flood and dynamic storage method when the maximum level for flood

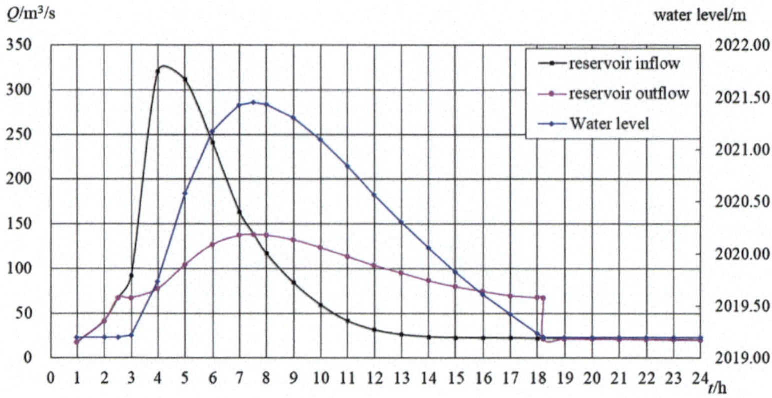

Fig. 6.5 Flood regulation process line of reservoir

routing included in dynamic storage is higher than the calculation result of flood routing of static storage. When dam-site flood is used for the designed flood, the static capacity method should be still used [18, 19].

Reservoir flood regulation calculations require the establishment of water balance equations based on changes in incoming flow, outgoing flow and reservoir storage. As there are two unknown variables in the equation, a single equation cannot be solved, so it needs to be solved jointly using the flood storage boundary conditions. V and q are both functions of the reservoir level H, so the flood storage boundary conditions can be established from the reservoir capacity curve $V = f(H)$ and the outflow curve $q = f(H)$. List trial algorithm is often used in flood control calculation (Fig. 6.5). The water balance equation is:

$$(Q_1 + Q_2)\Delta t/2 - (q_1 + q_2)\Delta t/2 = V_2 - V_1 \tag{6.4.7}$$

where

Q_1, Q_2 the input flow at the beginning and end of the calculation period, m^3/s;
q_1, q_2 the discharge flow at the beginning and end of the calculation period is calculated, m^3/s;
V_1, V_2 reservoir capacity at the beginning and end of calculation period, m^3;
Δt calculating period, *h*.

6.4.4 Flood Regulation Calculation and Flood Forecast

At flood routing calculation, weather forecast should not be considered. However, for those reservoirs provided with better weather forecast conditions, perfect forecast plans and higher accuracy of forecast, flood forecast and pre-discharge operation can be considered as appropriate at flood routing calculation provided that margin of forecast error is expected to be considered.

If flood forecast is considered in the original designed flood routing calculation and the flood forecast plan is reliable after review, flood forecast can be considered in flood routing calculation for evaluation of flood safety, but with caution. Consideration of flood forecast in flood routing calculation is in general studied only when excellent conditions for forecasting are available in those grand rivers.

6.5 Evaluation on Dam Flood Control Ability

6.5.1 Evaluation of Dam Flood Control Capacity

The dam crest freeboard shall be added to the flood resistance characteristic level determined in the flood routing calculation in Sect. 7.4 to give the minimum dam crest elevation or wave wall top elevation that meets the requirements of the flood resistance standard. In addition, these should be compared with the current actual dam crest or wave wall top elevation to assess whether the current flood resistance of the dam meets the requirements.

According to requirements, for earth-rock dam, the dam crest elevation shall not be lower than the checked flood level of reservoir, and the top elevation of soil impervious body shall not be lower than the checked flood level. The crest elevation of upstream wave protection wall on dam crest should be higher than that of wave crest. The difference between the crest elevation and the design flood level or the checked flood level can be calculated by the following formula [20].

$$y = R + e + A \tag{6.5.1}$$

where

y the content of fine particles of soil, calculated by mass percentage, %;
R porosity of soil, %;
e the maximum backwater surface height, m;
A dam safety heightening value, m, see in Table 6.4.

For concrete dams, the dam crest of the water retaining dam section should not be lower than the calculated height of normal operation and very application of static water level plus wave. The superelevation of the top of the wave-break wall upstream of the dam crest above the reservoir static water level shall be determined according to the following formula.

$$\Delta h = h_b + h_z + h_c \tag{6.5.2}$$

where

Table 6.4 Dam safety heightening value, unit: m

Dam grade	Earth rock dam			Concrete dam, masonry dam	
	Design flood level	Check flood level		Normal pool level	Check flood level
		Mountainous and hilly areas	Plain area and coastal area		
1	1.5	0.7	1.0	0.7	0.5
2	1.0	0.5	0.7	0.5	0.4
3	0.7	0.4	0.5	0.4	0.3
4 and 5	0.5	0.3	0.3	0.3	0.2

Δh dam crest superelevation, m;
h_b wave height, m;
h_z height difference from wave centerline to normal storage level or check flood level, m;
h_c dam safety heightening value, m, see in Table 6.4.

When the wave wall of the dam crest has the function of freeboard of the dam, the wave walls shall be inspected and reviewed to make sure if they are stable, solid and enclosed and if they are closely jointed with the dam impervious bodies.

6.5.2 *Evaluation of Flood Control Capacity of Flood Releasing Structures*

The dam flood resistance ability shall be reviewed in accordance with the current state of the dam and actual dimensions of the structure. Therefore, the designed flood level and check flood level and the corresponding maximum discharged volume to ensure a safe flood season under the current dam conditions shall be determined in accordance with related codes and specifications and the result of construction quality evaluation.

To ensure a safe discharging under the designed and check flood conditions, the following aspects of the discharging structures shall be reviewed at flood control safety evaluation.

- The safety of the discharging structures themselves: Including whether the water flow cross sections of the discharging structures are in accordance with the design requirements, whether energy dissipation facility is complete and sound, whether the operation state of gate and hoist is good and whether they can be safely operated in the period of higher water level.
- Impact of discharge to the dam safety: Review shall be made to see whether the discharged flood scours the dam toe and whether atomization influences the dam stability.

- Impact to the safety of downstream river channels and embankments by maximum discharged flow.

For the large sized reservoirs and those medium sized reservoirs that are the key ones for flood control, the discharged design flood, checked flood and range of impact by dam breach, and loss of lives and economic loss to be possibly caused should be estimated.

6.5.3 Impact Scope of Dam Flood Control

The hazard potential of a dam pertains to the potential for loss of human life or property damage in the area downstream of the dam in the event of failure or incorrect operation of a dam. Hazard potential does not refer to the structural integrity of the dam itself, but rather the effects if a failure should occur.

The possible consequences resulting from a dam breach include loss of human life; economic, social, and environmental impacts; damage to national security installations; and political and legal ramifications [21]. Estimates of the potential for loss of human life and the economic impacts of damage resulting from dam breach are the usual bases for defining hazard potential. Social and environmental impacts, damage to national security installations, and political and legal ramifications are not easily evaluated, and are more susceptible to subjective or qualitative evaluation. Therefore, these other considerations do not usually affect decisions on hazard potential. Because their actual impacts cannot be clearly defined. The following factors should be evaluated regarding potential for loss of human life when estimating the potential for fatalities resulting from dam breach:

- The number and location of habitable structures within the potential area inundated by dam breach. The presence of public facilities within the potential area inundated by dam breach that would attract people on a temporary basis (e.g., improved campgrounds, organized or unorganized recreation areas, State or national parks, etc.) requires special consideration.
- Type of flow conditions based on water depths, temperatures and velocities, rate of rise of the flood wave, duration of flood flow, and special hazardous conditions such as the presence of surface waves, debris flow or terrain conditions which may increase potential for loss of lives.

The evaluation of the economic impacts of failure should consider damages to residences; commercial property; industrial property; public utilities and facilities including transmission lines and substations; transportation systems; agricultural buildings, lands, and equipment; dams; and loss of production and other benefits from project operation [22].

The qualitative classification of dam risk [23] can be expressed using a risk matrix based on the level of engineering safety of the reservoir dam and the level of consequences of a dam failure, see Fig. 6.6.

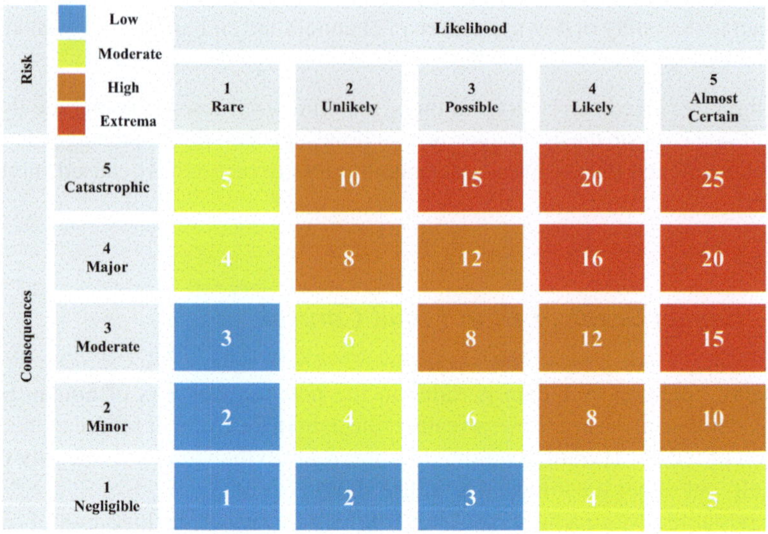

Fig. 6.6 The qualitative classification of dam risk

6.5.4 Exceeding Standard Flood Calculations Using Hydrodynamic Models

The core of the hydrodynamic model is the solution of the basic hydrodynamic equations. For different characteristics of the flow law of the water body, different control equations can be used to describe, such as the description of the general fluid motion of the most basic Navier–Stokes (NS) equations [24], based on this simplified water depth-averaged plane two-dimensional shallow water equations, a set of one-dimensional de Saint–Venant system of equations, and so on. With the continuous development of computational methods and computer technology, as well as the increasing requirements for the accuracy of water motion simulation, hydrodynamic models based on numerical simulation techniques have become more widely used [25].

Hydrodynamic models can be divided into river hydrodynamic models, lake hydrodynamic models and offshore hydrodynamic models according to their application; By model dimension, they can be divided into one-dimensional hydrodynamic models, two-dimensional hydrodynamic models and three-dimensional hydrodynamic models; By common functions, they can also be divided into hydrodynamic models that simulate only water movement and hydrodynamic models that simulate both water movement and convective diffusion transformation of water quality components.

The dam-break flood is a transient, rapid flow, and the sudden release of flood water at the dam-break results in significant inertia and dramatic changes in the flow regime. In addition, the large volume and rapid evolution of the dam-break flood will easily break out of the river channel and inundate both banks to form an extensive floodplain, therefore at least a two-dimensional hydrodynamic model should be used to simulate the evolution of the flood inundation. The two-dimensional hydrodynamic models currently in use at home and abroad include HEC-RAS, MIKE21, Delft3D, etc. [26].

6.5.4.1 Dam Breach Models

Before introducing the two 2D hydrodynamic computational models, HEC-RAS and MIKE21, a brief introduction to the dam breach model is given. The 2D hydrodynamic simulation and the dam breach model are inextricably linked. The dam-break flood downstream of a reservoir is mainly derived from the breach discharge obtained from the dam-break simulation.

Dam breach modelling is a technique for estimating the parameters of a breach (shape, height, width, final shape time, etc.) and thus predicting the peak flow and outflow process of the breach [27]. Concrete dams are often instantaneous or partial breaches, and based on the open channel flow principle, the prediction of peak breach flow and outflow processes is well established and can be simulated using the segmental method.

Earth-rock dam breach is a gradual process with complex physical mechanisms of water-sand equilibrium that have not yet been fully resolved. Earth-rock dam breach modelling is an important area of research and has produced a number of results. This article focuses on the process of earth-rock dam breach. There are currently two main types of earth-rock dam breach models, as follows:

(1) Parametric dam failure models. In the past, scientists collected historical dam breach parameters to build up a database and used regression analysis to derive parametric equations for the final failure parameters in relation to reservoir capacity, dam height, reservoir level and other characteristics, often accompanied by prediction equations for the peak flow rate of the failure, but without the spillway process of the failure. Some of the better known equations are MacDonald, Froehlich (1995), Froehlich (2008), Von Thun-Gillette, and Xu-Zhang [28].
(2) Physical modelling. Based on historical dam breach cases and laboratory scale tests, the physical mechanism of the dam breach process is established, often involving hydraulics, sediment dynamics, geotechnics, etc. Based on the physical mechanism, a numerical model of the dam breach is developed to solve for the failure propagation and simultaneously calculate the failure flow, ultimately yielding the failure parameters and the spillway process. Models such as NWS-BREACH, WINDAM C, DLBREACH and EMBREA are now widely used.

Most dam failure triggers can be attributed to two mechanisms of dam failure, i.e. overtopping and piping, which are also the main mechanisms of most physical dam failure models [29]. The overtopping refers to the process of dam failure in which the reservoir water spreads over the top of the dam, scouring the downstream face of the dam to form an erosion that gradually expands and progresses towards the top of the dam. The erosion reaches the top of the dam and pulls a small rectangular breach from the top of the dam. The water scour carries away the dam material causing the rectangular breach to expand downstream and on both sides of the dam until the scouring action of the water is balanced by the resistance of the dam material or the reservoir is drained.

Piping are rectangular or prototypical paths (initial breaches) formed by the infiltration of water into the dam from the upstream face to the downstream face. As the water passes through the initial breach, it carries away the dam material, causing the breach to widen and creating a cavity in the downstream face. As the spillway expands to a certain point, the upper part of the spillway collapses, creating an open breach, which then follows the pattern of a diffuse breach (Figs. 6.7 and 6.8) [30].

Von Thun and Gillette's equation to define the average breach width and breach formation are [31]:

$$B_{ave} = 2.5h_w + C_b \tag{6.5.3}$$

$$t_f = 0.02h_w + 0.25 \text{ (erosion resistance)} \tag{6.5.4}$$

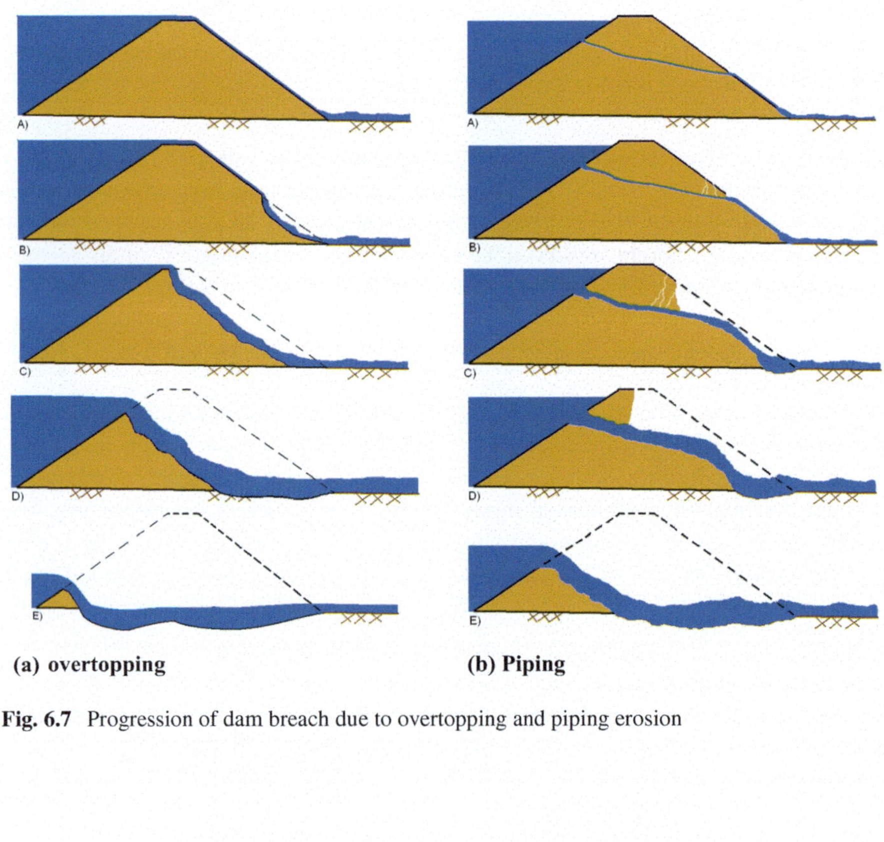

(a) overtopping **(b) Piping**

Fig. 6.7 Progression of dam breach due to overtopping and piping erosion

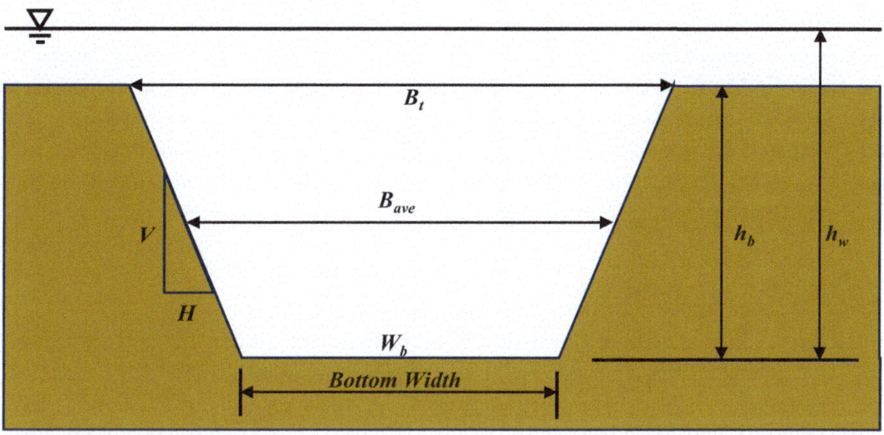

Fig. 6.8 Type diagram of break

$$t_f = 0.015h_w \text{ (easily eroded)} \tag{6.5.5}$$

where

B_{ave} average breach width in meter, m;
V_w depth of water above the bottom of the breach in meter;
C_b coefficient, which is defined by the volume of the reservoir;
t_f breach formation time in hours.

Froehlich's regression equations to define the average breach width and breach formation time are:

$$B_{ave} = 0.1803K_o V_w^{0.32} h_b^{0.19} \tag{6.5.6}$$

$$t_f = 0.00254V_w^{0.53} h_b^{-0.90} \tag{6.5.7}$$

where

V_w the reservoir volume at the time of failure in a cubic meter;
h_b the height of the final breach in meter.

DLBreach is a simplified physically-based Dam/Levee Breach model developed by Wu [32, 33]. It can simulate the breaching processes of homogenous and composite earthen embankments due to overtopping and piping in inland and coastal contexts. The model uses a one-way breach for inland dam and levee breaching by unidirectional flows, and a two-way breach for coastal and estuarine levee and barrier breaching, in which flow may reverse.

DLBreach divides the overtopping breaching process into two stages. The first stage is the intensive breaching or erosion stage, in which the breach flow is supercritical, controlled by upstream. The second stage is the general breach or inlet evolution stage, in which the flow is subcritical, controlled by downstream or both upstream and downstream. In the first stage the breach flow is calculated using the weir flow equation, and in the second stage the breach flow is calculated using the Keulegan equation. The Keulegan equation is the simplified energy equation for steady nonuniform flow with local head loss due to channel contraction and expansion, added the wind driving force to consider the effect of wind. In the case of piping, the breach flow is determined using the orifice flow equation.

DLBreach has been extensively tested by using 50 sets of dam breach data and 4 sets of levee/barrier breach data from laboratory experiments and field case studies. The calculated peak breach discharges, breach widths, and breach characteristic times agree generally well with the measured data. It has been also compared with other embankment breach models, such as NWS BREACH, HR BREACH and WinDAM B, and demonstrated to have more capabilities.

6.5.4.2 HEC-RAS

Two-dimensional flood routing simulation uses the river hydraulic calculation software HEC-RAS developed by the Hydrological Center of the U.S. Engineering Corps. HEC-RAS supports two-dimensional hydrodynamic model, which can simulate dam break/embankment, exceed the standard river type flood, and can meet the calculation requirements of this time. The HEC-RAS 2D non-constant flow model can be used to obtain instantaneous values and process lines of water depth, flow velocity and water surface elevation for the entire computational domain; it also supports the output of distribution layers of water depth, flow velocity and water surface elevation in raster format. In addition, a one-dimensional slice of the computational domain can be created, allowing the flow process, velocity and water depth to be viewed in the same way as in a one-dimensional model. Compared to the large 2D computational results, the 1D section results are more intuitive and easier to use [34].

HEC-RAS provides results in the form of gridded data such as velocity, depth, water surface elevation and hydraulic hazards. Assuming the fluid is incompressible and neglecting vertical velocities, the non-deterministic differential mass conservation equation applies:

Continuous equation:

$$\frac{\partial H}{\partial t} + \nabla \cdot hV + q = 0 \tag{6.5.8}$$

Momentum equation:

$$\frac{\partial V}{\partial t} + V \cdot \nabla V = -g\nabla H + v_t \nabla^2 V - c_f V + fk \times V \tag{6.5.9}$$

$$\frac{\partial u}{\partial t} + u\frac{\partial u}{\partial x} + v\frac{\partial u}{\partial y} = -g\frac{\partial H}{\partial x} + v_t\left(\frac{\partial^2 u}{\partial x^2} + \frac{\partial^2 u}{\partial y^2}\right) - c_f u + fv \tag{6.5.10}$$

$$\frac{\partial v}{\partial t} + u\frac{\partial v}{\partial x} + v\frac{\partial v}{\partial y} = -g\frac{\partial H}{\partial x} + v_t\left(\frac{\partial^2 v}{\partial x^2} + \frac{\partial^2 v}{\partial y^2}\right) - c_f v + fu \tag{6.5.11}$$

where

H water level elevation, m;
h water depth, m;
V current speed, m/s;
q lateral inflow, m^2/s;
g gravity acceleration, 9.81 m/s^2;
υ_t horizontal kinematic viscosity, m^2/s;
c_f riverbed bottom roughness;
f Coriolis parameters;
k unit vector in vertical direction;
t time;
u, v the components of the flow velocity V in the x and y directions;
R hydraulic radius, m.

The 2D model construction mainly includes the division of the simulation area and the definition of the grid size selection. In the process of selecting suitable areas for 2D modelling, the 2D grid should be constructed by selecting as many full flood evolution areas as possible based on the increased computational efficiency. During the 2D modelling process, different areas are selected to study the different effects of the dam breach flood. At the same time, the actual lower boundary is placed in such a way that the boundary error disturbance does not propagate far enough to the lower boundary of the study area.

The numerical calculation of HEC-RAS two-dimensional hydrodynamic model combines the finite volume method with the finite difference method [35]. The computational grid is unstructured, the non-boundary grid is square and the boundary grid is irregular polygon, as shown in Fig. 6.9. The edges of each grid are similar to river sections and the topography can be extracted. Therefore, at low grid density, enough terrain details can be extracted to ensure the accuracy of the model [34, 36]. Figure 6.10 shows the Hecras operating interface and the calculated flood inundation map.

The HEC-RAS breakdown module is used to solve the problem, which is a semi-parameter semi-physical model. It mainly calculates the change of outflow and water level of earth-rock dam break with time under the condition of overtopping or pipe

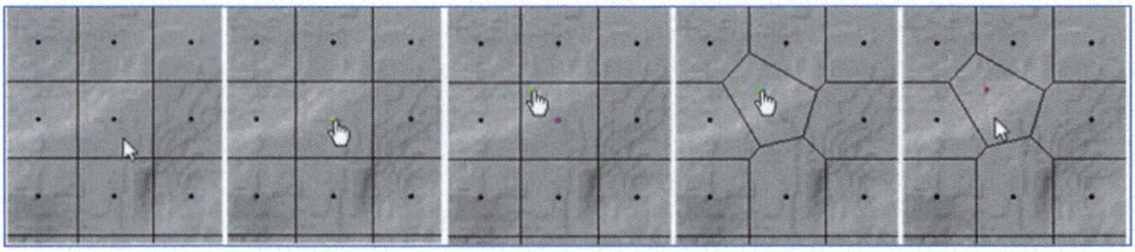

Fig. 6.9 Unstructured grid

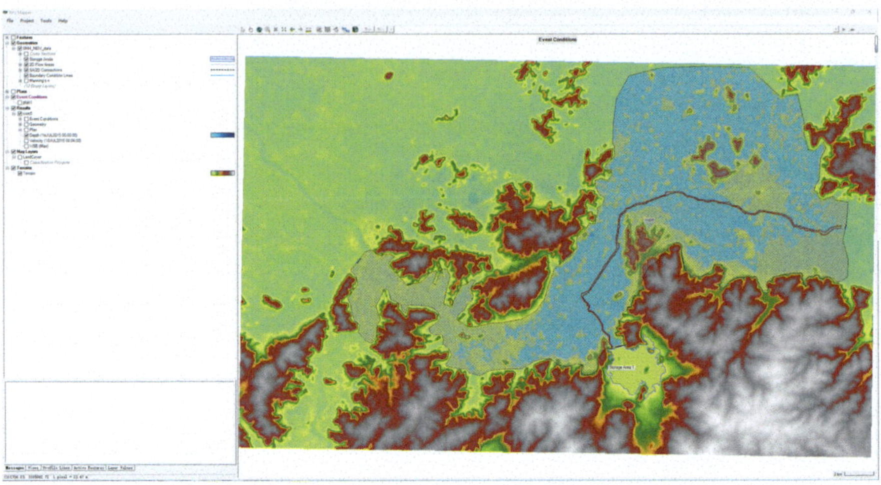

Fig. 6.10 Mapping results

inrush failure. When modeling a break, a model can be built which combines reservoir, dam body and river, in which the reservoir needs to input elevation-reservoir capacity curve. For earth-rock dams, the dam body is considered as a linear structure in the break simulation, but as concrete dams are analyzed, they generally break instantaneously, so the break curve is set as a curve of complete break from the beginning.

(1) Relationship between water balance and reservoir water level change

During dam break, the water level of upstream reservoir is a dynamic process, including the inflow and outflow of upstream channel. When calculating the change of upstream reservoir water elevation, the inflow and outflow should be considered simultaneously. The whole process follows the water balance equation:

$$Q_{out} = Q_{in} - \frac{\Delta V}{\Delta t} \tag{6.5.12}$$

where

Q_{Out} discharge from dam breach, m^3/s;
Q_{in} reservoir inflow, m^3/s;
ΔV variation of reservoir capacity in time interval Δt, m^3.

(2) Burst flow process

When the HEC-RAS breakdown module simulates the outflow process of the break, the break flow rate is assumed to be the outflow form of the wide-crested weir.

$$Q = \sigma_s \varepsilon m B' \sqrt{2g}(Z_i - Z_{i+1})^{\frac{3}{2}} \tag{6.5.13}$$

where

Q dam break flow, m^3/s;
σ_s submergence coefficient;
ε weir flow side collection coefficient;
m discharge coefficient;
B' width of broad-crested weir, m;
Z_i, Z_{i+1} water level before and after weir, m.

6.5.4.3 MIKE 21

MIKE21 is a numerical simulation software developed in Denmark for the calculation of hydraulic properties. MIKE21 can simulate the flow of water under different conditions, such as in rivers, seas, harbours, reservoirs and wetlands. MIKE21 is a numerical simulation software developed in Denmark for the calculation of hydraulic properties. MIKE21 can simulate the flow of water under different conditions, such as water in rivers, seas, harbours, reservoirs and wetlands. The hydrodynamic module is the core module of the MIKE21 numerical simulation software and provides the hydrodynamic basis for simulating the evolution of water bodies and the dispersion of pollutants. Figure 6.13 shows the flood inundation map calculated using MIKE21.

MIKE21 offers four meshes to simulate the evolution of a body of water: a single mesh, a nested mesh, a curved mesh and a finite element mesh. Each mesh has different application scenarios, with the finite element mesh (triangular mesh) being a good simulation of the flow field around curves or buildings on the water and suitable for simulating this complex area of the city.

1. Model calculation principle: The calculation of MIKE21 is based on the three-way incompressible equation and the Reynolds averaged Navier–Stokes equation [24], obeying the hydrostatic pressure and Boussincsq equation [37]. The FM model of MIKE21 is the MIKE21 hydrodynamic model, which is used to simulate the non-constant two-dimensional flow of the fluid, this model has been used in a large number of the calculation in the hydrodynamic model uses the shallow water equation, whose basic equations include the continuity equation and the momentum equation, and the mass and momentum equations in the x-direction are mainly considered in the calculation. Since the mass equation in the

y-direction affects the centroid of the mass equation in the x-direction, a differential approximation of this equation is also considered. The momentum equation in the y-direction is similar to the momentum equation in the x-direction.

Equation of mass in the *x*-direction:

$$\frac{\partial \zeta}{\partial t}+\frac{\partial p}{\partial x}+\frac{\partial q}{\partial y}=\frac{\partial d}{\partial t} \tag{6.5.14}$$

Equation of mass in the *y*-direction:

$$2\cdot\left(\frac{\zeta^{n+1}-\zeta^{n+\frac{1}{2}}}{\Delta t}\right)_{j,k}+\frac{1}{2}\cdot\left\{\left(\frac{p_j-p_{j-1}}{\Delta x}\right)^{n+1}+\left(\frac{p_j-p_{j-1}}{\Delta x}\right)^{n}\right\}_k$$
$$+\frac{1}{2}\cdot\left\{\left(\frac{q_k-q_{k-1}}{\Delta y}\right)^{n+\frac{3}{2}}+\left(\frac{q_k-q_{k-1}}{\Delta y}\right)^{n+\frac{1}{2}}\right\}_j=2\cdot\left(\frac{d^{n+1}-d^{n}}{\Delta t}\right)_{j,k} \tag{6.5.15}$$

Equation of momentum in the *x*-direction:

$$\frac{\partial p}{\partial t}+\frac{\partial}{\partial x}\left(\frac{p^2}{h}\right)+\frac{\partial}{\partial y}\left(\frac{pq}{h}\right)+gh\frac{\partial \zeta}{\partial x}+\frac{gp\sqrt{p^2+q^2}}{C^2\cdot h^2}$$
$$-\frac{1}{\rho_w}\left[\frac{\partial}{\partial x}(h\tau_{xx})+\frac{\partial}{\partial y}(h\tau_{xy})\right]-\Omega_q-fVV_x+\frac{h}{\rho_w}\frac{\partial}{\partial x}(p_a)=0 \tag{6.5.16}$$

Equation of momentum in the *y*-direction:

$$\frac{\partial q}{\partial t}+\frac{\partial}{\partial y}\left(\frac{q^2}{h}\right)+\frac{\partial}{\partial x}\left(\frac{pq}{h}\right)+gh\frac{\partial \zeta}{\partial y}+\frac{gp\sqrt{p^2+q^2}}{C^2\cdot h^2}$$
$$-\frac{1}{\rho_w}\left[\frac{\partial}{\partial y}(h\tau_{yy})+\frac{\partial}{\partial x}(h\tau_{xy})\right]+\Omega p-fVV_y+\frac{h}{\rho_w}\frac{\partial}{\partial xy}(p_\alpha)=0 \tag{6.5.17}$$

where

h	water depth, m;
d	dead water depth, m/s;
ξ	free surface elevation, m;
p, q	the flow per unit width in the x and y directions, m^2/s;
C	Chezy coefficient, $m^{0.5}/s$;
g	gravity acceleration, 9.81 m/s^2;
V, V_x, V_y	wind speed and its components in x direction and y direction, m/s;
Ω	Coriolis parameter, which elated to latitude, s^{-1};
f	coefficient of friction resistance;
p_a	atmospheric pressure, kg s^2/m;

ρ_w	the density of water, kg/m^3;
x, y	space coordinates, m;
t	time, s;
τ_{xx}, τ_{xy}, τ_{yy}	the component of effective shear stress.

The numerical method used for the MIKE21FM unstructured grid model is the Cell-Centred Finite Volume method. The two-dimensional calculation is discretised by dividing the computational domain into independent unstructured grid cells (quadrilateral or then triangular) as shown in Fig. 6.11: when the control equations are discretised, the resultant variables are located in the centre and the boundary fluxes are perpendicular to the cell sides. The calculation of the normal flux in the finite volume element method is obtained by constructing new hydrodynamic model cells along the outer normal direction and solving the Riemann problem [38] for these one-dimensional normal fluxes.

The equations are solved in one dimension by means of equations in the x and y directions respectively, for which the ζ term iterates from n to $n + 1/2$ and the p term from n to $n + 1$. While the q term is calculated iteratively using values known from previous calculations, i.e. $n - 1/2$ and $n + 1/2$. For the y direction, the ζ term iterates from $n + 1/2$ to $n + 1$, the q term iterates from $n + 1/2$ to $n + 3/2$, and the p term uses the values of n and $n + 1$ previously calculated for the x direction. The time difference of the $n + 1/2$ steps is obtained by iterating over the two calculated values (Fig. 6.12).

6.6 Climate Change Impact

In recent years, the world has been experiencing climate change characterised by global warming due to the combined effects of human activities and natural factors [11, 39]. Global warming has had a profound impact on human survival and development, and the international community has become increasingly aware of the serious threat and challenge posed by climate warming to the present and future human

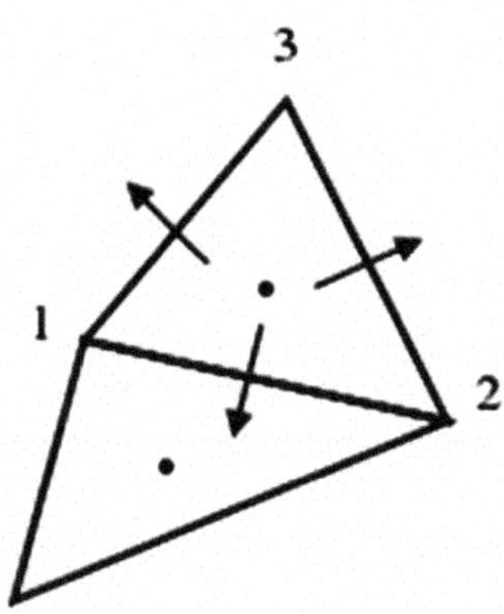

Fig. 6.11 Diagram of the numerical solution of the unstructured grid model

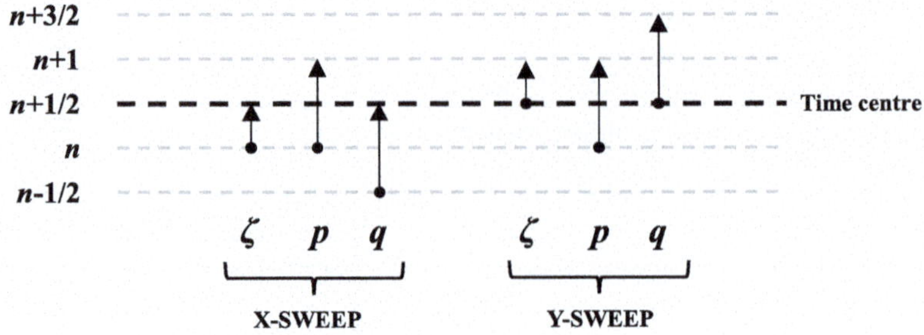

Fig. 6.12 Schematic diagram of the time difference format

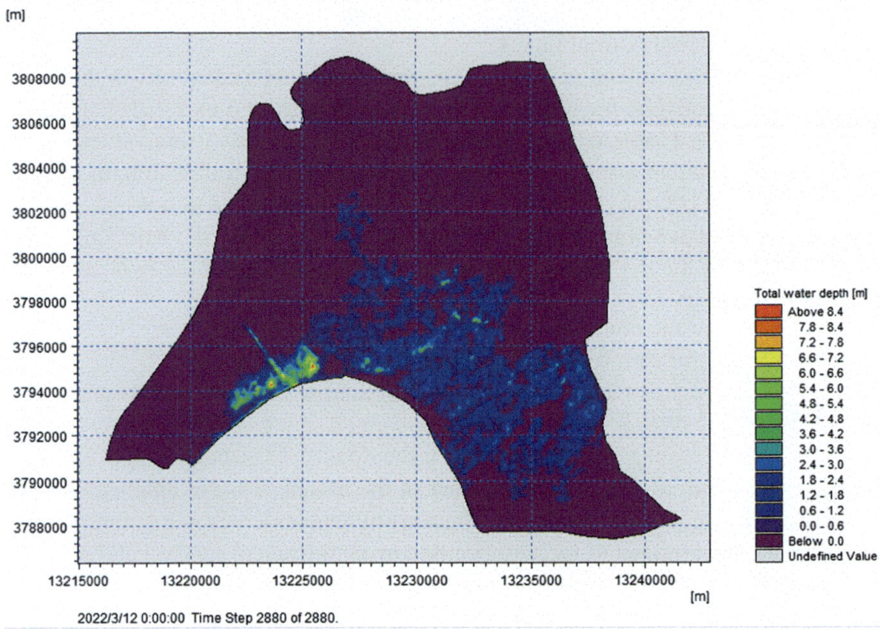

Fig. 6.13 Final calculation result chart

habitat, as well as the importance and urgency of taking collective action to reduce and prevent climate risks. Natural disasters are becoming more frequent around the world, and the intensity and frequency of extreme events caused by climate change, such as high temperatures, floods, heavy rainfall and sea level rise, are increasing [40].

As environmental conditions change due to climate change, the frequency and intensity of various types of sudden and extreme events are increasing, posing many challenges to the operation and management of reservoir dam projects [41]. Under the influence of climate change, global warming, extreme weather frequently, the global river bank breach, reservoir dam, dam breach almost every year, the United

States, China, Germany, India, Japan, Brazil and other major countries in the world are not spared, the global dams and reservoirs engineering safety and emergency management face more serious new challenges [42–44] (Fig. 6.14).

Climate change is causing changes in hydrological extremes such as heavy rainfall and flooding, which directly alter the design flood series of dam projects [45]. The risk posed by the changing environment to the future operation of reservoir dam projects built using existing hydrological design standards will continue to increase.

In addition, China's natural disasters are frequent and complex, and there is a dynamic chain of logical relationships between disasters due to their associated effects. In the reservoir area, the disaster chain of natural disasters mainly includes heavy rainfall, floods, strong earthquakes, various types of geological hazards, and the failure of dams in neighbouring upstream reservoirs. Climate change causes a series of changes in the thresholds, spatial and temporal distributions, and patterns of changes in global temperature, humidity, and precipitation, which trigger extreme events such as high temperatures, floods, heavy rainfall, sea level rise, and various geological hazards, thus forming a chain reaction to dams [11, 46]. The current lack of identification and understanding of the various catastrophic factors acting on dams in the context of climate change makes it difficult to apply targeted prevention and control measures in the early stages of dam breach. For example, extreme rainfall during prolonged droughts can lead to dam seepage damage and slope instability, increasing the likelihood of dam breach.

The current lack of knowledge about the driving factors, impact mechanisms and severity of consequences of each extreme event on reservoir xdams in the context of climate change makes it difficult to effectively quantitatively assess the impact of climate change on the safe operation of reservoir dams, and greatly increases the difficulty and complexity of controlling extreme events caused by climate change.

6.6.1 Climate Change Impacts on Hydrology

Climate change destroys the representativeness of hydrological data series and increases the uncertainty in the calculation of the size of the design flood of reservoir dams, which makes the reservoir dam operation and planning scheme developed by the existing engineering hydrological calculation method face the risk brought by the changing environment [47].

In addition, climate change causes changes in hydrological extremes such as heavy rainfall, floods and droughts, which will certainly affect the size of the design flood and subsequently the design of water conservancy projects [48]. A very large flood exceeding the design standard may result in excessive water loading on the dam or dam breach. Reservoir dams that fail to meet flood control standards will be at increased risk of failure during their future operation (Fig. 6.15).

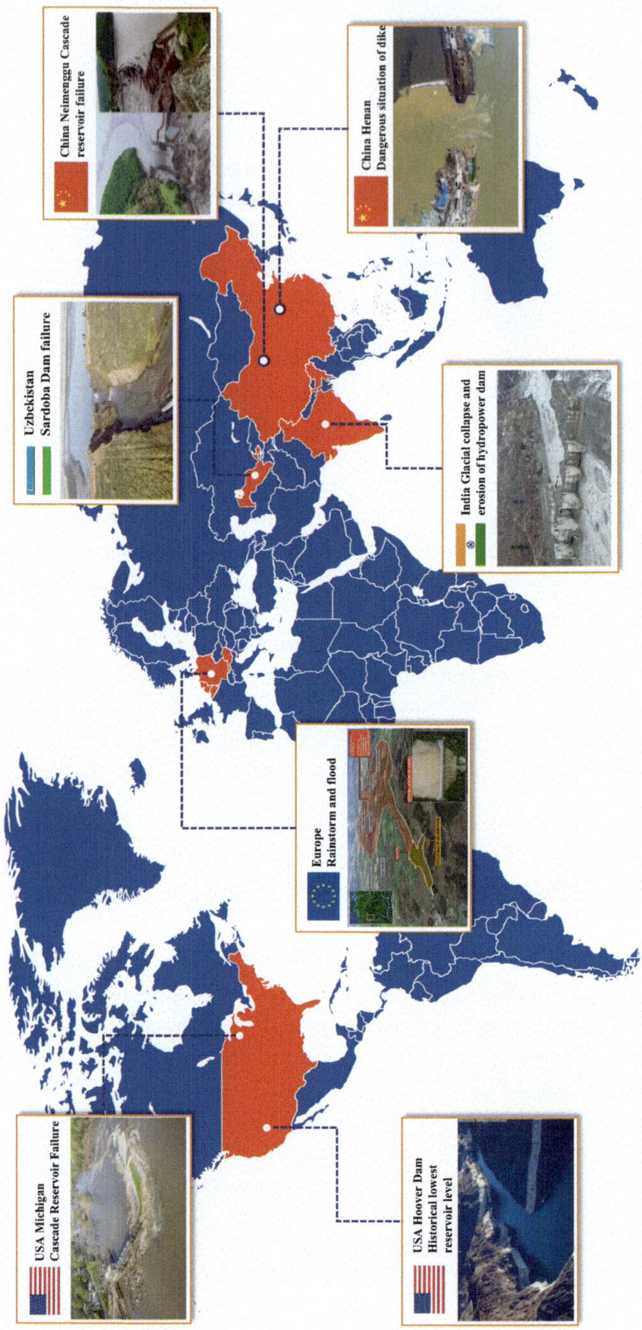

Fig. 6.14 Global water engineering outages caused by climate change

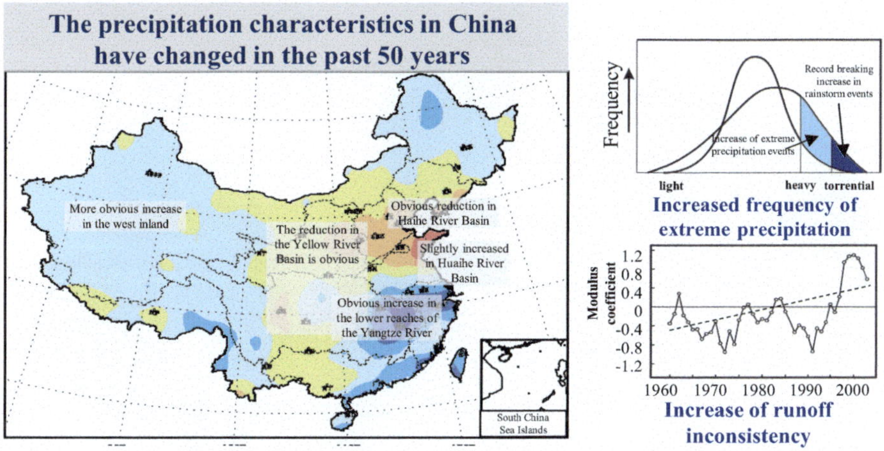

Fig. 6.15 The precipitation characteristics in China have changed in the past 50 years

6.6.2 *Climate Change Impact on Dams*

The possibility of extreme temperature events associated with climate change is increasing, directly affecting the performance of hydraulic materials and the structural stability of dams. Low temperatures have a greater impact on the safety of concrete dams during construction and operation. In addition, climate change will lead to more unfavourable operating conditions, such as prolonged droughts and rapid droughts and floods [49] (Fig. 6.16).

Extreme drought and high temperature mostly occur in summer, during the critical period of flood control, it is easy to form a dangerous situation of drought and flood, which seriously affects the operation and management of water conservancy projects, and even threatens the safety of the project. Prolonged drought will lead to dry shrinkage cracks in reservoir dams, which will destroy the reservoir sealing system. Studies have shown that sudden changes in droughts and floods are important factors contributing to reservoir dam breaches, firstly because managers do not pay attention to droughts and floods after prolonged droughts, and secondly because reservoir emergency management measures after rapid water level rise are not effective in dealing with extreme operating conditions caused by climate change. Extreme temperatures lead to cracks in the dam body, which can cause a series of adverse effects such as leakage, concrete carbonation and reduced durability of the dam in the incoming water environment, while the material properties of concrete materials such as compressive strength, modulus of elasticity, creep deformation and shrinkage deformation can change under high temperature and drought conditions, which can have a great impact on the safety of concrete dams [50].

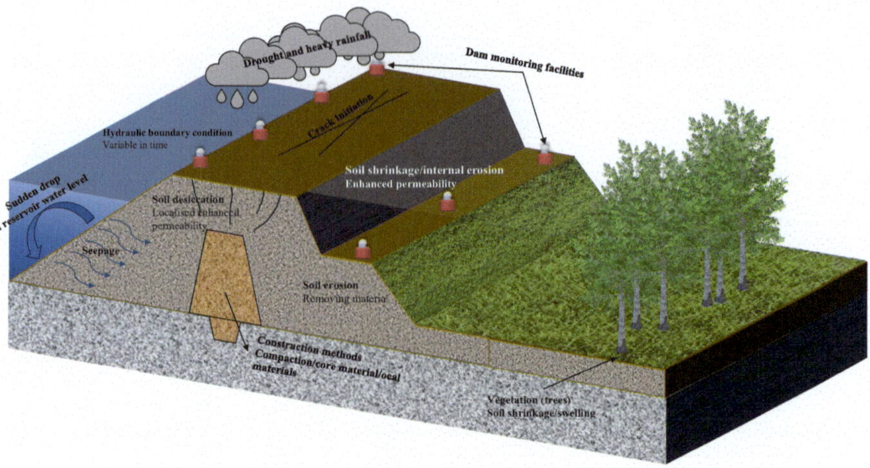

Fig. 6.16 Potential climate interactions and their impact on the earth and rock dam project

6.6.3 U.S. Flood Control Standards Construction

The United States was one of the first countries to recognise the importance of emergency management and has advanced experience in social emergency management [51].

The classification of reservoir dams and the selection of flood control standards in the United States are based on risk analysis, which quantifies the potential hazards of reservoir dam breach, and this approach fully reflects the concept of human-centred development, which belongs to the category of dam risk management.

The selection of the inlet design flood is the first step in the evaluation and design of a dam. It refers to the process of inundation into the reservoir and is used to determine the size of spillways and spillway buildings, estimate maximum capacity, dam height and superelevation requirements. The US Federal Agencies FEMA, USACE and USBR have established technical criteria for the selection of dam inflow design floods, with FEMA recommending a specified approach, site-specific maximum probable rainfall studies, incremental consequence analysis and risk-based strategies for inflow flood selection [52] (Table 6.5).

6.6.4 China's Flood Control Capacity Improved

China is one of the countries with the most frequent and severe flood disasters in the world. Historically, major flood disasters have caused severe damage and affected economic and social development. After years of construction, China's major river basins have gradually improved their flood prevention and mitigation systems with a

Table 6.5 IDF requirements for dams using a prescriptive approach [53]

Hazard potential classification	Definition of hazard potential classification	Inflow design flood
High	Probable loss of life due to dam breach or misoperation (economic loss, environmental damage, or disruption of lifeline facilities may also be probable, but are not necessary for this classification)	PMF
Significant	No probable loss of human life but can cause economic loss, environmental damage, or disruption of lifeline facilities due to dam breach or misoperation	0.1% annual chance exceedance flood (1000-year flood)
Low	No probable loss of human life and low economic and/or environmental losses due to dam breach or misoperation	1% annual chance exceedance flood (100-year flood) or a smaller flood justified by rationale

(1) Incremental consequence analysis or risk-informed decision making may be used to evaluate the potential for selecting an IDF lower than the prescribed standard. An IDF less than the 0.2% annual chance exceedance flood (500-year flood) is not recommended
(2) Incremental consequence analysis or risk-informed decision making studies may be used to evaluate the potential for selecting an IDF lower than the prescribed standard. An IDF less than the 1% annual chance exceedance flood (100-year flood) is not recommended

combination of engineering and non-engineering measures, such as reservoirs, dams and flood storage areas, and have overcome many major basin-based regional floods [54].

Flood management in river basins should reduce the impact of heavy rainfall and flooding, while increasing the adaptability and resilience of economic and social systems [55] (Fig. 6.17).

(1) Improve the ability to expand river flood discharge. For northern rivers, restore the flood discharge capacity of rivers, reduce river siltation and shrinkage, and ensure the smooth flow of floods; for southern rivers, maintain the stability

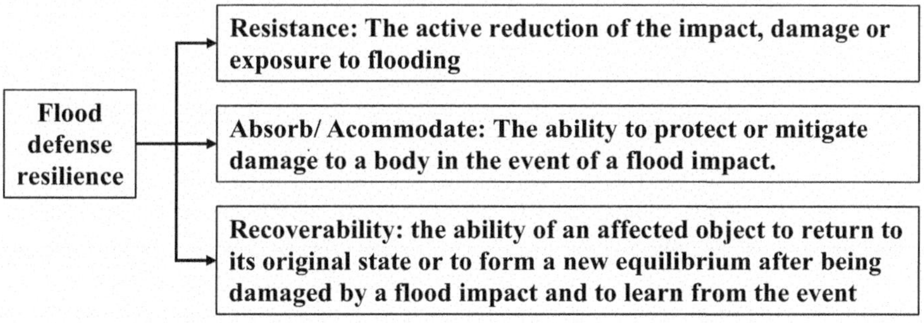

Fig. 6.17 Basin flood resilience interpretation

of rivers, coordinate the relationship between main rivers and tributaries, and reduce the pressure of main river flood control.

(2) Improve flood storage capacity. To improve the overall flood control capacity of the basin and flood control capacity of key areas for the purpose of accelerating the implementation of the basin control reservoir project construction, give full play to its role in effective flood control and peak reduction, reduce the downstream flood control pressure, mainly through the removal and reinforcement of dangerous reservoirs, consolidate and improve the existing control reservoir project on the flood control capacity.

(3) Ensure the flood storage function of flood storage areas. Accelerate the layout optimization and construction of flood storage areas, with emphasis on the Yangtze River, Huaihe River, Haihe River and other watersheds, and promote the construction of flood storage area engineering and safety, to ensure that the flood storage area "can be divided, stored and retired" [56].

(4) Digital empowerment to improve the level of information management of small and medium-sized reservoirs. Strengthen the leadership of digital technology, rely on the Internet, big data, edge computing and other information technology means to solve "the last mile" problem of safety management of small and medium-sized reservoirs. Use big data technology to collect data related to small and medium-sized reservoirs, explore the potential information value of the data, and empower the data in management to achieve scientific and effective use of data. Based on cloud computing, big data, Internet of Things, artificial intelligence and mobile Internet technology, we will deeply integrate the standard management content of reservoirs with information technology tools to improve the information management level of small and medium-sized reservoirs in terms of rainwater monitoring, engineering safety monitoring, flood forecasting and scheduling, emergency scheduling and planning, reservoir inspection and other work, and realise the transformation from rough management to refined management and from traditional management to digital management.

6.7 Conclusions on Review of Flood Control Ability

For review of flood control ability, clear conclusions shall be made on: Whether the original designed flood control standard of the reservoir meets the table requirements and whether adjustment is required; Whether the result of the original designed flood needs to be adjusted following the extension of hydrological series; Whether the discharging ability of the discharging structures can satisfy the requirement for safe flood discharge; Whether the operation mode for flood regulation is in conformity with the characteristics of the reservoir and safe operation of the dam; whether modification is needed; Whether elevation of current dam crest or wave wall top elevation and the impervious body top elevation meet the requirements of specifications.

The following three aspects shall be included for safe discharging of flood:

- The structure safety, dimension of water flow cross section, energy dissipation facility of the discharging structures all meet the requirements of specifications and design and the operation state of gate and hoist is sound and reliable.
- The discharged flow does not scour the dam toe and atomization for discharging does not influence the stability of the dam shoulder.
- The flood passage downstream the discharging structure is unobstructed and the discharge capacity meets the requirements of the standard for flood control downstream.

When any one aspect of the discharging structure is not as required in those three aspects, it means that the flood cannot be safely discharged. When both the reservoir flood control standard and the dam's flood resistance ability meets the requirements of specifications, and the flood can be discharged safely, the safety of flood control of the dam shall be rated as **Class A**. The safety of flood control of the dam shall be rated as **Class B** when the reservoir flood control standard and the dam's flood resistance ability fail to meet the requirements of specifications, but meet the requirements of the standard for short-term and emergency operation of flood, or the reservoir flood control standard and the dam's flood resistance ability meet the requirements of specifications, but flood cannot be discharged safely. The safety of flood control of the dam shall be rated as **Class C** when the reservoir flood control standard and the dam's flood resistance ability fail to meet the requirements of the standard for short-term and emergency operation of flood.

References

1. Rui XF (2004) Principles of hydrology. China Water & Power Press
2. Angelino CV, Cicala L, Fiscante N et al (2016) Estimated post-flood effects through free, full and open satellite remote sensing data to support civil protection. In: 2016 IEEE international geoscience and remote sensing symposium (IGARSS), pp 4215–4218. https://doi.org/10.1109/IGARSS.2016.7730098
3. Fischer S (2018) A seasonal mixed-POT model to estimate high flood quantiles from different event types and seasons. J Appl Stat 45:2831–2847. https://doi.org/10.1080/02664763.2018.1441385
4. Hilly G, Vojinovic Z, Weesakul S et al (2018) Methodological framework for analysing cascading effects from flood events: the case of Sukhumvit area, Bangkok, Thailand. Water 10:81. https://doi.org/10.3390/w10010081
5. Sheng J, Li H, Sheng T (2023) Statistical analysis of dam failure and its loss of life in China. Hydro-Sci Eng 1–15. https://doi.org/10.12170/20220919001
6. Gu C, Su H, Liu H. (2018) Review on service risk analysis of dam engineering. Shuili Xuebao/Journal of Hydraulic Engineering 49:26–35. https://doi.org/10.13243/j.cnki.slxb.20170721
7. Ministry of Housing and Urban Rural Development, General Administration of Quality Supervision, Inspection and Quarantine of the People's Republic of China (2015) Standard for flood control: GB 50210—2014. Standards Press of China, Beijing
8. Ministry of Water Resources of the People's Republic of China (2017) Standard for classification and flood control of water resources and hydroelectric project: SL 252-2017. China Water Conservancy and Hydropower Press, Beijing
9. Guo S, Liu Z, Xiong L (2016) Advances and assessment on design flood estimation methods. J Hydraul Eng 47:302–314. https://doi.org/10.13243/j.cnki.slxb.20150913

10. Yi X, Zhou F, Wang X et al (2014) Classification and runoff simulation of data-scarce basins based on self-organizing maps. Prog Geogr 33:1109–1116
11. Wisser D, Frolking S, Hagen S, Bierkens MFP (2013) Beyond peak reservoir storage? A global estimate of declining water storage capacity in large reservoirs, Water Resour Res 49:5732–5739. https://doi.org/10.1002/wrcr.20452
12. Lane PNJ, Best AE, Hickel K, Zhang L (2005) The response of flow duration curves to afforestation. J Hydrol 310:253–265. https://doi.org/10.1016/j.jhydrol.2005.01.006
13. Zheng H, Zhang L, Zhu R et al (2009) Responses of streamflow to climate and land surface change in the headwaters of the Yellow River Basin. Water Resour Res 45. https://doi.org/10.1029/2007wr006665
14. Batalla RJ, Gómez CM, Kondolf GM (2004) Reservoir-induced hydrological changes in the Ebro River basin (NE Spain). J Hydrol 290:117–136. https://doi.org/10.1016/j.jhydrol.2003.12.002
15. Amador A, Sánchez-Juny M, Pomares J et al (2018) Feasibility study of stepped spillways in RCC dams controlled by a Tainter gate, pp 555–560. https://doi.org/10.1201/9780203741276-67
16. Singh D, Kumar M (2023) Effect of the inlet-to-outlet key width ratio of Piano Key Weir on its hydraulic behaviour. Flow Meas Instrum 91:102342. https://doi.org/10.1016/j.flowmeasinst.2023.102342
17. Design specification for spillway: SL 253-2018 (2018) China Water Conservancy and Hydropower Press, Beijing
18. John D Fenton (2019) Flood routing methods. J Hydrol 570:251–264. https://doi.org/10.1016/j.jhydrol.2019.01.006
19. Rui X, Zhang C (2014) Research on design flood calculation. Adv Sci Technol Water Resour 34:20–26. https://doi.org/10.3880/j.issn.1006-7647.2014.01.004
20. Zhao J, Lu S, Liu L et al (2011) Points of flood-prevention evaluation of tailings ponds. South-to-North Water Transf Water Sci Technol 9(6):58–60
21. Figueiredo R, Merz B, Weiss-Motz A et al (2018) Multi-model ensembles for assessment of flood losses and associated uncertainty. Nat Hazards Earth Syst Sci 18:1297–1314. https://doi.org/10.5194/nhess-18-1297-2018
22. Zhang L, Peng M, Chang D, Xu Y (2017) Dam failure mechanisms and risk assessment. Wiley
23. Peng X, Cai Y, Sheng J et al (2015) Study of risk criteria for reservoirs and dams in China. China Water & Power Press. ISBN: 978-7-5170-3705-7
24. Chorin AJ (1968) Numerical solution of the Navier–Stokes equations. Math Comput 22:745. https://doi.org/10.1090/s0025-5718-1968-0242392-2
25. Zhang T, Xu B, Kang A et al (2021) Advances in joint application of hydrology, hydrodynamics and water environment models in river basins. Adv Sci Technol Water Resour 41:11–19. https://doi.org/10.3880/j.issn.1006-7647.2021.03.002
26. Shrestha Alen, Bhattacharjee Linkon, Baral Sudip (2020) Understanding Suitability of MIKE 21 and HEC-RAS for 2D Floodplain Modeling. 237–253. https://doi.org/10.1061/9780784482971.024
27. Azmi M, Thomson K, Dam breach parameters: from data-driven-based estimates to 2-dimensional modeling. Nat Hazards 120:4423–4461 (2024). https://doi.org/10.1007/s11069-023-06382-3
28. Zhong Q, Wu W, Chen S, Wang M (2016) Comparison of simplified physically based dam breach models. Nat Hazards 84:1385–1418. https://doi.org/10.1007/s11069-016-2492-9
29. Dong K, Li Z, Lu X et al (2021) Analysis of dam overtopping failure risks caused by landslide-induced surges considering spatial variability of material parameters. Front Earth Sci 9. https://doi.org/10.3389/feart.2021.675900
30. ICOLD (2013) Internal erosion of existing dams, levees and dikes, and their foundations. Bulletin n. 164, vol. 1: "internal erosion processes and engineering assessment"
31. Von L, Gillette DR (1990) Guidance on breach parameters. Unpublished internal document, U.S.

32. Wu W (2013) Simplified physically based model of earthen embankment breaching. J Hydraul Eng 139:837–851. https://doi.org/10.1061/(asce)hy.1943-7900.0000741
33. Wu W, Li H (2017) A simplified physically-based model for coastal dike and barrier breaching by overtopping flow and waves. Coast Eng 130:1–13. https://doi.org/10.1016/j.coastaleng.2017.09.007
34. Pathan AI, Agnihotri PG (2020) Application of new HEC-RAS version 5 for 1D hydrodynamic flood modeling with special reference through geospatial techniques: a case of River Purna at Navsari, Gujarat, India. Model Earth Syst Environ 7:1133–1144. https://doi.org/10.1007/s40808-020-00961-0
35. Aung Pyae Phyo, Helmut Yabar, Delmaria Richards (2023) Managing dam breach and flood inundation by HEC-RAS modeling and GIS mapping for disaster risk management, Case Studies in Chemical and Environmental Engineering. 100487:2666-0164. https://doi.org/10.1016/j.cscee.2023.100487
36. Brunner G (2016) HEC-RAS, river analysis system hydraulic reference manual. U.S. Army Corps of Engineers, Davis, CA
37. Zhang YG, Li YC (1999) A new form of Boussinesq equation with improved linear dispersion characteristics. China Ocean Eng 13:11–20
38. Toro EF (2011) Riemann solvers and numerical methods for fluid dynamics: a practical introduction. Springer, Berlin
39. Merz B, Blöschl G, Vorogushyn S et al (2021) Causes, impacts and patterns of disastrous river floods. Nat Rev Earth Environ 2:592–609. https://doi.org/10.1038/s43017-021-00195-3
40. China Meteorological Administration Center (2022) Blue book on climate change in China 2022. Science Press
41. Chernet HH, Alfredsen K, Midttømme GH (2014) Safety of hydropower dams in a changing climate. J Hydrol Eng 19:569–582. https://doi.org/10.1061/(asce)he.1943-5584.0000836
42. Stanford University (2020) Global warming influence on extreme weather events has been frequently underestimated. ScienceDaily, 18 March 2020
43. Chernet HH, Alfredsen K, Midttømme GH (2014) Safety of hydropower dams in a changing climate. J Hydrol Eng 19:569–582
44. Fluixá-Sanmartín J, Altarejos-García L, Morales-Torres A, Escuder-Bueno I (2018) Review article: climate change impacts on dam safety. Nat Hazards Earth Syst Sci 18:2471–2488. https://doi.org/10.5194/nhess-18-2471-2018
45. Song X, Zhang J, Zhan C et al (2013) Review for impacts of climate change and human activities on water cycle. J Hydraul Eng 44:779–790
46. Calvin K, Dasgupta D, Krinner G et al (2023) IPCC, 2023: climate change 2023: synthesis report, summary for policymakers. Contribution of working groups I, II and III to the sixth assessment report of the intergovernmental panel on climate change [Core Writing Team, H. Lee and J. Romero (eds.)]. IPCC, Geneva, Switzerland, pp 1–34. https://doi.org/10.59327/ipcc/ar6-9789291691647.001
47. Shirzaei M, Bürgmann R (2018) Global climate change and local land subsidence exacerbate inundation risk to the San Francisco Bay Area. Sci Adv 4:eaap9234. https://doi.org/10.1126/sciadv.aap9234
48. Qin P, Xu H, Liu M et al (2020) Climate change impacts on Three Gorges Reservoir impoundment and hydropower generation. J Hydrol 580:123922. https://doi.org/10.1016/j.jhydrol.2019.123922
49. Ayyub BM, Committee on Adaptation to a Changing Climate (2018) Climate-resilient infrastructure: adaptive design and risk management. American Society of Civil Engineers, Reston, VA. https://doi.org/10.1061/9780784415191
50. Vardon PJ (2015) Climatic influence on geotechnical infrastructure: a review. Environ Geotech 2:166–174. https://doi.org/10.1680/envgeo.13.00055
51. Zou Y (2021) A comparative study of flood control standards for small and medium-sized reservoirs and recommendation. Hydro-Sci Eng 1–8. https://doi.org/10.12170/20200503003
52. FEMA (2013) The federal guidelines for inundation mapping of flood risks associated with dam incidents and failures. Federal Emergency Management Agency, Washington, DC

53. FEMA (2004) Federal guidelines for dam safety: selecting and accommodating inflow design floods for dams. Federal Emergency Management Agency, Washington, DC
54. Wu A, Zhou H, Wu Q (2019) Levees in Europe and United States: characteristics and comparison with China. J Yangtze River Sci Res Inst 36:11–18. https://doi.org/10.11988/ckyyb.20191016
55. Wang Lihong, Cui, Shenghui, Li Yuanzheng, Huang, Hongjie (2022) A review of the flood management: from flood control to flood resilience. Heliyon 8:e11763. https://doi.org/10.1016/j.heliyon.2022.e11763
56. Quansen, Wang (2022) Concept of digital twin construction scheme for flood storage space in mid-lower Yangtze River. Metaverse 3:12. https://doi.org/10.54517/m.v3i2.2125

Chapter 7
Evaluation on Seepage Safety

7.1 Introduction

Seepage behavior is a key factor affecting the safety and stability of earth-rock dams and concrete dams [1, 2]. It is necessary to fully grasp the spatial and temporal distribution characteristics of seepage in order to reduce its operational safety risks and improve its safe operation guarantee ability. Seepage control methods for dams include upstream seepage prevention measures (blanket, seepage cut-off wall, curtain grouting, etc.), downstream seepage guidance measures (pressure-reducing shaft, seepage guidance trench, blanket drainage, chimney drainage, etc.) and filter drainage measures at the seepage outlet (slope filter and prism drainage). The actual seepage behaviour means the actual seepage flow distribution of the dam, the seepage flow and its variation pattern, and the seepage stability of the critical points (impermeable body, contact surfaces of different structures, seepage outlet, etc.).

The evaluation of seepage resistance involves a comprehensive assessment of various aspects related to seepage control, filter drainage, and the overall safety of the dam. The following contents shall be included in the evaluation of seepage resistance:

1. Inspection of Seepage and Filter Drainage Facilities: Check the condition of seepage prevention measures, cut-off walls, curtain grouting, blanket, etc. The evaluate the quality of design, construction, and foundation treatment in accordance with requirements.
2. Identification of Seepage Anomalies: Identify any instances of seepage anomalies that have occurred during the operation of the dam. Assess whether these anomalies have impacted the safety of the dam and its surrounding environment.
3. Analysis of Seepage Control and Filter Drainage Facilities: Analyze the operational behavior of seepage control measures and filter drainage facilities. Evaluate the safety performance of the dam's seepage behavior under different operational conditions.

Y. Xiang et al., *On-site Inspection and Dam Safety Evaluation*, Hydroscience and Engineering, https://doi.org/10.1007/978-981-97-3987-5_7

4. Seepage Stability Analysis: Ensure that the seepage stability of the dam meets the required safety standards. Analyze how well the dam's seepage control measures are maintaining the desired stability [3].
5. Analysis of Existing Seepage Safety Problems: Investigate any existing seepage-related safety issues observed in the dam. Identify the root causes of these problems and assess potential hazards they may pose.

The purpose of the seepage safety assessment is to re-examine the seepage control measures and the current actual seepage behaviour to ensure that they meet the requirements for safe operation of the dam under the designed conditions. By addressing issues related to design, construction, anomalies, safety, and problem analysis, this evaluation contributes to the overall safety and performance of the dam structure.

7.2 Methods of Seepage Safety Evaluation

For evaluation on seepage safety, on-site inspection method, monitoring data analysis method [4], computation analysis method and empirical analogy method can be used. In the evaluation of seepage safety [5], a combination of different methods is essential to gather comprehensive and accurate information about the dam's behavior and its potential for seepage-related risks.

With on-site inspection and monitoring data analysis, the current dam behavior is revealed and reflected. Generally, these methods can not accurately predict seepage safety under higher water level in the future but can invert real seepage parameters and seepage boundary conditions through current status. After that, the dam seepage safety under the designed and check operation conditions is predicted through calculation analysis method. It's important to note that the calculation analysis method should be based on the information gathered from on-site inspections and monitoring data to ensure that the models and simulations accurately reflect the real conditions of the dam.

The dam seepage safety shall be assessed following the on-site safety inspection in accordance with geological survey data, seepage monitoring data, safety test information, comprehensive monitoring data analysis and seepage calculation. For dams where seepage monitoring data is available, analysis of monitoring data shall be carried out first. Identify correlations between different monitoring parameters that could provide insights into seepage patterns. Selective analysis shall be made to the abnormal seepage revealed during operation. For earth-rock fill dams with structures built in and running through, emphasize shall be on the analysis of the stability of contact seepage between the structures crossing the dam and the dam body to make sure if it meets the requirement.

7.2.1 On-site Inspection Method

On-site inspection method involves direct visual inspection of the dam and its seepage control measures, can help identify physical conditions, structural integrity, and any signs of seepage anomalies, and provides valuable information about the current state of the dam. Abnormal seepage can be found through routine walk around inspection. This is the most intuitive and effective means for evaluation of seepage safety. For those small-sized reservoirs with few monitoring facilities, on-site inspection is of particular importance. Visual inspection of seepage consists of visually detecting or observing wet spots, depressions, or other evidence of seepage. While instrumentation is very important in assessing the overall seepage stability of a dam, nothing is as important as effective visual inspection. See Chap. 3 for details.

When the following phenomena exist in the project, it can be preliminarily thought that the dam seepage state is not safe or serious potential seepage risk exists. Then further analysis and argumentation are necessary.

- Seepage flow continues to increase under the same condition. The leaked water becomes turbid or suspect substances appear. The position of water leakage is elevated or shifted.
- The upstream dam slope of the earth-rock fill dam collapses; leaching occurs at the downstream dam slope and the humid and soft range increasingly expands. Water and sand come out of the dam toe and sponging and uplift or collapse occurs. Whirlpools and water leakage occur inside the reservoir and serious collapse pits or cracks appear on the blanket.
- Water is leaking at the junction parts between the dam body and bank slopes at the two ends of the dam and water conveyance tunnels. The adjacent dam surface collapses and the leaked water becomes turbid.
- Seepage pressure and flow increase simultaneously or suddenly change its previous relationship with the reservoir water level and remarkably increase under the same condition [6].

Therefore, the dam must be monitored (particularly by visual inspection) and monitored (by the use of instruments) to detect potential vulnerabilities, and the frequency and intensity of monitoring and monitoring should be set to allow early detection of seepage damage before failure occurs. Through necessary monitoring and monitoring, ensure sufficient time to react to the seepage failure and adopt certain measures to reduce or eliminate the serious consequences of seepage failure [7].

7.2.2 Monitoring Data Analysis Method

The monitoring data analysis method, which collected utilizes data from various monitoring systems installed on the dam, can help track changes in seepage patterns, water levels, and structural behavior over time, and provide insights into how the dam responds to different loading conditions [8]. Judgment shall be made to see if the dam seepage behavior is normal through analysis of the correlation between seepage pressure and flow and the water level of the reservoir. Meanwhile, the degree of dam's safety and danger shall be ascertained by comparing measured values or calculated values by mathematical models of seepage pressure and flow with the designed, tested values or allowed values given by specifications [9]. See Chap. 4 for details.

In the analysis of monitoring data of seepage pressure, the permissive anti-seepage gradient [J] is a threshold value that defines the maximum allowable gradient for seepage under normal conditions [10]. It acts as a reference point to gauge whether the seepage behavior of the dam is within safe limits. If, in actual operation of the dam, seepage damage occurs under the condition of the actual anti-seepage gradient $J \leq [J]$, or the seepage behavior is normal under the condition of $J \leq [J]$, it is suggested that the determination of original allowed anti-seepage gradient is not in agreement with reality. The allowed anti-seepage gradient shall be corrected through analysis or test when necessary [11].

In the analysis of seepage monitoring data, emphasis is placed on the analysis of the relative changes of current and historical seepage measurements. Design parameters often differ significantly from actual conditions. The seepage resistance of different impervious bodies is different, and therefore the absolute value of the seepage flow is different. There is no strict standard for its evaluation. Taking the measured absolute value of seepage flow as the only criterion for judgment in evaluating dam seepage safety should be avoided. Taking seepage hysteresis effect into account, the seepage flow monitoring data corresponding to the reservoir water level that is present most frequently and for longer duration shall be selected for analysis. For earth-rock fill dam, the principle is whether a stable seepage field can be formed.

Care shall be taken when statistical model or correlation diagrams are used for calculation of seepage safety under the condition of future higher water level because statistical model or correlation diagrams are set up in accordance with historical monitored values and it is unable to consider the randomness of future factors.

7.2.3 Calculation Analysis Method

The computational analysis method can predict seepage patterns under different scenarios, including different water levels, and requires accurate input parameters and boundary conditions for realistic results, which involves the use of mathematical models and numerical simulations to analyze seepage behavior [12, 13].

Through theoretical methods or numerical models, such hydraulic elements as dam seepage flow, water head, seepage pressure, seepage gradient and their distributions shall be calculated, flow net chart shall be draught to evaluate anti-seepage effect of impervious body, to make sure whether the seepage gradient for the critical location is smaller than the allowed seepage gradient, whether the seepage line (plane) is lower than the designed value and whether the elevation of egress point is within the range of slope filter protection. Special attention is given to critical locations such as weak zones, contact interfaces between structures, and areas prone to erosion or high seepage rates.

As a commonly used numerical approach for seepage calculations, the seepage finite element method is more often used for calculation of commonly used numerical values [14–16]. This method accounts for the complex geometries and material properties of the dam and foundation, providing a more accurate representation of seepage behavior.

The seepage parameters used in calculation analysis method shall be selected by integrating geological survey data and general engineering experience. When seepage monitoring data is available, engineers can perform inverse analysis to calibrate and adjust the parameters used in the numerical models. Inverse analysis helps refine the accuracy of the calculations by adjusting seepage parameters to match real-world observations. Similarly, the deterministic forecast model can become more reliable only after continuous inverse correction against the future higher water level.

7.2.4 *Empirical Analogy Method*

While monitoring data and seepage test parameters are ideal for accurate assessments, the absence of such data doesn't mean that seepage safety cannot be evaluated. For small and medium-sized reservoirs where seepage monitoring data and test parameters are not available, evaluating dam seepage safety can be approached based on the specific project's conditions, dam structure, and engineering geological characteristics. In such cases, practical methods include relying on project experience and drawing comparisons with similar projects. The empirical analogy method draws upon historical data, experience, and analogies to assess seepage behavior. may be used to estimate potential risks based on similarities with other projects providing a qualitative understanding of seepage safety. The empirical analogy method supplements this information with lessons learned from similar cases.

7.3 Evaluation on Seepage Safety of Earth-Rock Fill Dam

7.3.1 *Statistics of Embankment Dam Failures*

Based on the records of dam incidents and the dam register in ICOLD analyzed [17] the statistics of dam failures for large dams constructed between 1800 and 1986. This is summarized in Table 7.1.

It can be seen that internal erosion has been responsible for about half of dam failures where the mode of failure is known. It is approximately equal in importance to failures by overtopping in floods due to inadequate spillway capacity and malfunction

Table 7.1 Statistics of dam failures

Failure mechanism	Erosion		Embankment sliding	
Mode of failure	External erosion (overtopping) (%)	Internal erosion (%)	Static instability (%)	Seismic instability (%)
% over the world	48	46	4	2
% over the world	94		6	

of gates and other outlets leading to overtopping. By comparison dam slides and failures due to earthquakes account for only 4% and 1.7% of dam failures in operation respectively.

In China, as shown in Table 7.2, a total of 3356 earth-rock dams (including concrete and rockfill dams) have collapsed, accounting for 94.32% of the total number of dams that have collapsed, exceeding their share of the total number of

Table 7.2 Statistics of dam failures in China

NO.	Classification	1954~1979		1980~1999		2000~2021	
		Number of dam breaks	%	Number of dam breaks	%	Number of dam breaks	%
1	Earth-rock dam	2781	95.44	485	88.99	90	90.91
1.1	Homogeneous earth dam	2550	87.51	446	81.83	69	69.70
1.2	Inclined-wall dam	5	0.17	18	3.30		
1.3	Core wall dam	162	5.56	8	1.47	2	2.02
1.4	Concrete face rockfill dams	13	0.45	5	0.92	1	1.01
1.5	Mixed earth-rock dam	50	1.72	8	1.47	4	4.04
1.6	Other	1	0.03			3	3.03
1.7	Unknown					11	11.11
2	rigid dam	28	0.96	17	3.12	7	7.07
2.1	gravity dam	1	0.03	3	0.55	2	2.02
2.2	arch dam	5	0.17	4	0.73	2	2.02
2.3	Stone masonry dam	22	0.75	10	1.83	3	3.03
3	Other	12	0.41	1	0.18		
4	Unknown	93	3.19	42	7.71	2	2.02
Summary		2914	100	545	100	99	100

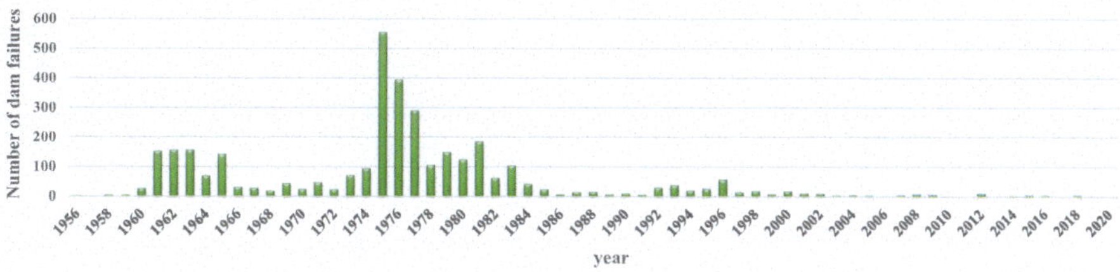

Fig. 7.1 The fig about dam failures in record in China

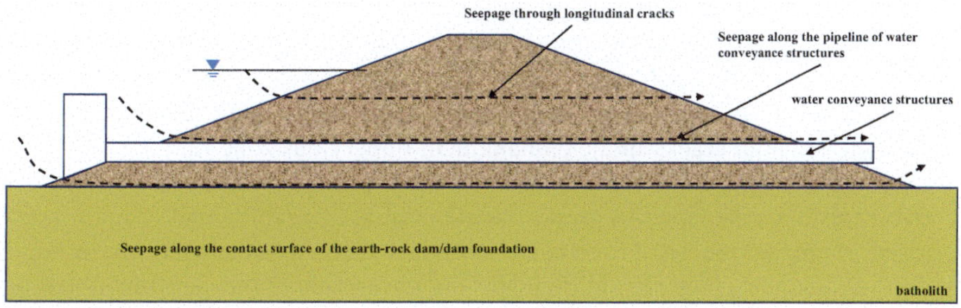

Fig. 7.2 Internal erosion

dams (91.80%). This is mainly due to the inability of earth-rock dams to withstand flooding as rigid dams (concrete dams and masonry dams) do [18]. In the 1950s and 1970s, as a local material dam, earth rock dams were used on a large scale because they could obtain local materials, which was convenient for construction and had low technical requirements. They were suitable for local people to invest in labor and local construction methods; Due to the inherent deficiencies of engineering construction and the inability to resist flood overtopping, the proportion of earth rock dam failure at this stage is higher, accounting for 95.44%. From 1980 to 1999, the proportion of earth rock dam failure decreased significantly, accounting for 88.99%, which is lower than its proportion in the total number of dams, indicating that dam type is not the absolute factor affecting dam failure, and design standards, engineering construction quality and operation management level are the decisive factors affecting dam failure. Since 2000, the proportion of earth rock dam break is about 90.91%, which is slightly higher than that of the previous stage, which is related to the obvious increase in the frequency and intensity of extreme weather events, and has a more significant impact on the safety of earth rock dams (Fig. 7.1).

7.3.2 *Erosion Control Measures*

All earth-rock fill dam are subject to some seepage passing through, under, and around them. If uncontrolled, seepage may be detrimental to the stability of the structure as a result of excessive internal pore water pressures or by piping (Fig. 7.2). For existing dams, records or evidence that seepage flows have removed any significant degree of fine grained material must be evaluated. Any such records requires further field investigation [19]. In the evaluation of seepage reduction or seepage control measures as they pertain to dam safety, one should review and evaluate the following:

1. Protective control measures such as relief wells, weighted graded filters, horizontal drains, or chimney drains which prevent seepage forces from endangering the stability of the downstream slope [20, 21].
2. Filters and transition zones designed to prevent movement of soil particles that could clog drains or result in piping [22].
3. Drainage blankets, chimney drains, and toe drains designed to ensure that they control and safely discharge seepage for all conditions. The design of these features must also provide sufficient flow capacity to safely control seepage through potential cracks in the embankment impervious zone [23].
4. Contacts of seepage control features with the foundation, abutments, embedded structures, etc., designed to prevent the occurrence of piping and/or hydrofracturing of embankment and foundation materials. If conduits or pipes exist through the embankment, they should be inspected to insure that they are functional or have been properly sealed. If there is the slightest doubt about through going conduits, they should be decommissioned and replaced.
5. Grouting, cut-off trenches, and impervious blankets. The use of impervious blankets or new structures in place of a cutoff is discouraged because of piping of the blanket materials into the foundation on recent projects. Blankets should only be used as a last resort element; and, then the use of a filter between the blanket and the foundation must be evaluated, not immediately omitted.
6. Construction records for foundation shaping, treatment and grouting at the contact between the impervious core and foundation [24].
7. Measures such as compaction requirements, seepage collars, placement of special materials, or other similar features to prevent internal erosion from seepage at the interface with concrete structures. If seepage collars are present, special attention should be given to compaction requirements around them. The use of seepage collars is not recommended in new construction.
8. For existing dams, all seepage records compiled during the existence of the structure should be reviewed for significant trends or abnormal changes. The causes of any abnormalities should be determined as accurately as possible.

7.3.3 Key Points of Earth-Rock Dam Seepage Safety Evaluation

7.3.3.1 Safety Evaluation of Dam Foundation Seepage

The foundations of earth-rock fill dams are mostly made of quaternary loose sediments, and a few have rock foundations, or only part of the impermeable bodies (such as core wall and spillway) are in contact with the rock foundation. The dam foundation safety problems are mostly piping, soil flow, or contact scouring, which have a direct impact on the overall safety of the dam. The impervious bodies and seepage outlets are important points for checking the seepage safety of the dam foundation. The stability of the seepage outlet is closely related to the availability of filter protection, soil mass material, and construction quality. For example, qualified filter protection at the seepage outlet will significantly improve the seepage stability [25]. The evaluation of dam foundation seepage safety shall include the following essential points:

(1) To evaluate the seepage stability of the gravel layer, the seepage deformation pattern shall be determined according to the soil type and grain size distribution to verify the allowable seepage gradient [J]. By comparing with the actual seepage gradient J, it shall be judged whether there is any piping at the seepage outlet or the possibility of destruction of flowing soil, and whether there is any potential seepage hazard such as piping and contact scour within the seepage field.
(2) If the cover layer consists of soil with relatively high permeability, its resistance to buoyancy should be reevaluated. The permissible seepage gradient must be established either through permeability testing or by referencing the flowing soil index. In cases where a filter cover has weight, calculations should be conducted to verify whether the cover's thickness and extent satisfy the specified requirements.
(3) The seepage stability of the contact surface must be evaluated based on the following two conditions:

   - Examination for Potential Contact Scour and Soil Flow: An assessment should be conducted to determine if there is any possibility of contact scour occurring at the interfaces between coarse and fine soil layers. Additionally, the likelihood of soil flow at vertical interfaces in the direction from fine to coarse particles needs to be considered. Furthermore, it should be verified whether the coarse particle layer can effectively shield and safeguard the underlying fine particle layer.
   - Evaluation of Seepage Stability at Contact Interfaces with Rigid Structures: The seepage stability at interfaces between particulate soil masses and rigid structures, such as concrete anti-seepage walls, tunnels, and rocks, should be thoroughly reviewed. This involves analyzing the level of compactness in the connection between the particulates and the rigid surface. Moreover, it's important to ensure the presence of proper filter protection at outlets and to address any contact surfaces with fault fracture zones, limestone dissolution zones, and extensive fractures. A comprehensive analysis of the anti-seepage stability is also necessary.

(4) The anti-seepage performance and seepage stability of the impervious body in the foundation shall be analyzed [26].

7.3.3.2 Safety Evaluation of Dam Seepage

When conducting an assessment of dam seepage safety, the primary areas of focus should include the seepage stability of impervious structures, the seepage sections on the downstream dam slope, the contact seepage at transition layer, the contact zone of dam-crossing structure, and the construction interfaces [27] The following essential points shall be covered in the evaluation of dam seepage safety:

(1) For homogeneous earth dam, it is crucial to conduct a thorough review of the dam's anti-seepage capabilities to confirm whether they meet the specified criteria. The elevations of the actual seepage line (plane) of the dam and the seepage section along the downstream dam slope should be compared against the initially designed values. This assessment helps ensure that the dam's seepage performance aligns with the intended design. Particular attention should be given to the identification of any transverse or horizontal cracks within the dam structure. Additionally, the presence of soft binding zones or potential seepage pathways inside the dam should be examined.

(2) For zoned dam, the following requirements shall be met:

- It is essential to review the anti-seepage performance and seepage stability of key components such as core walls, sloping walls, blankets, and face plates. Pay close attention to the presence of qualified transition layers on both the upstream and downstream sides of the core walls or sloping walls. These transition layers play a vital role in ensuring proper seepage control. The evaluate whether the horizontal impervious blanket's bottom bedding layer, which could be natural gravel or another suitable material, functions as a protective layer to prevent seepage. Verify the presence of a qualified bedding layer for the face plate. This layer is essential to support and stabilize the face plate's structure and seepage prevention performance.
- Assess the anti-slide stability of the upstream dam slope, particularly when there's a rapid drawdown of the reservoir water level. This evaluation aims to ensure that the slope can withstand the changing water conditions without experiencing slides or instability. Examine the seepage stability of the egress section (zone) along the downstream dam slope. This assessment is essential to verify that the slope remains stable and resistant to potential erosion caused by seepage. Pay special attention to the embankment slope layer at the seepage section of the downstream dam slope. Confirm whether this slope layer aligns with the design specifications for a proper filter layer. A well-designed filter layer prevents erosion and maintains slope stability by managing water movement.
- Review the transition areas between coarse and fine fillings within the dam structure. Confirm whether the soil conservation conditions and drainage conditions align with the design requirements for the filter layer. This assessment ensures that the filter layer's effectiveness in preventing soil erosion

and facilitating drainage is maintained. Evaluate the drainage systems in place, including prism drainage, blanket drainage, and embankment slope drainage. Ensure that the design criteria for the filter layer's soil conservation and drainage capacity are met. Additionally, examine the presence of any concentrated seepage or abnormal occurrences, such as the discharge of a large quantity of solid particles.

7.3.3.3 Overage Operation of the Dam and the Under Dam Culvert

Among the reservoirs constructed in the early years, many were developed without proper survey, design, or quality control during construction. These projects often lacked synchronized coordination between design, construction, and operation phases. Geological conditions were not adequately understood, and foundation cleaning was insufficient. The soil used in dam construction lacked strict control, leading to non-homogeneous and inadequately compacted fill materials. The complex construction interfaces and connections further compounded the issue, making the dam structure intricate and sometimes unclear.

In addition, many early earth-rock dams featured structures integrated within or passing through the dam, such as water conveyance tunnels beneath the dam and open spillways on the dam's surface. These structures, often made of masonry, presented weak points in construction quality where they contacted the dam. They were prone to deformations due to contact seepage, which posed significant risks to dam safety. Recent dam failure incidents, even in cases of dam reinforcement projects, have frequently resulted from contact seepage issues with these integrated structures. As a result, it is imperative to give special attention to evaluating the stability of contact seepage associated with structures integrated within and running through the dam during dam seepage safety assessments.

7.3.4 Determination of Seepage Deformation of Earth-Rock Dam

The discrimination of soil seepage deformation stability is an important part of seepage safety evaluation of earth-rock dam. The discrimination of seepage deformation stability of earth-rock dam includes: the discrimination of seepage deformation stability of dam body, the discrimination of seepage deformation stability of dam foundation, the discrimination of seepage deformation stability of contact surface between dam body and dam foundation, and the discrimination of seepage deformation stability of contact surface between them and adjacent buildings.

7.3.4.1 Types and Characteristics of Seepage Deformation

The forms and developmental processes of seepage-induced deformations in earth-rock dams are influenced by various factors, including geological conditions, soil particle gradation, hydraulic conditions, and anti-seepage and drainage measures. These phenomena can be broadly classified into the following four types. The first two types of seepage deformation mainly occur in a single soil layer foundation, and the latter two types of seepage deformation mostly occur in multi-layer structure foundation (Figs. 7.3, 7.4 and 7.5).

(1) **Piping**: under the action of certain seepage, the fine particles in the soil move along the pore pipe formed by the skeleton particles or are taken away by seepage. This phenomenon is called piping, which occurs in gravel material [28].
(2) **Flowing soil**: under the action of certain seepage, the particle group in the soil starts at the same time and loses. It can occur in both cohesionless soil and cohesive soil. The deformation of flowing soil in cohesionless soil shows that

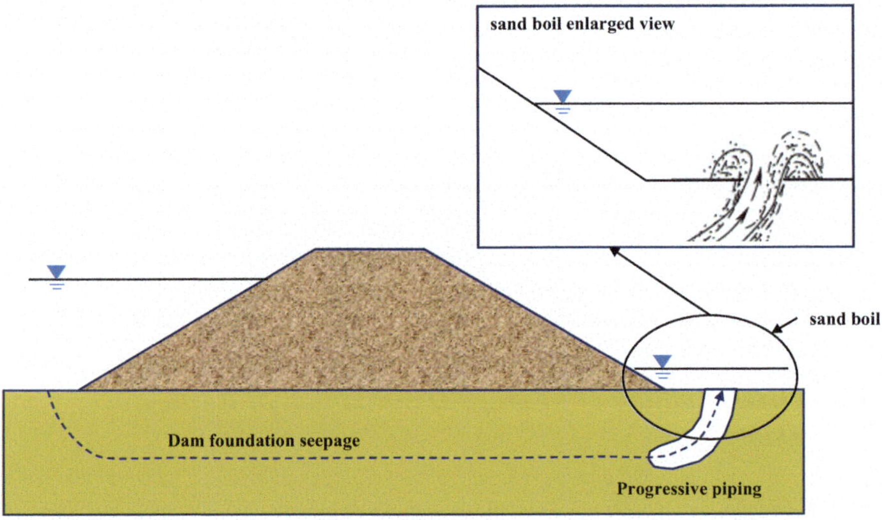

Fig. 7.3 The piping

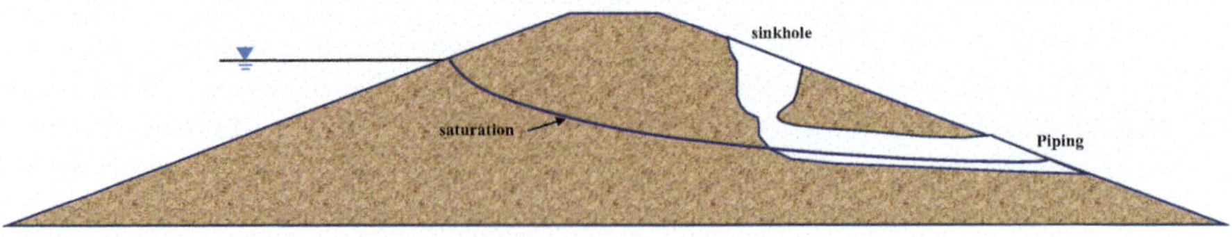

Fig. 7.4 The formation of sinkholes

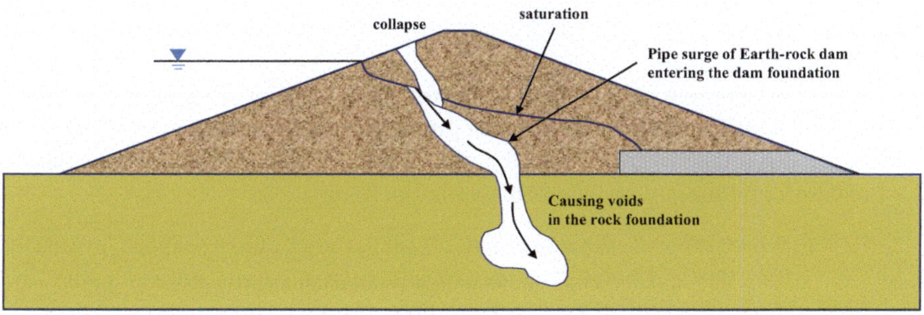

Fig. 7.5 Pores of earth-rock dam pipes entering the rock foundation

the particle group starts at the same time, such as spring group, sand boiling, and soil rolling is finally lifted by seepage. The deformation of flowing soil in cohesive soil is manifested as soil uplift, expansion, floating, and fracture.

(3) **Contact erosion**: When seepage flows along the contact surface of two different media, the fine particles of the particle layer are taken away. The direction of the contact surface is arbitrary. See Fig. 7.6.
(4) **Contact loss**: When the seepage flows perpendicular to the contact surface of two different media, the fine particles of one layer are moved into another layer. For example, the clogging of the filter layer is this deformation. See Fig. 7.6.

Piping occurs in the gravel foundation. Piping can occur at the outlet of seepage, or in the dam foundation or inside the dam.

- The external piping, which occurs at the downstream seepage outlet, can be protected by properly designed and installed filter layers to prevent soil particles from being carried away by seepage.
- The internal piping, which can be particularly problematic as it can compromise the stability and integrity of the dam structure, happens when seepage erodes soil

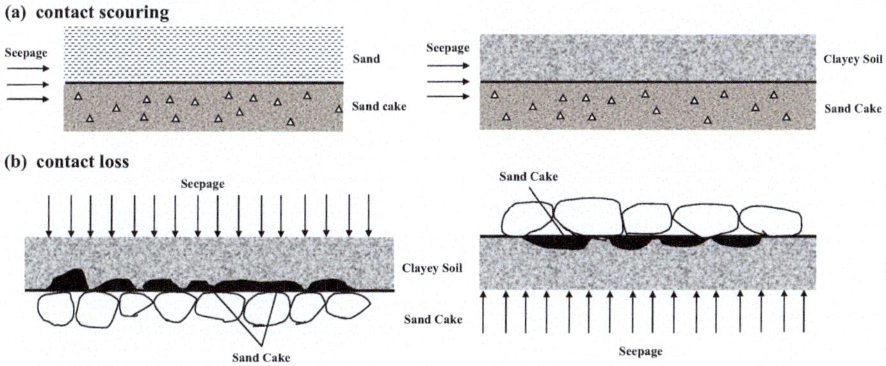

Fig. 7.6 Seepage deformation at contact surface of soil layer

particles within the dam foundation, creating voids and channels that can lead to the formation of pipes.
- The horizontal piping refers to piping that occurs along horizontal layers or planes within the dam foundation. This type of piping can occur in layered soils or where there are varying permeabilities within the foundation. It can lead to the formation of underground tunnels or channels within the foundation.

Flow soil, which involves the movement of soil particles due to seepage, primarily occurs in various types of homogeneous sands (coarse, medium, fine) and silty fine sand foundations. In addition to dispersive clay, the seepage deformation type of clay is mainly flowing soil. Flow soil occurs downstream of the dam seepage outlet. This type of soil movement and erosion happens when the filter layer protection or filter layer design at the downstream outlet is inadequate or absent. The flowing soil cannot occur inside the foundation, and only when the flowing soil occurs at the downstream outlet, it will extend to the inside of the foundation.

7.3.4.2 Permeability Deformation Discrimination Method

Identifying the type of seepage deformation in piping and flowing soils is complex and involves the following three steps:

- Identify the type of seepage deformation;
- Determining the critical hydraulic gradient and allowable hydraulic gradient;
- Assessing the safety of the seepage deformation.
- According to the fine particle content of soil (P) to identify piping and flowing soil:

Piping:

$$P_c < \frac{1}{4(1-n)} \times 100 \tag{7.3.1}$$

Flowing soil:

$$P_c \geq \frac{1}{4(1-n)} \times 100 \tag{7.3.2}$$

where

P_c The content of fine particles of soil, calculated by mass percentage, %;
n Porosity of soil, %

The fine particle content P of soil can be determined by the following method:

For soil with discontinuous gradation, there is at least one smooth section in the gradation curve where the particle content of the particle size class is less than or equal to 3%. The discriminating particle size d of coarse and fine particles is distinguished by the average particle size of the maximum and minimum particle sizes of the particle size class in the smooth portion, or the minimum particle size is the discriminating particle size and the content corresponding to this particle size is the fine particle content P. For continuously graded soil, The discriminating particle size d that distinguishes coarse and fine particle sizes is calculated as follows:

$$d_f = \sqrt{d_{70}d_{10}} \tag{7.3.3}$$

where

- d_f Distinguishing particle size of coarse and fine particles, mm;
- d_{10} The particle size of the content less than this particle size accounting for 70% of the total soil weight, mm;
- d_{70} The particle size of particles smaller than this size accounting for 10% of the total soil weight, mm.

(2) For discontinuous graded soil with nonuniformity coefficient greater than 5, the following methods can be used for discrimination:

Piping:

$$P_c < 25\% \tag{7.3.4}$$

The transition type depends on the density, particle size and shape of the soil:

$$25\% \le P_c < 35\% \tag{7.3.5}$$

Flowing soil:

$$P_c \ge 35\% \tag{7.3.6}$$

The non-uniform coefficient of soil can be calculated by the following formula:

$$C_M = \frac{d_{60}}{d_{10}} \times 100 \tag{7.3.7}$$

where

- C_M Uneven coefficient of soil;
- d_{60} The particle size of the content less than this particle size accounting for 60% of the total soil weight, mm;
- d_{10} The particle size of particles smaller than this size accounting for 10% of the total soil weight, mm.

(3) Critical hydraulic gradient of seepage deformation. The critical hydraulic gradient of seepage deformation is calculated as follows:

Piping or transition type:

The following formula should be used to calculate the critical hydraulic gradient of bottom-up seepage piping or transitional type:

$$J_{cr} = 2.2(G_s - 1)(1 - n)^2 \frac{d_5}{d_{20}} \tag{7.3.8}$$

where

d_5, d20 The soil particle size accounting for 5% and 20% of the total soil weight, mm.

Piping type can also be calculated by the following formula:

$$J_{cr} = \frac{42d_3}{\sqrt{\frac{k}{n^3}}} \tag{7.3.9}$$

where

k Permeability coefficient of soil, cm/s;
d_3 The soil particle size accounting for 3% of the total soil weight, mm.

The critical hydraulic gradient of the bottom-up seepage flow soil type is calculated by the following formula:

$$J_{cr} = (G_s - 1)(1 - n) \tag{7.3.10}$$

where

J_{cr} Critical hydraulic gradient of soil;
G_s The ratio of soil particle density to water density;
n Porosity of soil, %.

(4) Allowable soil gradient. The allowable hydraulic gradient of seepage deformation [J] is obtained by dividing the critical hydraulic gradient of the soil by the safety factor. The allowable gradient of the pipe is the hydraulic gradient when the soil particles begin to move in the pores and are removed. Generally, the soil has a certain potential to withstand the hydraulic gradient under this hydraulic gradient, so the safety factor K of 1.5 is used. Usually flow soil damage is the total failure of the soil, the greater the damage to hydraulic structures, safety factor of 2, for particularly important projects can also be used 2.5.In the absence of test data, the allowable hydraulic gradient of cohesionless soil can be selected according to Table 7.3.

Table 7.3 Allowable hydraulic gradient of non viscosity

Allowable hydraulic gradient	Type of seepage deformation					
	Fluvial soil form			Transitional type	Piping type	
	$Cu \leq 3$	$3 < Cu \leq 5$	$Cu \geq 5$		Grading continuity	Grading discontinuity
[J]	0.25~0.35	0.35~0.50	0.50~0.80	0.25~0.40	0.15~0.25	0.10~0.20

7.3.5 Seepage Around the Dam and Contact Seepage

The relationship between seepage flow and water level should be thoroughly analyzed. Particular attention should be paid to the presence of contact seepage [29]. In addition, the quality of the seepage water, including its turbidity or presence of suspicious substances, should be investigated. Seepage around the dam can have a significant impact on its stability. The stability of contact seepage at the interface between the embankment at each end of the dam and the embankment slope must be evaluated. The potential impact of subsurface seepage from the embankment ridges on the seepage stability of the natural bedrock layer and the slope stability of the embankment should be evaluated. If a grout curtain is provided along the embankment shoulder, its seepage performance and seepage stability should be analyzed [30]. The effectiveness of the grout curtain in preventing seepage and ensuring its long-term stability should be investigated.

7.3.6 Seepage Calculation and Analysis

The seepage calculation shall obtain the seepage field elements of the dam and its foundation abutment under different operation conditions:

- Determine the location of the dam body phreatic line and its downstream escape point, and draw the equipotential line distribution map or flow network map in the dam body and dam foundation.
- Determine the single width seepage flow and total seepage flow of the dam body and dam foundation.
- Determine the seepage gradient between the downstream dam embankment and the dam foundation, the escape gradient of the escape section of the dam slope, and the seepage gradient between different soil layers.
- Determine the position of the phreatic line or pore pressure in the upstream dam slope when the reservoir water level drops.
- Determine the equipotential line, seepage flow and seepage gradient of the dam abutment.
- The seepage calculation should include the steady seepage under each condition combination. The unstable seepage calculation should be carried out under the

water level drop condition of the first stage dam, the second stage dam and the high dam reservoir below the third stage. The seepage calculation shall include the following water level combinations:

- the normal upstream water level and the corresponding lowest downstream water level;
- The upstream design flood level corresponds to the downstream water level.
- Check flood level upstream and corresponding water level downstream;
- Upper and lower water levels before and after reservoir water level drops.
- The values of permeability coefficient shall comply with the following provisions: the anisotropy of permeability coefficient should be considered for dam body and dam foundation materials, and different permeability coefficients should be used for different directions of anisotropic materials; The average value of large permeability coefficient should be used when calculating the seepage flow rate. The average value of small permeability coefficient should be used to calculate the wetting line when water level drops. Material values of dam foundation shall take into account the influence of rock characteristics, weathering degree and geological structure.

1 **Hydraulics Formula Method**

The calculation of the hydraulic formula method is relatively simple, which can be used for seepage calculation and analysis of earth-rock dams in medium and small projects [31]. The calculation of the hydraulic formula method can simplify the more complex actual conditions as follows: the adjacent soil layer with a permeability coefficient difference of less than 5 times can be regarded as one layer, and its permeability coefficient adopts the weighted average parameter; for two-layer dam foundation, if the lower layer is thick and its permeability coefficient is less than 1 / 100 of the permeability coefficient of the upper layer, the lower layer can be regarded as a relatively impermeable layer. If the depth of the permeable embankment is more than 1.5 times the length of the impermeable bottom of the structure, it can be estimated according to the infinitely deep permeable embankment.

When using the hydraulic method for seepage analysis of earth and rock dams, the following assumptions are often made. (1) The soil of the embankment is homogeneous, and the permeability coefficient of each point in the embankment is the same in all directions. (2) The infiltration flow is a binary steady laminar flow state according to Darcy's law. (3) The infiltration flow is asymptotic, and the infiltration gradient and flow velocity are equal at each point in any straight section. Figure 7.7 shows a rectangular soil on an impermeable foundation with a permeability coefficient of k. Applying Darcy's Law, the average flow velocity over the entire section is:

Fig. 7.7 Diagram of infiltration calculation of rectangular soil on impermeable foundation

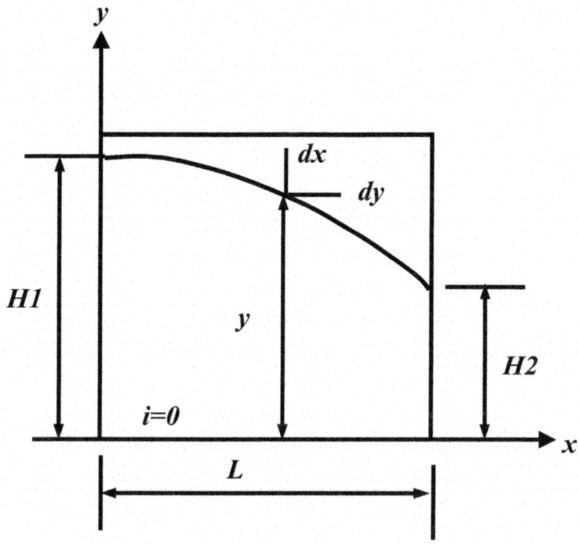

$$v = -k\frac{dy}{dx} \tag{7.3.11}$$

Let the single-width seepage volume be q, then

$$q = vy = -ky\frac{dy}{dx} \tag{7.3.12}$$

After separating the variables in the above equation and dividing the area from the upstream surface ($x = 0$, $y = H_1$) to the downstream surface ($x = L$, $y = H_2$), we obtain the following equation:

$$H_1^2 - H_2^2 = \frac{2q}{k}L \tag{7.3.13}$$

$$q = \frac{k(H_1^2 - H_2^2)}{2L} \tag{7.3.14}$$

Transforming the integral of Eqs. (7.3.6–7.3.9), the dip line equation is obtained as

$$q = \frac{k(H_1^2 - y^2)}{2x} \tag{7.3.15}$$

$$y = \sqrt{H_1^2 - \frac{2q}{k}x} \tag{7.3.16}$$

2 **Numerical Simulation**

The numerical simulation analysis method is a numerical solution that discretizes the actual physical model and divides it into small units to approximate the actual physical model [32]. The commonly used methods are finite element method (FEM), difference method, etc. The numerical simulation analysis method can effectively use the permeability parameters of various media partitions and various complex boundary conditions to obtain the dam seepage field data including potential, seepage gradient and seepage discharge through iterative calculation, so as to judge the seepage safety of the dam more comprehensively and accurately.

The numerical analysis method can simulate the three-dimensional seepage characteristics of the dam and its foundation, which greatly improves the calculation accuracy of the dam seepage analysis. The numerical analysis method can also be used to analyze the seepage parameters and leakage causes of the dam material by using the measured seepage field of the current water level. With the rapid development of computer seepage calculation software, the seepage calculation and analysis of large and medium-sized reservoir dams generally adopt numerical analysis methods [33–36] (Figs. 7.8, 7.9 and 7.10).

7.4 Evaluation on Seepage Safety of Concrete Dam and Masonry Dam

7.4.1 Characteristics of Concrete Seepage Diseases

Seepage deformation of the concrete dam body does not generally occur, but cracks can easily occur during construction and operation, causing the dam body to leak. The seepage of the dam body may increase the seepage pressure of the dam body and cause the stress of the dam body structure not to meet the requirements. More importantly, the seepage water will dissolve and remove the calcium hydroxide in the concrete and form white calcium carbonate on the outside of the concrete. This will destroy the equilibrium between cement and other hydration products, leading to the decomposition of hydration products and the deterioration of concrete performance. On the other hand, leakage can cause and accelerate the occurrence and development of other diseases. When the environmental water has the effect of erosion on concrete, the leakage will cause the environmental water erosion to develop toward the inside of the concrete, thus increasing the depth and breadth of damage. In cold areas, leakage will increase the water content of concrete and promote freeze–thaw damage of concrete. For hydraulically reinforced concrete structures, leakage will also accelerate the corrosion of reinforcement, etc. According to the characteristics of concrete dams and the conditions of concrete dams in China. The seepage diseases of concrete dams are classified into the following categories:

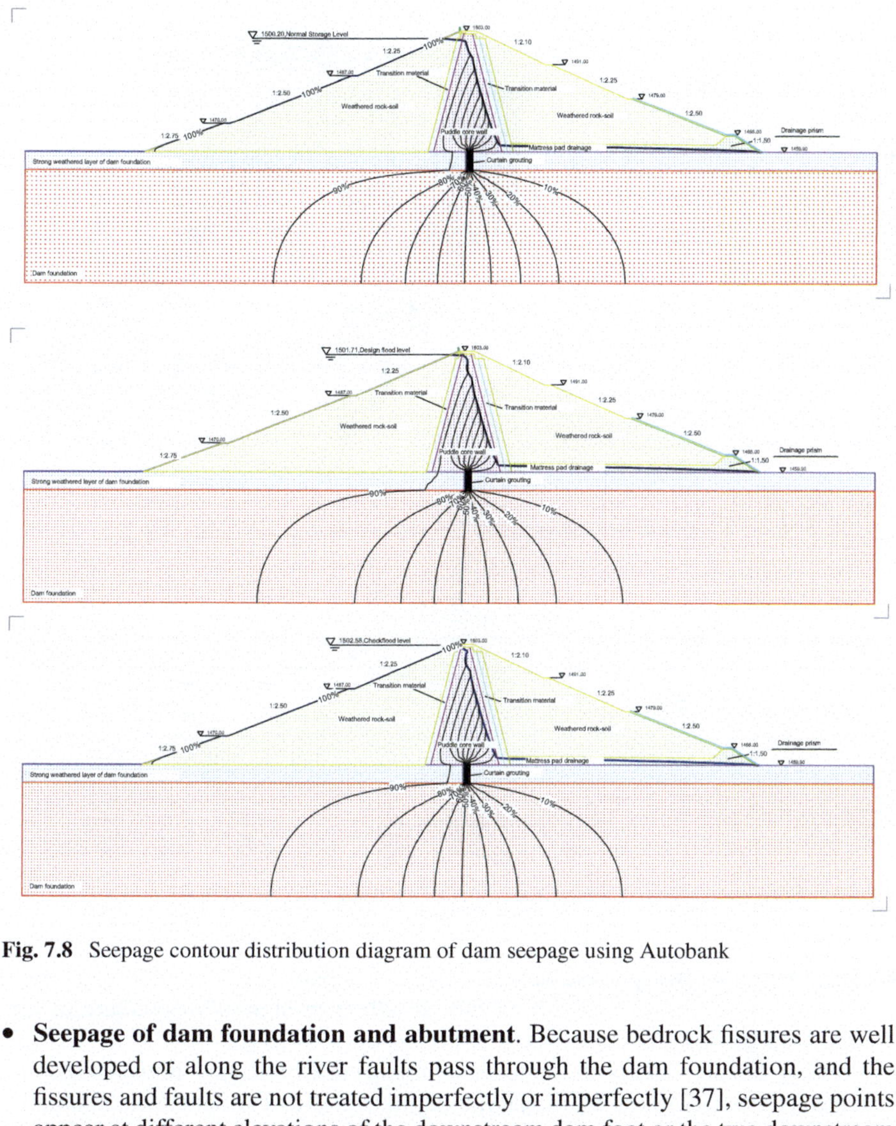

Fig. 7.8 Seepage contour distribution diagram of dam seepage using Autobank

- **Seepage of dam foundation and abutment**. Because bedrock fissures are well developed or along the river faults pass through the dam foundation, and the fissures and faults are not treated imperfectly or imperfectly [37], seepage points appear at different elevations of the downstream dam foot or the two downstream dam abutment slopes, or the uplift pressure of the dam foundation is high.
- **Leakage of the dam body**. Due to the poor construction quality of the dam, honeycomb or cold joint occurs in concrete, and mortar is used as a seepage passage to make concrete permeable or layer joint seepage. Temperature crack occurs in the dam body or dam foundation and abutment deformation cause cracking of the dam body, which results in seepage of downstream dam slopes, such as stone masonry joint seepage and cracks seepage, concrete crack seepage, layer joint seepage of RCC. This causes the dam body uplift pressure to rise.

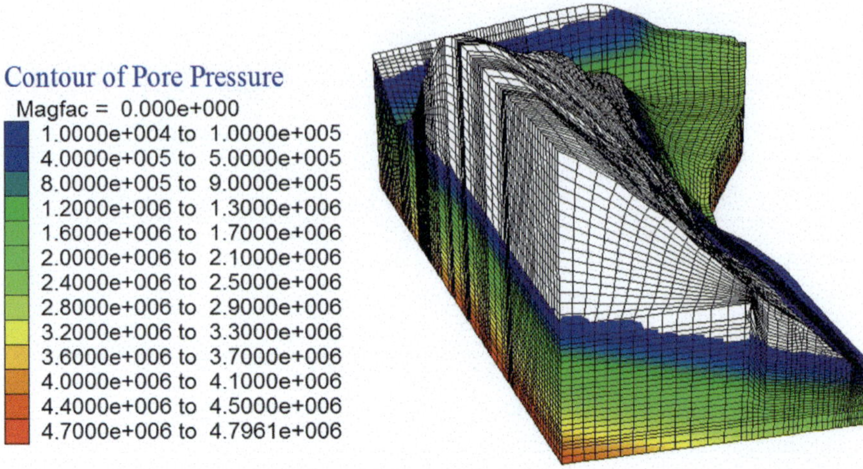

Fig. 7.9 Seepage analysis of homogeneous earth-rock dams using FLAC 3D

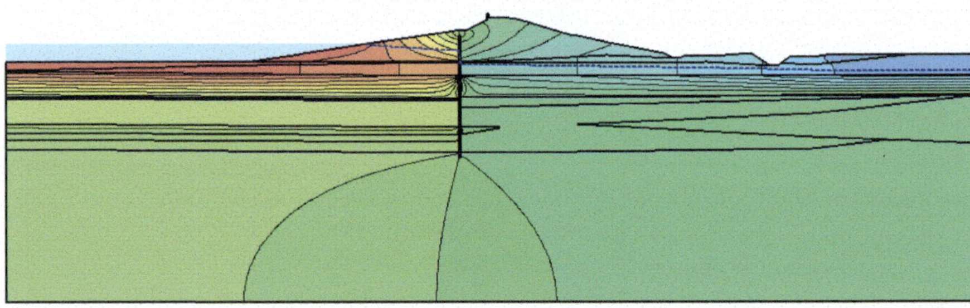

Fig. 7.10 Seepage analysis of homogeneous earth-rock dams using Geo-Studio

- **Karst leakage**. Because the karst around and under the reservoir is not treated imperfectly, large karst leakage occurs, resulting in high uplift pressure of the dam foundation or difficulty in water storage of the reservoir.
- **Etching hazard**. There are soluble components in the dam foundation. The erosion of groundwater increases the permeability of the dam foundation and forms leakage, or the erosion of groundwater causes erosion damage to the concrete.

7.4.2 Safety Evaluation of Dam Foundation Seepage

Dam foundation seepage is a critical factor in the safety assessment of both concrete dams and masonry dams. The effectiveness of dam foundation seepage prevention directly impacts not only the seepage stability of the dam, but also the overall stability of the dam against sliding. The seepage safety evaluation of the dam foundation covers the following essential elements:

(1) The analysis of the grouting curtain's seepage stability and anti-seepage performance, as well as the effectiveness of dam foundation drain holes, is necessary. Reviewing the dam uplift pressure coefficient is crucial using monitoring data to ensure it complies with design specifications. Additionally, its impact on the dam's anti-slide stability should be assessed.
(2) When fault fracture zones, weak and soft intercalated layers, and crack fillings are present on the dam contact face, a review of the substances' anti-seepage stability is necessary. The allowed anti-seepage gradient should be determined through specialized testing. If drain holes are present in the weak stratum, a review should be conducted to determine if the inverse filter material is adequately protected.
(3) The analysis of the anti-seepage performance and seepage stability of vertical impervious bodies, such as grouting curtains and anti-seepage walls, in non-rock dam foundations, is necessary. Additionally, a review of the seepage stability of the horizontal seepage flow and seepage outlet of the corresponding soil type at the point where the dam foundation contacts the soil should be conducted.

7.4.3 Calculation of Uplift Pressure

Uplift pressure, which is the hydrostatic pressure exerted by water underneath the dam, plays a significant role in dam stability. The criteria for evaluating uplift pressure are specified in design guidelines. These criteria may be relaxed only if substantial field measurements of actual uplift pressures are available and support the proposed deviations.

Any stability analysis of the dam should seek to apply forces that are compatible with the failure mechanism being assumed. For this reason, it is less important to determine what the uplift pressures on a dam are at present than it is to determine what they would be during failure. Uplift should be assumed to exist between the dam and its foundation, and within the foundation below the contact plane and it should also be applied within any cracks within the dam. Uplift is an active force which must be included in the analysis of stability. Uplift shall be assumed to act over 100% of the area of any failure plane whether that plane is within the dam, at the contact with the foundation or at any plane within the foundation.

Uplift reduction can be achieved through a drainage system, a grout curtain, or sometimes simply by the accumulation of low permeability silt against the upstream face of the dam. If uplift reduction is assumed in analysis, it must be verified by instrumentation. There must also be a reasonable assurance that the uplift reduction effect that is measured under normal conditions will persist under unprecedented load conditions such as extreme floods or earthquakes. The uplift pressure exists in both the concrete dam body and foundation, and there are certain differences between the uplift pressure of gravity dam and arch dam. Uplift along failure planes within the body of the dam shall be assumed to vary from 100% of normal headwater at the upstream face to 100% of tailwater or zero, as the case may be, at the downstream face. When a vertical drainage system has been provided within the dam, the drain effectiveness and uplift assumptions should follow the knowledge provided in below, and should be verified by instrumentation [38].

7.4.3.1 The Gravity Dam

(1) The distribution patterns of uplift pressure on the bottom surface of various gravity dams on the rock foundation are determined according to the following three conditions:

- When the dam foundation is provided with anti-seepage curtain and drainage holes, the uplift pressure head at the upstream (dam heel) of the dam bottom surface is H_1, the center line of the drainage hole is $H_1 + a(H_1\text{-}H_2)$, and the downstream (dam toe) is H_2. The sections are connected in a straight line in turn, see Fig. 7.11
- When the dam foundation is equipped with an anti-seepage curtain, upstream main drainage hole, downstream auxiliary drainage hole and drainage system, the head of uplift pressure at the upstream of the dam bottom surface is H_1, the center lines of the main and auxiliary drainage holes are a_1H_1 and a_2H_2 respectively, and the downstream is H_2. The sections are connected in a straight line, as shown in Fig. 7.11d.
- When there is no anti-seepage curtain and upstream drainage hole in the dam foundation, the head of the uplift pressure upstream of the dam bottom is H_1, and the head downstream is H_2, which is connected in a straight line, see Fig. 7.11e.
- The seepage pressure reduction factor α, the uplift pressure reduction factor α_1, and the residual uplift pressure reduction factor α_2 can be used according to Table 7.4.

(2) The distribution of uplift pressure on the calculated section inside the dam can be determined according to Fig. 7.12 when the drainage pipe of the dam is

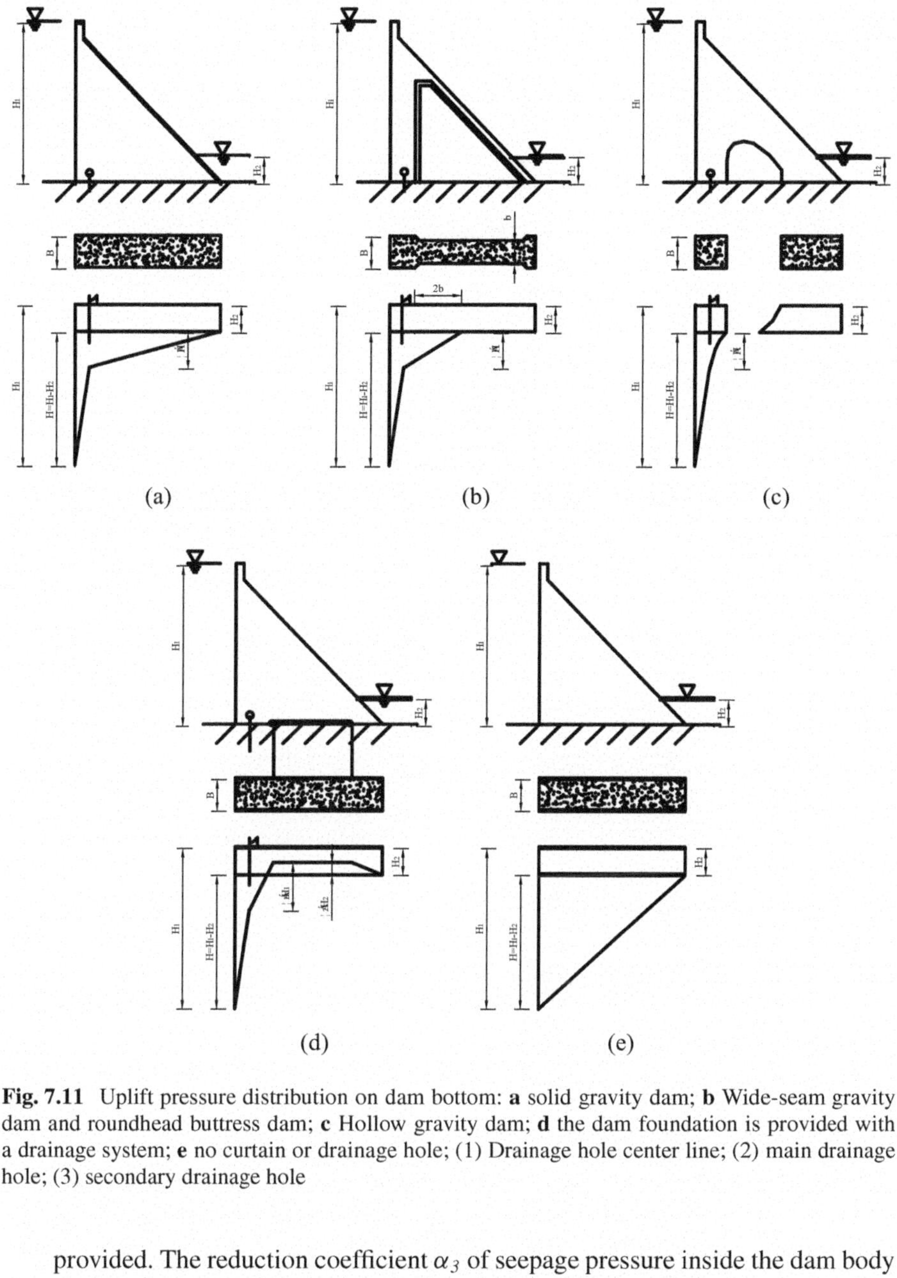

Fig. 7.11 Uplift pressure distribution on dam bottom: **a** solid gravity dam; **b** Wide-seam gravity dam and roundhead buttress dam; **c** Hollow gravity dam; **d** the dam foundation is provided with a drainage system; **e** no curtain or drainage hole; (1) Drainage hole center line; (2) main drainage hole; (3) secondary drainage hole

provided. The reduction coefficient α_3 of seepage pressure inside the dam body at the center line of the drainage pipe can be adopted as follows:

- The solid part α_3 of the solid gravity dam and the hollow gravity dam is 0.2.
- The α_3 of the non-wide seam part of the wide seam gravity dam and the large head buttress dam is 0.2, and the α_3 of the wide seam part is 0.15.

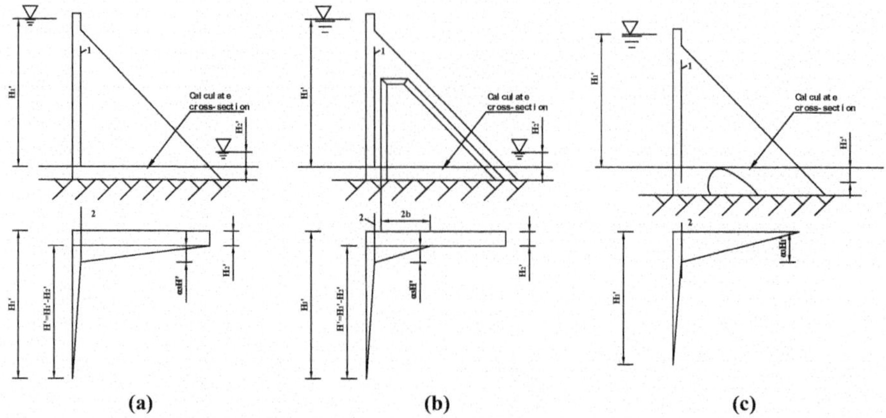

Fig. 7.12 Uplift pressure distribution of dam body calculation section: **a** solid gravity dam; **b** wide joint gravity dam; **c** hollow gravity dam; (1) drainage pipe inside the dam; (2) Drainage pipe centerline

Table 7.4 Reduction coefficient of seepage pressure and uplift pressure of dam bottom

Dam type and location		Dam foundation treatment		
		(A) provided with anti-seepage curtain and drainage holes	(B) equipped with an anti-seepage curtain, upstream main drainage hole, downstream auxiliary drainage hole and drainage system	
Location	Types	The seepage pressure reduction factor α	The uplift pressure reduction factor α_1,	The residual uplift pressure reduction factor α_2
River bed dam section	Solid gravity dam	0.25	0.20	0.50
	Wide-seam gravity dam	0.20	0.15	0.50
	Roundhead buttress dam	0.20	0.15	0.50
	Hollow gravity dam	0.25	–	–
Bank slope dam section	Solid gravity dam	0.35	–	–
	Wide-seam gravity dam	0.30	–	–
	Roundhead buttress dam	0.30	–	–
	Hollow gravity dam	0.35	–	–

(3) The distribution graph of uplift pressure on the calculated section inside the dam body, when there is no drainage pipe in the dam body, the water head of uplift pressure on the upstream dam surface is H_1, and that on the downstream dam surface is H_2, which are connected in a straight line.

7.4.3.2 The Arch Dam

The following standards are put forward for the calculation of uplift pressure of dam foundation and dam body:

(1) The calculation of uplift pressure acting on dam foundation surface should meet the following provisions:

- When the dam foundation is equipped with anti-seepage curtain and drainage holes, the head of the uplift pressure at the dam heel is H_1 (upstream water depth), the center line of the drainage hole is $H_2 + a(H_1 - H_2)$, and the dam toe is H_2 (downstream water depth), which is connected in a straight line in turn. The reduction factor α can be 0.25, see Fig. 7.13.
- When the anti-seepage curtain is not set in the dam foundation, only the drainage hole is set. The head of the uplift pressure at the dam heel is H_1, the center line of the drainage hole is $H_2 + a(H_1 - H_2)$, and the dam toe is H_2, which are connected in a straight line. The reduction coefficient α can be used from 0.30 to 0.45, as shown in Fig. 7.14.
- When only anti-seepage curtain without drainage holes is set in the dam foundation, the water head of the uplift pressure at the dam heel is H_1, the center line of the curtain is $H_2 + a_1(H_1 - H_2)$, and the dam toe is H_2, which are connected in a straight line. The reduction factor a_1 can be 0.50 ~ 0.70, as shown in Fig. 7.15.
- When there is no anti-seepage curtain and drainage hole in the dam foundation, the head of the uplift pressure at the dam heel is H_1, and the head at the dam toe is H_2, which is connected in a straight line, see Fig. 7.16.

(2) The calculation of seepage pressure at the side of bank slope arch seat in arch seat stability analysis shall comply with the following provisions:

- When anti-seepage curtain and drain hole are set on the side of arch seat, they are $H_1 - H_2$ at the upstream point of arch end, the reduction coefficient a at the centerline of drain hole can be 0.35, and 0 at the downstream point of arch end, which are connected in a straight line. For important projects with complex geological conditions, it shall be determined by three-dimensional seepage calculation or test.
- When there is no impervious curtain and drainage hole at the side of the arch seat or drainage hole is easily invalid, the upper and lower ends of the side of the arch seat are calculated by straight line connection from $H_1 - H_2$ to 0.

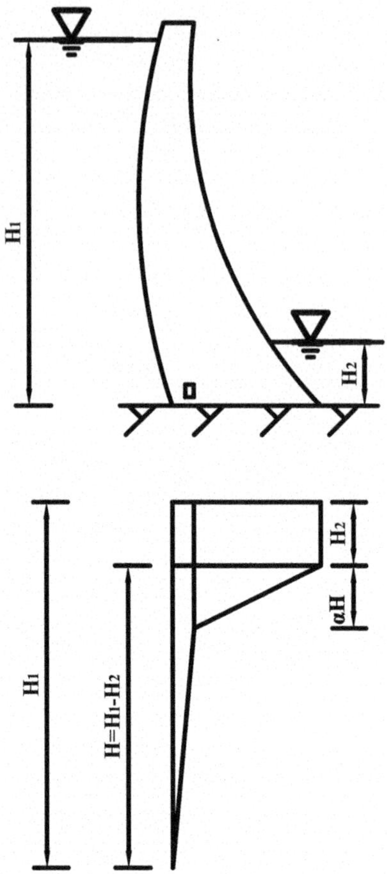

Fig. 7.13 Distribution of uplift pressure of dam foundation with impervious curtain and drainage hole

- When no anti-seepage curtain is installed at the side of the arch seat only with drainage holes or only with no drainage holes in the anti-seepage curtain, due to the influence of groundwater and three-way seepage on the bank slope, the reduction coefficient should be increased appropriately.

(3) The calculation of uplift pressure inside thick and medium-thick arch dams shall comply with the following requirements:

- When a drainage pipe is installed in the dam, the head under uplift pressure at the upstream dam face is H'_1 on the center line of the drainage pipe and $H'_2 + a_1(H'_1 - H'_2)$ on the upstream water depth above the calculated section. (When the calculated section is above the downstream water level, take $H'_2 = 0$). The head of uplift pressure acting on the downstream dam surface is H'_2, which is connected in a straight line and the reduction coefficient a can be 0.20, see Fig. 7.17.

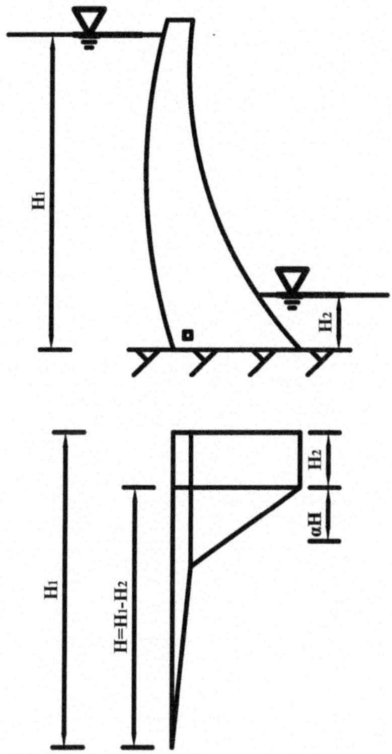

Fig. 7.14 Uplift pressure distribution of dam foundation without anti-seepage curtain and only with drainage holes

- When there is no drainage pipe in the dam, the head of uplift pressure acting on the upstream dam face is H'_1, and the downstream dam face is H'_2, which is connected in a straight line, see Fig. 7.18.

7.4.4 Other Situations in Calculation of Uplift Pressure

In the context of gravity dams on rock foundations, the consideration of potential failure planes is essential for ensuring the dam's stability. When analyzing the stability of a gravity dam on a rock foundation, engineers need to assume a failure plane between the dam and the foundation. Beyond the dam-foundation interface, the analysis must also take into account the potential for failure planes within the rock itself beneath the dam.

Engineers are advised to adopt a conservative approach when reviewing assumptions related to uplift pressure reduction. This means erring on the side of caution and ensuring that safety margins are maintained in the analysis. Any assumptions made regarding uplift reduction should be supported by instrumentation data collected

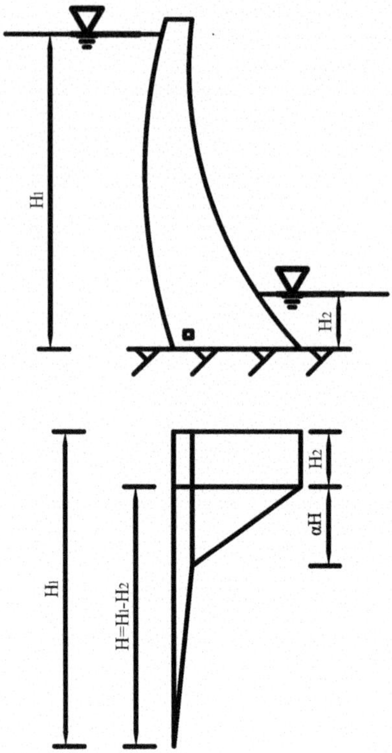

Fig. 7.15 The uplift pressure distribution of dam foundation when the anti-seepage curtain is set without drainage hole

from the dam and its foundation. Even when instrumentation data indicates that uplift reduction is occurring under certain conditions, engineers must question whether the factors influencing headwater, tailwater, and foundation stresses will remain constant or predictable under more severe or extreme conditions.

- **Uplift Assumptions**

For structures equipped with an open and verifiable drainage system, the uplift at the concrete/rock interface should be considered to change linearly. It starts with the full headwater pressure at the heel or theoretical crack tip, gradually decreases at the drain, and ultimately reaches the full tailwater pressure at the toe. The effectiveness of the drain must be confirmed through instrumentation, and an efficient maintenance plan must be implemented as outlined. Additionally, it's important to note that if heads are measured from a datum other than the dam base, the dam base elevation must be subtracted from the absolute heads to determine the actual uplift pressure. Furthermore, it is assumed that the gallery is designed for free drainage.

The character of foundation stresses significantly impacts crack extent and dimensions, as these are influenced by the imposed stresses. When analysis suggests that foundation stresses will vary considerably, this can alter crack geometry, thereby potentially affecting drain efficiency. Moreover, there may be a discrepancy between

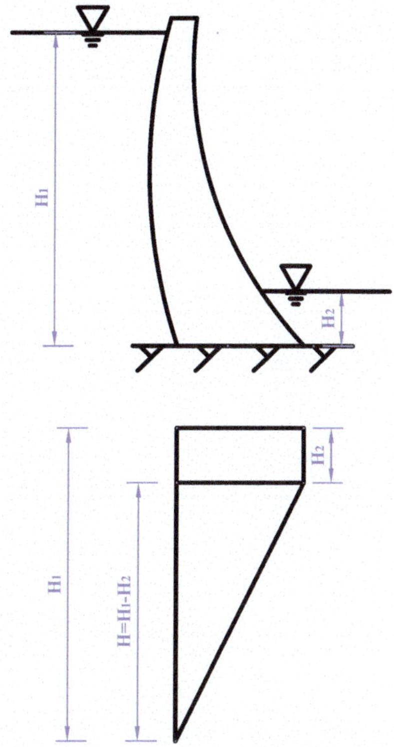

Fig. 7.16 Distribution of uplift pressure of dam foundation without anti-seepage curtain and drainage hole

the drain efficiency assumed during design and the measured efficiency. If a certain degree of error tolerance is permissible, extrapolation becomes more justifiable. The level of comprehension regarding the dam's foundation geology is also crucial. Reducing uncertainties in the selection of design parameters can lead to a corresponding decrease in required safety factors. This principle applies to the extrapolation of drain efficiencies as well. A more precise definition of the geologic characteristics that affect seepage parameters can minimize the uncertainties surrounding drain efficiency extrapolation. Lastly, the stability drain effectiveness assumptions must be carefully considered. If drain efficiency is crucial in preventing the theoretical base crack from penetrating the entire dam, extrapolating these assumptions into unprecedented loading conditions should be approached with skepticism.

When crack does not extend beyond the drain line (Fig. 7.19):

$$H3 = K * [(HW - TW) * \frac{(L-X)}{(L-T)} + TW - H4] + H4 \quad (\text{H4} > \text{TW}) \tag{7.4.1}$$

$$H3 = K * [(HW - TW) * \frac{(L-X)}{(L-T)}] + TW \quad (\text{H4} < \text{TW}) \tag{7.4.2}$$

where

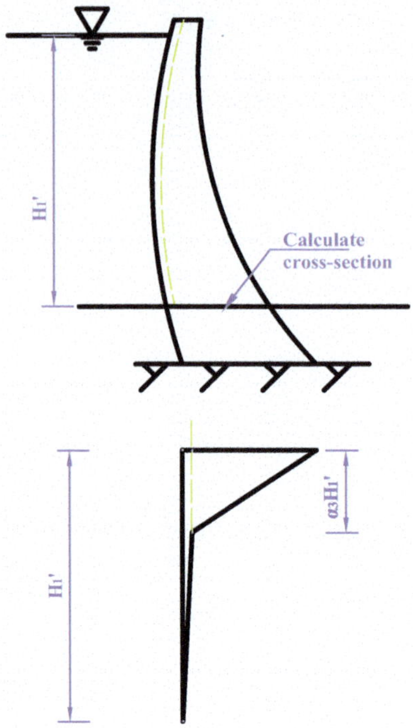

Fig. 7.17 Uplift pressure distribution in dam with drainage pipe

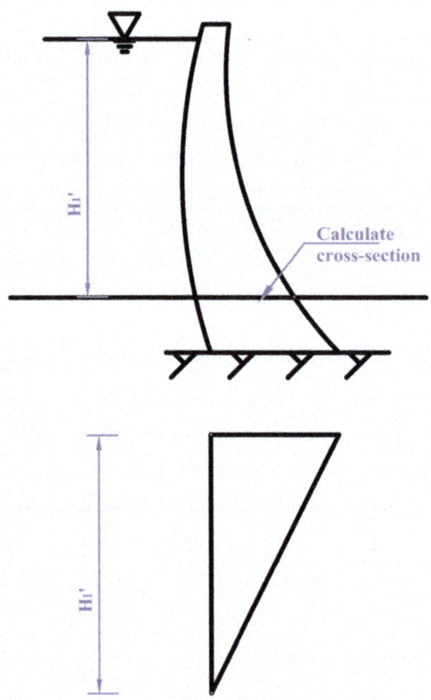

Fig. 7.18 Uplift pressure distribution in dam without drainage pipe

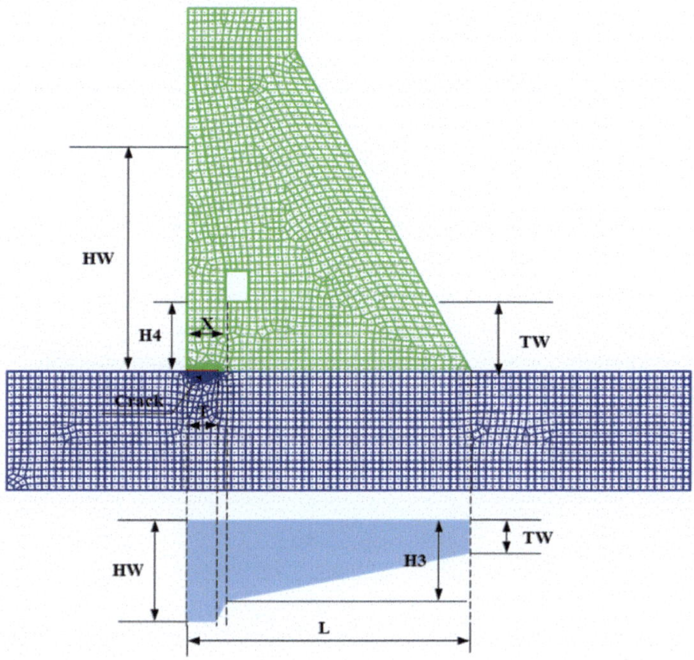

Fig. 7.19 When crack does not extend beyond the drain line

$K - K = 1 - E, E =$ Drain effectiveness
When crack does extend beyond the drain line (Fig. 7.20):

$$H3 = K * (HW - H4) + H4 \quad \text{(H4 > TW)} \tag{7.4.3}$$

$$H3 = K^*(HW - TW) + TW \quad \text{(H4 < TW)} \tag{7.4.4}$$

When analysis indicates that a theoretical crack propagates beyond the drains for an unprecedented load condition such as the PMF, the amount of drain efficiency that can exist is limited by certain physical constraints. For cases in which the theoretical base crack extends beyond the drains, the resulting uplift force should not be assumed to be less than that calculated by the idealization shown below , where $\nabla^2 \cdot P = 0$. and the boundary conditions are those depicted in Fig. 7.21.

- **Grouting**: Although a grout curtain may initially slow down foundation flows, as the dam ages, the curtain's effectiveness may deteriorate, resulting in a diminished reduction in uplift. To ensure the long-term integrity and safety of the dam, it is crucial to have a drainage system operating downstream of the grout curtains. This system will assist in managing any residual seepage and help maintain the structural stability of the dam. As to dam body, the anti-seepage performance of the dam body, the upstream impervious panel or core wall shall be reviewed to

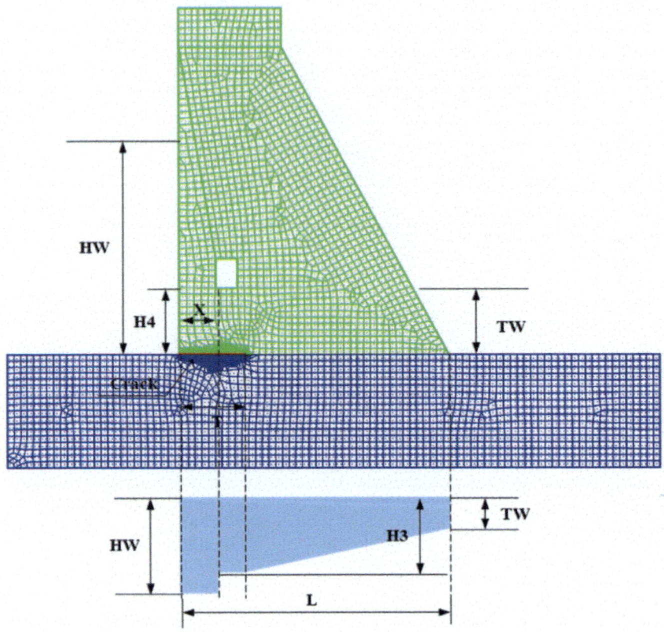

Fig. 7.20 When crack does extend beyond the drain line

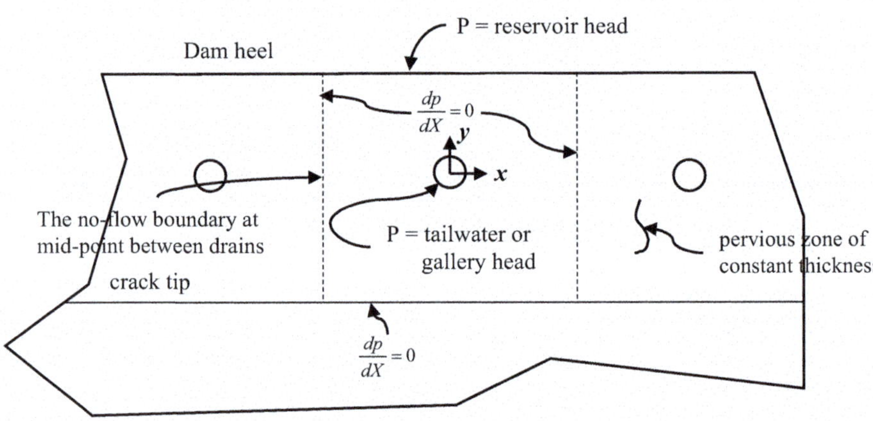

Fig. 7.21 The boundary conditions of the theoretical crack

see if it meets the requirements. For masonry dam where seepage exists, the variation of strength and anti-seepage performance of the masonry mortar shall be inspected and measured and the dam body strength and anti-slide stability shall be reviewed. For masonry dam provided with impervious panels or core walls, review shall be also made on whether the impervious body and the foundation anti-seepage curtain form a continuous and closed anti-seepage system. Furthermore, the implementation of a monitoring system is essential. This system should be designed to measure actual uplift pressures and detect any decrease in drain efficiency caused by clogging or other factors. By continuously monitoring these key parameters, dam operators can promptly identify any issues and take appropriate measures to address them, thus ensuring the continued safe operation of the dam.

- **Aprons**: Upstream and downstream aprons have the effect of increasing the seepage path under the dam. For an upstream apron properly sealed to prevent leakage, the effect is to reduce the uplift under the dam. The effectiveness of upstream aprons in reducing uplift is compromised if cracks and joints in the apron permit leakage. Conversely, downstream aprons such as stilling basins have the effect of increasing uplift under the dam. (See Fig. 7.22) Uplift reduction should be justified by instrumentation. In the case of downstream aprons, it may be assumed that uplift is limited to that which would float the apron.
- **Reservoir Silt:** Reservoir silt can indeed contribute to reducing uplift forces under a dam, operating in a manner akin to an upstream apron. (See Fig. 7.22) However, it is crucial to justify this uplift reduction through instrumentation and careful monitoring. The reason for this is the potential for liquefaction of the silt during seismic events. In the event of liquefaction, the pore pressure within the silt increases significantly. This elevated pore pressure condition can persist for a considerable period even after the seismic event has subsided. As a result, the uplift reduction attributed to the silt may be temporarily or even permanently lost. Therefore, when assessing the post-earthquake stability of a dam, it is not advisable to rely solely on the uplift reduction provided by the silt. Instead, a comprehensive analysis should be conducted, considering various factors such as

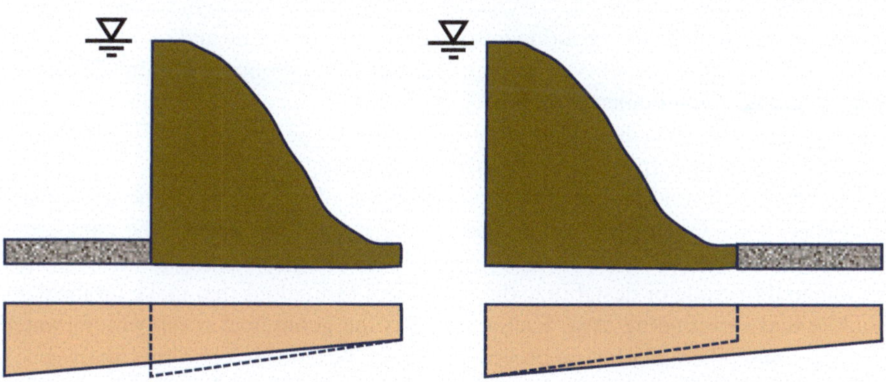

Reduce the uplift pressure on the dam through the upstream apron

Increase the uplift pressure on the dam through downstream apron

Fig. 7.22 Different layout forms of aprons

the dam's structural integrity, the characteristics of the foundation and reservoir, and the potential for liquefaction and other seismic-related hazards.

- **Earthquake**: Uplift pressures beneath a dam during earthquake loading should indeed be initially assumed to be those existing under normal conditions. However, it is crucial to recognize that seismic events can significantly alter these pressures due to various factors such as silt liquefaction, apron cracking, or potential offsets.
- **Flood Loading**: It is indeed important to avoid making overly optimistic assumptions when evaluating uplift reductions under a dam. Relying on the assumption that Intensity-Duration-Frequency (IDF) flood events will be brief and that the foundation's permeability will be so low that elevated headwater and tailwater pressures will not penetrate the base of the dam is not a conservative approach. This assumption can be problematic because extreme design floods can persist for hours or even days, resulting in prolonged exposure to elevated water levels. Furthermore, in a saturated rigid system such as a rock foundation with joints, even minimal volume changes can lead to significant pressure transmission. This means that changes in water levels can have a direct impact on uplift forces beneath the dam. In the absence of corroborative evidence, it is prudent to assume that uplift will vary directly with changes in headwater and tailwater levels. This approach ensures that the design considerations are based on realistic and conservative assumptions, thereby enhancing the safety of the dam structure.

7.4.5 Seepage Around the Dam and for Bank Slope Underground Water Seepage

The grouting curtain is an essential part of dam construction to control seepage through the foundation [39, 40]. Dynamic analysis of underground water seepage involves assessing how the grouting curtain performs over time and under changing conditions. This analysis aims to determine whether the grouting curtain effectively reduces seepage, maintains its integrity, and prevents excessive water flow. This can involve simulations and calculations to assess the performance of the grouting materials and the effectiveness of the grouting process.

The analysis of underground water seepage out of the ridges of the two banks should also consider the potential impact on the seepage stability of the geological structure zone of the dam shoulder. This involves evaluating whether the seepage has the potential to erode or destabilize the geological formations that make up the dam shoulder and surrounding area. Factors such as the geological composition, water flow rates, and potential erosion processes need to be considered in this analysis.

The underground water seepage out of the ridges can also influence the anti-slide stability of the dam shoulder. The seepage can potentially weaken the material and increase the risk of sliding along the dam shoulder. Analysis needs to be conducted to determine whether the seepage-induced changes in soil properties or water pressure could compromise the dam's stability against sliding. This may involve evaluating the strength of the materials, pore pressure conditions, and the potential for erosion or soil loss.

7.4.6 Analysis of Leakage Water Precipitates and Chemical Composition

Precipitates and chemical compositions must undergo analysis for seepage water. These measurements will be compared with the chemical compositions of the reservoir to verify if they cause any corrosive chemical reactions to concrete structures or natural foundations [41].

7.4.7 Temperature Effect

In the period when water level is relatively stable or water level drops, seepage flow and uplift pressure rapidly increase or decrease individually or simultaneously and the increase and decrease is much related with the temperature. At this time, structure deformation analysis shall be made in accordance with the temperature and deformation monitoring data [42].

Seepage of some concrete dams is related to temperature (but lagging behind atmospheric temperature, as in winter and spring). This is because the openness of some junction parts caused by cold shrinkage of dam or rock mass (such as dam block joints, dam heel contact face, dam cracks, rock fractures, fault joints of anti-seepage curtain) increases when temperature is reduced to a certain level. Once this happens, analysis on structures shall be conducted in accordance with monitoring data related to deformation and temperature. If trend of development is found, remedy measures shall be taken.

7.5 Evaluation on Seepage Safety of Discharging and Conveyance Structures

The seepage prevention and drainage design of spillway foundation shall be determined according to engineering geological and hydrogeological conditions, the importance and function of buildings, structural safety and stability, and the relationship between seepage prevention and drainage. The layout and design of seepage prevention and drainage for flood channel near dam abutment should be considered as a whole with respect to the seepage prevention and drainage system of the dam.

According to the engineering geology, hydrogeology, lining type and other design conditions of surrounding rocks along the tunnel, as well as the requirements of environmental protection, water protection and operation, the seepage prevention and drainage measures of the tunnel are determined by comprehensive analysis. For seepage safety evaluation of conveyance tunnels(culverts), the tunnels(culverts) shall checked to see if there is water leakage, if there is soil pClause deposit in the culvert, if there is water seepage at the junction between rock(soil) body and tunnel(culvert) and if filter protection is provided at the outlet. Based on the above, analysis shall be conducted to see if there is any seepage stability problem, like contact scour at the peripheral junction zone.

When the gate base is installed on the rock foundation with cement grout curtain and drainage holes, the head of infiltration pressure at the upstream end of the gate bottom plate is H - h_s, the center line of the drainage hole is $a(H - h_s)$, the downstream end is zero, and the sections are connected in a straight line in turn (see Fig. 7.23). The infiltration pressure acting on the bottom surface of the bottom plate of the gate can be calculated from Eq. (7.5.1).

$$U = \frac{1}{2}\gamma(H - h_s)(L_1 + dL) \tag{7.5.1}$$

where

U Infiltrate pressure acting on the bottom surface of gate bottom plate, kN/m;
L_1 The horizontal distance between the centerline of the drain hole and the upstream end of the bottom of the gate bottom plate, m.
a Permeability pressure strength coefficient, can be used 0.25 and 0.5;
L The length of the horizontal projection of bottom surface of the gate bottom plate, m.

When the sluice gate base is placed on the rock base with cement grout curtain and no drainage holes or drainage curtain failure, the infiltration pressure head at

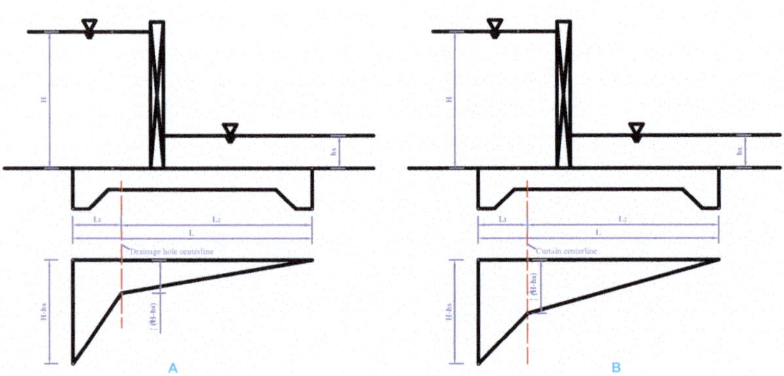

Fig. 7.23 Infiltrate pressure acting on the bottom surface of gate bottom plate

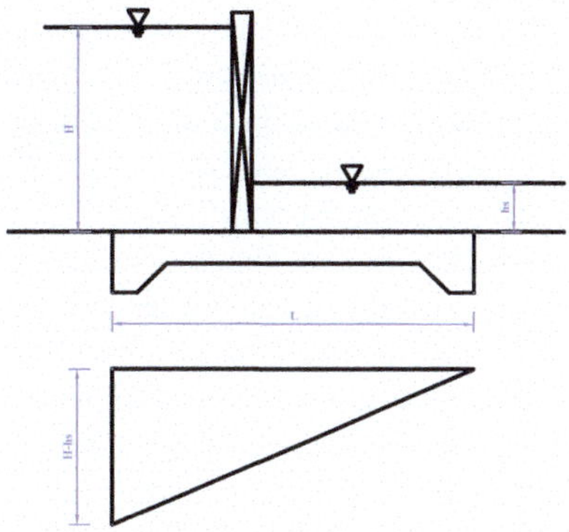

Fig. 7.24 Not equipped with a cement grout curtain and drainage holes

the upstream end of the bottom of the gate plate is $H - h_s$, the curtain centerline is $a(H - h_s)$, the downstream end is 0, and the sections are sequentially connected in a straight line (see Fig. 7.23). The infiltration pressure acting on the bottom surface of the gate bottom plate can be calculated from Eq. (7.5.2).

$$U = \frac{1}{2}\gamma(H - h_s)(L_1 + aL) \tag{7.5.2}$$

If the gate base on the rock foundation is not equipped with a cement grout curtain and drainage holes, the head of the infiltration pressure acting on the upstream end of the gate bottom plate is $H - h_s$, and the downstream end is zero, connected by a straight line (see Fig. 7.24). The infiltration pressure acting on the bottom surface of the bottom plate of the gate can be calculated from Eq. (7.5.3).

$$U = \frac{1}{2}\gamma(H - h_s)L \tag{7.5.3}$$

The analysis shall focus on the stability of contact seepage at the junction zones between the dam and discharging and conveyance structures that are built in and ruining through the dam and on whether the structures themselves break to leakage (including water seal damage) causing contact scouring.

7.6 Conclusions on Seepage Safety Evaluation

The safety state of engineering seepage is analyzed and evaluated. This part is the main part of seepage safety evaluation report. The analysis idea is to 'evaluate the current situation first, and then evaluate the future', that is, according to various measured data, the seepage pressure distribution and safety status obtained under the engineering conditions and common operating conditions (reservoir water level is generally not high) are compared (including the anti-seepage strength of the materials in the dam body and the dam foundation, the uplift pressure of the dam foundation and the reference seepage flow, etc.) to evaluate the seepage safety of the project under the current conditions.

On this basis, the extended forecast should be made according to the present data to calculate the seepage pressure distribution and seepage safety state of the high water level in the future (generally refers to the characteristic water level that the reservoir has not reached, such as the design flood level, check the flood level, etc.), so as to evaluate the seepage safety degree of the project under the condition of high water level in the future. After review of dam seepage safety, engineers shall come to clear conclusions on:

- Whether the dam anti-seepage and filter drainage facilities are sound and complete;
- Whether the change pattern of dam seepage pressure and seepage flow is normal and whether the dam seepage line(plane) or dam foundation uplift pressure is lower than the designed value;
- Whether the seepage stability of various types of rocks and soils and impervious bodies meet the requirements;
- Whether there is any abnormal seepage during operation.

1. The dam seepage state can be regarded as being safe and can be rated as **Class A** provided that the dam anti-seepage and filter drainage facilities are sound and complete and the design and construction quality meet the requirements of specifications; that through analysis by monitoring data and calculation, the dam seepage pressure and seepage flow changes in normal pattern, dam seepage line (plane) or dam foundation uplift pressure is lower than designed value and the seepage gradient of various kinds of rocks and soils and the impervious bodies is smaller than the allowed seepage gradient and that there is no abnormal seepage during operation.
2. The dam seepage state can be regarded as being basically safe and can be rated as **Class B** provided that the dam anti-seepage and filter drainage facilities are fairly sound and complete, the design and construction quality meet the requirements of specifications; that through analysis by monitoring data and calculation, the dam seepage pressure and seepage flow changes basically in normal pattern and dam seepage line (plane) or dam foundation uplift pressure is not higher than designed value; that local abnormal seepage comes out during operation but it does not seriously influence the safety of the dam. (Partial leaching and sponging at the downstream dam slope or on the ground in the rear of the dam can be regarded as local abnormal seepage.)
3. The dam seepage state can be regarded as being unsafe and can be rated as **Class C** under the following conditions. The dam anti-seepage and filter drainage facilities are NOT sound and complete, or serious quality defects exist; through analysis by monitoring data and calculation, the dam seepage pressure and seepage flow

changes but the previous change pattern is not followed, they obviously increase under the same condition and the seepage gradient at the critical location is greater than the allowed seepage gradient, or the seepage egress point is higher than the top elevation of filter drainage facility, or the dam foundation uplift pressure is higher than the designed value. Serious abnormal seepage appeared during operation. Serious abnormal seepage includes:

- Concentrated seepage that suddenly comes out at dam foundation, downstream slope and near the outlet of dam-crossing structures;
- Collapse pits that suddenly occur on the dam face near the dam-crossing structures;
- Seepage flow keeps increasing under the same conditions; seepage water becomes turbid or suspect substances appear; the level of seepage water is elevated or shifted;
- The upstream dam slope of the earth-rock fill dam collapses; leaching occurs at the downstream dam slope and the wet and soft range increasingly expands. Water and sand comes out of the dam toe and sponging and uplift or collapse occurs. Whirlpools and water leakage occur inside the reservoir and serious collapse pits or cracks appear on the blanket.
- Water is leaking at the junction parts between the dam body and bank slopes at the two ends of the dam and water conveyance tunnels. The adjacent dam surface collapses and the leaked water becomes turbid.
- Seepage pressure and flow increases simultaneously or suddenly changes its previous relationship with the reservoir water level and remarkably increases under the same condition.

References

1. Wang X, Yu H, Lv P et al (2019) Seepage safety assessment of concrete gravity dam based on matter-element extension model and FDA. Energies 12:502. https://doi.org/10.3390/en12030502
2. He L, Wang S, Gu Y et al (2021) Seepage behavior assessment of earth-rock dams based on Bayesian network. Int J Distrib Sens Netw 17:1881–1888. https://doi.org/10.1177/15501477211058672
3. Wei H, Shen Z-Z (2008) Reliability analysis on seepage stability of earth dams and its application. Chin J Geotech Eng 30:1404–1409
4. Bauer A, Strobl T (2017) Monitoring of existing dams—Criteria for the necessity of seepage measurement. Wasserwirtschaft 107:24–28
5. Wei B-W, Gu M-H, Li H-K et al (2018) Modeling method for predicting seepage of RCC dams considering time-varying and lag effect. Struct Control Health Monit 25:e2081. https://doi.org/10.1002/stc.2081
6. Cita Sari U, Prabandiyani Retno Wardani S, Suharyanto PW (2017) Influence of pore water pressure to seepage and stability of embankment dam (case study of Sermo Dam Yogyakarta, Indonesia). MATEC Web Conf 101:05007. https://doi.org/10.1051/matecconf/201710105007
7. ICOLD (2015) Internal erosion of existing dams, levees, and dikes, and their foundations

8. Li Bin,Yang Jie, Hu Dexiu (2019) Dam monitoring data analysis methods: a literature review. Struct Control Health Monit 27:e2501. https://doi.org/10.1002/stc.2501
9. Li Shengxi (2022) The factors affecting dam safety and seepage control measures. Highl Sci Eng Technol 10:135–140. https://doi.org/10.54097/hset.v10i.1241
10. Zhang C, Chai J, Cao J et al (2020) Numerical simulation of seepage and stability of tailings dams: a case study in Lixi, China. Water 12:742. https://doi.org/10.3390/w12030742
11. Jin J-F, Fang X-S (2017) Observed data of seepage of Xianlin Reservoir dam in Hangzhou during trial operation. Chin J Geotech Eng 39:224–227. https://doi.org/10.11779/CJGE2017S1044
12. Dong K, Yang DW, Yan JH, et al (2022) Anomaly identification of monitoring data and safety evaluation method of tailings dam. Frontiers Earth Sci 10: https://doi.org/10.3389/feart.2022.1016458
13. Chaiyo K, Rattanadecho P, Chantasiriwan S (2011) The method of fundamental solutions for solving free boundary saturated seepage problem. Int Commun Heat Mass Transfer 38:249–254. https://doi.org/10.1016/j.icheatmasstransfer.2010.11.022
14. Feng X-M, Chen C-H (2003) Research on h-version adaptive finite element method for 3-D seepage problem 24:653–656
15. Zhaokui H, Baoyu S, Jinchang S, Yulong L (2023) Seepage analysis based on weak finite element method. Chin J Geotech Eng 45:1526–1532. https://doi.org/10.11779/CJGE20220635
16. Li FZ, Tu Q (2012) The scaled boundary finite element analysis of seepage problems in multimaterial regions. Int J Comput Methods 09:1240008. https://doi.org/10.1142/s0219876212400087
17. United States, & United States (2015) Evaluation and monitoring of seepage and internal erosion
18. Jinbao S, Hongen L, Taozhen S (2023) Statistical analysis of dam failure and its loss of life in China. Hydro-Sci Eng 1–15. https://doi.org/10.12170/20220919001
19. Zhang X, Wang C-Y, Wong H et al (2022) Modeling dam deformation in the early stage of internal seepage erosion—Application to the Teton Dam, Idaho, before the 1976 incident. J Hydrol 605:127378. https://doi.org/10.1016/j.jhydrol.2021.127378
20. Caldeira L (2018) Internal Erosion in dams: studies and rehabilitation. Int J Civil Eng 17:457–471. https://doi.org/10.1007/s40999-018-0329-5
21. Flores Berrones JR, Li Liu X (2008) Analysis of the filter installed in the El Batán Dam, Querétaro. Tecnología y ciencias del Agua 23:165–178
22. Jiang N-J, Soga K, Kuo M (2017) Microbially Induced Carbonate Precipitation for Seepage-Induced Internal Erosion Control in Sand–Clay Mixtures. J Geotechnical Geoenvironmental Engi 143: https://doi.org/10.1061/(asce)gt.1943-5606.0001559
23. Rönnqvist H, Fannin J, Viklander P (2014) On the use of empirical methods for assessment of filters in embankment dams. Géotech Lett 4:272–282. https://doi.org/10.1680/geolett.14.00055
24. Razavi SK, Hajialilue Bonab M, Dabaghian A (2020) Investigation into the internal erosion and local settlement of Esfarayen earth-fill dam. J Geotech Geoenvironmental Eng 146:04020006. https://doi.org/10.1061/(asce)gt.1943-5606.0002216
25. Xiang Y, Sheng J, Liu C, Fang Z (2018) Research progress in long-term service and risk assessment of earth-rockfill dams. Adv Sci Technol Water Resour 38:86–94. https://doi.org/10.3880/j.issn.1006-7647.2018.05.016
26. Peng YF (2011) Calculation and control of seepage of hardfill dam with different foundation. Adv Mater Res 255–260:3666–3670. https://doi.org/10.4028/www.scientific.net/amr.255-260.3666
27. Wang S, Duan X (2002) Seepage safety evaluation for Fushui Dam. Hydro-Sci Eng 21–26
28. Diao Z, Wang C, He K, Guo L-P (2022) Principle of relationship between seepage movement and action. Sci Rep 12. https://doi.org/10.1038/s41598-022-21335-9
29. Van Beek VM (2015) Backward erosion piping: initiation and progression. Dissertation, Technische Universiteit Delft

30. Ganjalipour K, Fatemi Aghda SM (2018) Calibration of seepage analysis sections based on observation well monitoring data: a case study on Darian dam abutments–Kermanshah province, Iran. Model Earth Syst Environ 5:159–173. https://doi.org/10.1007/s40808-018-0525-y
31. Wu CG (2016) Hydraulics. Higher Education Press
32. Zhongkui Li, Hui Liu (2005) Application of numerical analysis principles and key technology for high fidelity simulation to 3-D physical model tests for underground caverns. Tunn Undergr Space Technol 0886-7798:390–399. https://doi.org/10.1016/j.tust.2005.01.004
33. Wei Y, Li J, Niu Y (2018) Application of AutoBank software in earth-rock dam seepage flow computation. IOP Conf Ser Mater Sci Eng 392:062069. https://doi.org/10.1088/1757-899x/392/6/062069
34. Su H, Li H, Zhang L, Wen Z (2019) Particle flow code method-based seepage behavior analysis and control effect evaluation for soil levee. Eng Comput 36:97–114. https://doi.org/10.1007/s00366-018-0687-2
35. Aldulaimi AA (2021) Safety evaluation of the chamrga earth dam. a seepage deformation, and stability analysis with GeoStudio. IOP Conf Ser Earth Environ Sci 664:012075. https://doi.org/10.1088/1755-1315/664/1/012075
36. Belew AZ, Belay SK, Wosenie MD, Alemie NA (2022) A comparative evaluation of seepage and stability of embankment dams using geostudio and plaxis models: the case of gomit dam in Amhara Region, Ethiopia. Water Conser Sci Eng 7:429–441. https://doi.org/10.1007/s41101-022-00152-1
37. Nourani V, Aminfar MH, Alami MT et al (2014) Unsteady 2-D seepage simulation using physical analog, case of Sattarkhan embankment dam. J Hydrol 519:177–189. https://doi.org/10.1016/j.jhydrol.2014.07.011
38. Salmasi F, Nourani B, Aghdam AT (2019) Investigation of relief wells performance in decreasing uplift pressure in foundation of embankment dams. J Dam Hydroelectric PowerPlant 6:11–25
39. Zhang S, Johansson F, Stille H (2022) Design methodology for grout curtains under dams founded on rock. Geotech Geol Eng 40:2167–2186. https://doi.org/10.1007/s10706-021-020 19-z
40. Dou J, Zhang G, Zhou M-X et al (2020) Curtain grouting experiment in a dam foundation: case study with the main focus on the Lugeon and grout take tests. Bull Eng Geol Env 79:4527–4547. https://doi.org/10.1007/s10064-020-01865-0
41. Bonotto DM, Garcia-Tenorio R (2019) Investigating the migration of pollutants at Barreiro area, Minas Gerais State, Brazil, by the 210Pb chronological method. J Geochem Explor 196:219–234. https://doi.org/10.1016/j.gexplo.2018.10.011
42. Yousefi S, Ghiassi R, Noorzad A et al (2013) Application of temperature simulation for seepage inspection in earth-fill dams. Građevinar 65:825–832. https://doi.org/10.14256/jce.878.2013

Chapter 8
Evaluation of Structure Safety

8.1 Introduction

As a man-made construction, the stability of the architectural structure of a reservoir dam is of utmost importance. The reliability and stability of hydraulic structures directly impact the secure and stable operation of the entire project. The safety assessment and maintenance of dam structures are particularly crucial. Structural safety evaluation mainly includes the review of structural strength, deformation and stability of the building. The purpose of structural safety evaluation is to review whether the deformation, strength, and stability of dams (including near-dam bank slopes) meet requirements under static conditions. The analysis of deformation and stability should be emphasized for earth-rock dams, while the analysis of strength and stability should be emphasized for concrete dams, masonry dams, water release structures, and water conveyance structures [1].

The dam structure safety shall be assessed based on the on-site inspection and conducted by integrating information and data collected from geological surveys, dam monitoring, safety inspections, along with the analysis of monitoring data and structural calculations, which assessed by way of special research and demonstration. For dams provided with information and data regarding deformation, stress, strain and temperature monitoring, the analysis of monitoring data shall be conducted first. The emphasis shall be laid on the analysis of cracks, holes, cavities, corrosion, collapse, sliding and other abnormal conditions that are revealed during operation and may produce impact to structure safety. The purpose of taking on-site inspection and analysis of monitoring data as the basis of evaluation on structure safety is to stress that great importance shall be attached to the current dam safety performance in structure safety evaluation and the evaluation work shall be carried out in accordance with the current dam state with targets and highlights so that the evaluation work would not become purely design review.

Y. Xiang et al., *On-site Inspection and Dam Safety Evaluation*, Hydroscience and Engineering, https://doi.org/10.1007/978-981-97-3987-5_8

The following is a brief description of the abnormal deformation of the structure.

The insufficient strength or abnormal deformation at the local level, which does not yet pose a serious threat to the project's safety, may encompass the following conditions:

- Longitudinal cracks or non-penetrating crosswise cracks on the surface of the earth-rock dam;
- Partial depressions observed on the face of the earth-rock dam;
- Local instances of damage or collapses on the upstream slope of the earth-rock dam's protection;
- Local superficial cracks observed on the concrete dam.

However, abnormal deformations that could potentially endanger the project's safety are more critical and encompass the following scenarios:

- Penetrating crosswise cracks and horizontal cracks appearing in the earth-rock dam;
- Landslides, cracks, or signs of instability occurring within the earth-rock dam, along with dam slope instability;
- Noticeable displacement of a gravity dam along its foundation surface or evident tilting of the dam body;
- Apparent deformation or displacement observed in the arch seats located at the two banks of an arch dam;
- Vertical penetrating cracks forming on the arch dam;
- Observable shear displacement occurring at the dam foundation or at fault lines along the dam's shoulders.

8.2 Evaluation on Structure Safety for Earth-Rock Fill Dam

A comprehensive review should be conducted to assess various critical aspects of the dam's structure during the evaluation of safety for earth-rock dams. This review aims to ensure the dam's overall stability, performance, and adherence to safety standards. In evaluation on structure safety for earth-rock dams, review shall be carried out on whether the change pattern of the dam body is normal, whether amplitude variation and settlement are within the range of safety/and whether stability of bank slope, elevation and width of dam crest and upstream slope protection meet the requirements.

The key areas to focus on include:

- Dam Body Change Pattern: Evaluate the pattern of changes in the dam body over time. Any deviations from the expected change pattern should be carefully analyzed to identify potential issues.
- Instrumentation Data: Utilize data from monitoring instruments (if available) to track changes in dam behavior, such as settlement, deformation, and water

pressure. Monitor and assess the amplitude variation and settlement of the dam. Ensure that these factors remain within the acceptable range of safety, as excessive settlement or variation could indicate structural instability.

- Stability of Bank Slope: Review the stability of the bank slope, both on the upstream and downstream sides of the dam. Ensure that the slopes are well-maintained and not prone to erosion, landslides, or other forms of instability.
- Elevation and Width of Dam Crest: Examine the elevation and width of the dam crest to verify if they adhere to the required specifications. Deviations from these specifications could impact the overall safety of the dam.
- Upstream Slope Protection: Evaluate the effectiveness of the upstream slope protection measures, such as erosion control and vegetation cover. Adequate protection is crucial to prevent erosion and maintain the integrity of the slope.
- Stability Analysis: Conduct thorough stability analyses to assess the dam's resistance against potential failure mechanisms, such as sliding, overturning, or slope instability.

The dam operators and engineers can identify potential safety risks and structural vulnerabilities. This information guides decision-making regarding necessary maintenance, repairs, or interventions to ensure the continued stability and safety of the earth-rock dam.

8.2.1 Failure Causes of Earth-Rock Dam

A failure for the purpose of this discussion is an uncontrolled release of the reservoir due to a breach of the embankment dam [2, 3]. The most common causes of earth-rock dam failures are [4]:

- Piping by internal erosion of fine-grained soils from the embankment dam. Piping through the foundations or abutments of the embankment dam. Piping along conduits constructed through the embankment dam [3, 5].
- Overtopping of the dam and subsequent breach by erosion due to overtopping. Overtopping has occurred due to inadequate spillway capacity and due to improper operation of spillway gates. Overtopping and a breach has also occurred on an upper reservoir of a pumped storage project without a spillway.
- Loss of shear strength due to high pore pressures and in some cases liquefaction of loose saturated granular materials in the foundation or embankment during earthquakes [6].
- High pore pressures in downstream shell due to inadequate drainage resulting in instability and failure of the downstream shell.
- Embankment failure due to sliding on clay-shale foundations.

Moreover, man-induced causes such as sabotage and terrorist attacks are gaining attention as another possible source of dam failure.

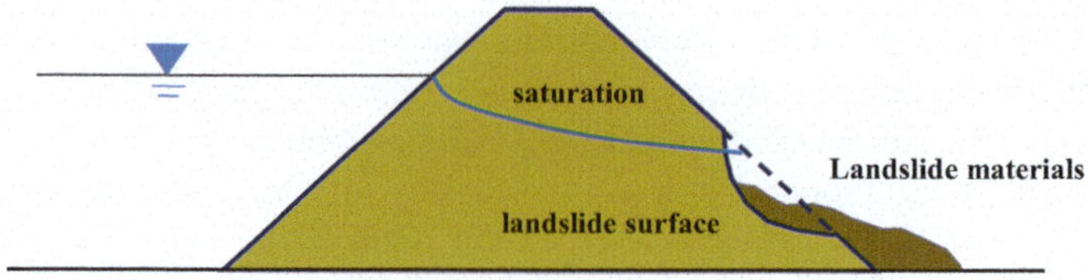

Fig. 8.1 Schematic diagram of shallow landslides

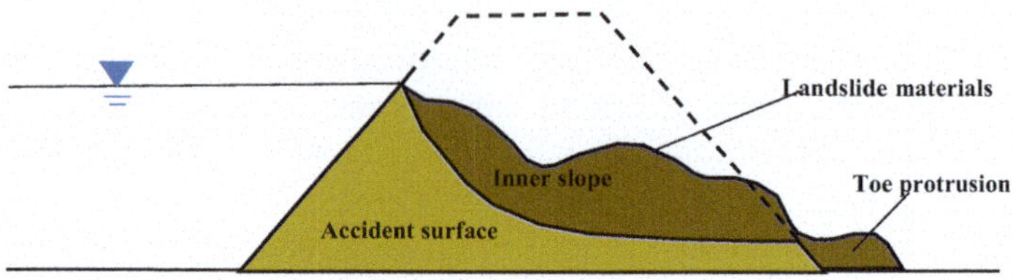

Fig. 8.2 Schematic diagram of deep landslide

The destabilization and failure state of earth-rock dam mainly includes landslide, liquefaction and plastic flow [7]. In order to ensure safe operation of earth-rock dam, three kinds of instability states are not allowed to occur. For earth-rock dams operating for a period of time, it is necessary to check whether the earth-rock dam in operation has sufficient stability and evaluate its safety condition by calculation.

- **Landslide**. Due to steep dam slope or insufficient shear strength of dam soil, the sliding force or moment of soil above a certain sliding surface exceeds the corresponding sliding force or moment of soil below it. Or, due to insufficient shear strength of the dam foundation soil, some dams slide along with the dam foundation, especially when there is a weak interlayer in the dam foundation, sliding always occurs along the interlayer. Sliding surface of dam slope may be similar to cylindrical surface, folded surface, plane or combination surface (Figs. 8.1 and 8.2).
- **Liquefaction**. It is easy to occur in the dam body or foundation composed of non-compact and uniform fine sand. Often, the volume shrinkage of saturated loose sand is caused by vibration or shearing, but the water in the pore can not be discharged immediately, and the effective stress of sand particles in contact with each other will change part or all into pore pressure, so the sand particles can liquefy with the flow of water. The specific and objective causes for liquefaction of saturated loose sand may be vibration caused by earthquake, explosion, etc., or vibration and shear applied during piling.
- **Plastic flow**. It refers to that the shear stress of the dam body or foundation soil exceeds the actual shear strength of the soil body and the deformation exceeds the elastic–plastic limit. Loss of bearing capacity, pressure or uplift of dam slope or foot soil, or serious cracks in dam body and foundation, excessive settlement, etc. The dam body or foundation is composed of soft clay, which can occur without proper treatment.

8.2.2 *Recheck of Dam Crest Elevation, Dam Crest Width and Upstream Slope Protection*

See Sect. 7.5 for recheck of dam crest elevation firstly.

The width of dam crest should be determined according to structure, construction, operation management and earthquake resistance. The width of dam crest for high dam can be selected as 10–15 m, for medium dam and low dam as 5–10 m. For seismic design intensities of level VII and level IX, it is recommended to widen the dam crest. The dam crest should not be used as a public transportation road and special demonstration should be made if necessary.

When the dam surface is composed of materials like soil, sand, gravel, soft rock, weathered material, etc., it is crucial to implement special slope protection measures to prevent erosion caused by factors such as water flow, rainfall, and wind and ensure the stability of the dam.

Coarse grains or oversized rocks from rockfill materials can be used for slope protection in the rockfill dam. These larger materials offer enhanced resistance to erosion and provide stability for the dam's slopes. By strategically placing and compacting these rocks on the surface, the dam's integrity can be preserved, especially in areas prone to erosive forces such as water flow, rainfall, or wind.

8.2.3 *Deformation Analysis of Earth Rock Dam Body*

Deformation analysis includes analysis of sedimentation (vertical displacement), analysis of horizontal displacement, analysis of cracks, and stress and strain analysis when necessary. The analysis methods or approaches shall include deformation monitoring data analysis and deformation calculation analysis and the two methods shall be mutually verified and supplemented [8, 9].

For dams with deformation monitoring data, monitoring data analysis shall be conducted first. Deformation calculation analysis shall be used when deformation monitoring data is not available or abnormal deformation and cracks have already taken place, or the landform and geological conditions along the dam axis change greatly and there is doubt of cracking. The deformation analysis shall include the following important points:

1. Deformation monitoring items of earth-rock dam mainly include surface and internal deformation of dam body (foundation), deformation of impervious body, deformation of interface, joint (crack) and void, deformation of bank slope near dam and deformation of surrounding rock of underground tunnel.
2. Deformation calculation analysis mainly includes crack analysis and stress and strain analysis. Inclination method based on sedimentation monitoring data can be used for crack analysis. And calculation formula for inclination method is as indicated below:

$$\gamma_{AB} = \frac{S_A - S_B}{\Delta l} \tag{8.2.1}$$

where

A, B Inclination in connection with the two measuring points A and B;
S_A, S_B Vertical displacement of measuring points A and B;
Δl Distance between two measuring points A and B.

If inclination exceeds critical inclination, dam cracks may possibly exist. Critical inclination can be obtained from deflection test on earth beam, or roughly taking 1% for critical inclination. The settlement after completion of the project shall be used for calculation of the probability of dam cracks when settlement rate is used for estimation.

When sedimentation monitoring data is not available, sedimentation calculation result can be used. Sedimentation can be calculated by using layer-wise summation method or using finite element method.

The settlement and final settlement of clay dam body and dam foundation at completion can be calculated by layered summation method according to formula (8.2.2). The settlement of dam crest after completion shall be the difference between the final settlement and the settlement at completion.

$$S_t = \sum_{i=1}^{n} \frac{e_{i0} - e_{it}}{1 + e_{i0}} h_i \tag{8.2.2}$$

where

S_t Total Settlement of Dam Body and Foundation at Completion or Final;
e_{i0} Vinitial porosity ratio of layer *I*;
e_{it} Porosity ratio of layer *i* corresponding to the effective vertical stress at completion or at the end;
h_i Thickness of soil layer *i*;
n Number of soil layers.

The final settlement of non cohesive earth dam body and dam foundation can be estimated by formula (8.2.3):

$$S_\infty = \sum_{i=1}^{n} \frac{p_i}{E_i} h_i \tag{8.2.3}$$

where

S_∞ Final settlement of dam body or dam foundation;
p_i Layer *i* calculates the vertical stress of the soil layer due to the load of the dam body;
E_i Calculation of Deformation Modulus of Soil Layer *I*.

The settlement of the crest of concrete face rockfill dam can be estimated according to the formula (8.2.4) by using the prototype observation results of the built dam with similar materials.

$$S_2 = \left(\frac{H_2}{H_1}\right)^2 \left(\frac{E_1}{E_2}\right) S_1 \tag{8.2.4}$$

where

S_2 Estimated settlement value of dam crest for dam to be built;
S_1 Settlement values observed by prototype dam crest of built dam;
H_2 Dam height of dam to be built;
H_1 Dam height of built dam;
E_2 Deformation modulus of the dam to be built; When calculating the settlement of dam crest in different periods of the dam to be built, E_2 should be the deformation modulus in corresponding period.
E_1 Deformation modulus of built dam.

For dams of Scale 1 and 2 and earth-rock fill dam with special requirements, stress and strain analysis shall be conducted and for this finite element method can be used.

3. In deformation analysis and evaluation, conclusions shall be drawn on: Whether the general deformation behavior of the dam and the dam sedimentation is stable; Whether cracks that threat dam safety are produced on the dam impervious body; Whether the dam deformation monitoring is in accordance with requirements.

8.2.4 Force Acting on Dam Body

Indeed, an earth-rock dam, as a water-retaining structure, is subjected to a variety of forces and factors that influence its stability and behavior. As a water retaining structure, earth rock dam will be subjected to self weight, water load, etc.; As a granular structure that can pass through seepage, it is also affected by seepage force, pore pressure, etc.; Earth rock dams in earthquake areas are also subject to earthquake action.

8.2.4.1 Self-weight and Water Load

The dam must withstand the self-weight of its materials and the additional load exerted by the water it retains. The combined weight of the dam and the water it holds can exert significant pressure on the structure.

Soil self-weight is usually the main load in stability analysis of earth-rock dam and may be the main factor on which shear strength on sliding surface is formed. It is usually calculated by natural wet gravity for soil above the dam wetting line and float gravity for soil below the wetting line.

As the upstream slope of earth-rock dam is gentle and the upstream slope seeps through porous granular accumulation surface (except reinforced concrete panel), the water load of this surface force is not unfavorable or even beneficial to the earth-rock dam in terms of stability analysis of dam slope.

8.2.4.2 Permeability

The hydrodynamic force acting on the soil body along the streamline within the seepage field is a critical consideration in soil mechanics and geotechnical engineering. This force arises from the flow of water through the soil and can significantly impact the stability of soil structures. Permeability per unit volume of soil is γJ (γ is the weight of water). Because of the change of seepage direction, it is not convenient to obtain the seepage force, which depends on the seepage elements at all points of the whole seepage field. When calculating the stability of the sliding surface method, an alternative method for calculating the seepage force is to take the saturated bulk density to calculate the effect on the sliding force for the soil above the downstream water level and below the saturation line, and take the floating bulk density to calculate the effect on the anti-sliding force [10, 11].

8.2.4.3 Pore Pressure

It mainly refers to the two-phase soil body of saturated water in pore which can not be drained away immediately and can not be compressed under the action of new loading. The new loading temporarily becomes the positive pressure borne by pore water and can not be transformed into the effective compressive stress of soil frame immediately, and of course does not produce shear strength. If the soil is a three-phase unsaturated soil with some air in the pore, this part of the positive pressure should be called pore pressure because the air can be compressed and a part of the newly loaded load is converted to the positive pressure borne by both water and air.

Pore pressure is a special force that must be considered in stability calculation of earth-rock dam under certain conditions. According to the formation mechanism of pore pressure, pore pressure is the stress condition and corresponding shear strength of the soil mass under sudden load. When pore pressure is u and effective stress of soil skeleton is $\overline{\sigma}$, the sum of the two is called total stress σ, [12] which can be expressed as:

$$\sigma = \overline{\sigma} + u \tag{8.2.5}$$

$$\overline{\tau} = \overline{\sigma} \tan \varphi' + c' = (\sigma - u) \tan \varphi' + c' \tag{8.2.6}$$

where

$\overline{\tau}$ Effective shear strength;
φ', c' Effective shear strength index, i.e. effective internal friction angle and cohesion.

It can be seen from Eqs. (8.2.5) and (8.2.6) that for earth dam construction, pore pressure will immediately arise when load is added to the dam. New load only increases sliding force, but can not (or only a small amount) increase anti-sliding force, thus greatly reducing the stability of the dam. Pore pressure may only occur in compressed soils.

Clearly, pore pressure is typically generated by clay with weak water permeability and high compressibility, combined with the possible conditions for a sudden increase in load in the soil. The sudden increase in load on soil is as follows:

- Continuous filling and heightening of dam body during construction period;
- The water level of reservoir drops and the wetting line of dam body changes;
- Additional pore pressure under earthquake.

Pore pressure is a process that changes with time from the beginning to the gradual dissipation. The falling of reservoir water level during construction period may bring adverse effects on the stability of dam slope upstream of earth-rock dam [13, 14].

Design conditions of earth-rock dams shall be divided into:

- **Normal operating conditions** include the following: stable seepage period of reservoir water level between normal water level and design flood level and dead water level; The reservoir water level regularly drops within the above range; Frequent change and fall of reservoir water level of pumped storage power station.
- **Extraordinary application condition I** includes the following contents: Construction period; Checking flood level may result in stable seepage; The water level of reservoir drops very much, such as from checked flood level drops, drops below dead water level, and rapid discharge of large flow, etc.
- **Extraordinary operating condition II** includes normal operating condition in case of earthquake. The adverse conditions of stability against sliding for upstream dam slope include rapid drawdown of water level, dead water level or water level of 1/3 dam height.

8.2.5 Shear Strength Determination

The shear strength is a fundamental property of a material that governs its ability to withstand internal forces caused by external loads or environmental factors. It is a measure of the material's resistance to sliding or deformation along a particular plane or surface. The shear strength is influenced by a combination of factors including the material's cohesion, internal friction (angle of internal friction), and pore pressure.

In geotechnical engineering, evaluating the shear strength along potential failure surfaces is critical for assessing the stability of soil and rock structures, such as embankment dams, retaining walls, and slopes. Different loading conditions, such as static loads, dynamic loads, changes in water levels, and seismic events, can influence the shear strength and contribute to potential failure [15].

- **Initial construction stage**: During the early stages of construction, the dam is being built. Shear stresses occur due to the weight of the materials being placed and compacted to form the dam structure.
- **Partial load stage**: As construction progresses, the dam starts to impound water and fill up, causing variations in water levels within the reservoir. These pool fluctuations exert additional shear stresses on the dam and its foundation.
- **End of construction stage**: As construction concludes, shear stresses emerge due to the weight of the dam and its structural elements. This phase marks the completion of the dam's initial construction.
- **Partial pool with steady seepage**: At a stable water level within the reservoir, shear stresses continue to be influenced by seepage forces, which are sustained as long as water continues to move through the dam and its foundation. During periods of reservoir filling, as water levels rise, the dam experiences shear stresses related to the rising water pressure and seepage forces. These forces contribute to the overall stability considerations.
- **Operational stage**: During the operational life of the dam, fluctuations in water levels due to seasonal variations, water demand, and other factors continue to impose varying shear stresses on the structure.
- **Sudden drawdown stage**: Once construction is complete, but before the reservoir is filled to capacity, the dam still experiences shear stresses from its weight and any remaining construction activities.
- **Earthquake loading stage**: In earthquake-prone regions, the dam and foundation are subjected to dynamic seismic forces that induce shear stresses and ground shaking. These forces can vary based on the magnitude and duration of the earthquake. Earthquake forces can impact the dam's structural integrity and stability, necessitating specialized design considerations to mitigate these effects.
- **Appropriate flood surcharge pool**: During flood events, the dam's design may include provisions for water surcharge to accommodate excess inflow. The additional water load results in shear stresses, and proper design is required to manage these conditions safely.

In all loading scenarios, it is essential to determine the shear strength along potential failure surfaces. The capacity of the material to resist failure along a specific failure surface depends on the applied loading conditions and the rate at which these conditions change. Throughout these stages and scenarios, engineers must meticulously analyze, design, and monitor the dam to ensure its structural integrity and safety. Adequate engineering measures, construction techniques, and ongoing evaluation are essential to manage the various shear stresses and loading conditions the dam and its foundation encounter, thereby guaranteeing the continued stability and performance of the structure over time.

From the time construction begins until the reservoir has been filled and a state of steady seepage has been established, and during reservoir operation subsequent to establishment of the steady-state seepage condition, three different static loading conditions may act on the embankment and foundation. These are the end-of construction, steady seepage, and sudden drawdown conditions. Shear strength values used in stability analyses for these loading conditions depend on consolidation conditions and on the shear-induced excess pore pressures generated by the loadings. Laboratory tests on specimens of embankment material which are compacted in the laboratory to the dry densities and water contents anticipated in a proposed dam, and on undisturbed samples of natural foundation soils or embankment materials from an existing dam, are conducted to simulate the conditions of consolidation and shear-induced pore pressure dissipation expected for the various loading conditions, in order to determine the appropriate shear strength values.

In general, static stability analysis requires several different shear strength values. The following describes the shear strength determined by three different types of laboratory triaxial tests [16, 17]:

1. Unconsolidated-Undrained Strength, determined by a test (referred to as UU-type) in which no consolidation is permitted under the initial confining pressure and in which no drainage is permitted during the shearing stage so that the water content is kept constant. In other words shear-induced pore pressures are not allowed to dissipate in this test. A special type of this test in which the initial confining pressure is zero is the Unconfined Compression Test (referred to as UC-type).
2. Consolidated-Undrained Strength, determined by a test (referred to as CU-type) in which the sample is consolidated under the initial confining pressure, and in which the drainage is not permitted during the shearing stage so that the water content is kept constant. In other words, the shearing stage of this test is the same as in the UU test, in that shear induced pore pressures are not allowed to dissipate. A variation often used in this type of test is to measure the shear-induced pore pressures during the test, in which case the test is referred to as a $\overline{CU}$-type.
3. Consolidated-Drained Strength, determined by a test (referred to as CD-type) in which consolidation is permitted under the initial confining pressure, and complete dissipation of shear-induced pore pressure is allowed for each increment of shear stress applied in the shearing stage of the test. The consolidated-drained strength can also be determined from the $\overline{CU}$-type test as explained in the following paragraphs.

The shear strength values described in (1) and (2) are commonly referred to as undrained shear strengths, abbreviated as S_U. Whether the undrained shear strength determined in (1) or (2) is appropriate for use in stability analyses depends on the initial consolidation conditions as-discussed in the following paragraphs. The shear strength described in (3) is commonly referred to as the drained shear strength. The undrained shear strength is used in stability analyses for loading conditions during which shear-induced pore pressures cannot dissipate, due to the rapidity of the application of the shear stresses by the loading with respect to the permeability

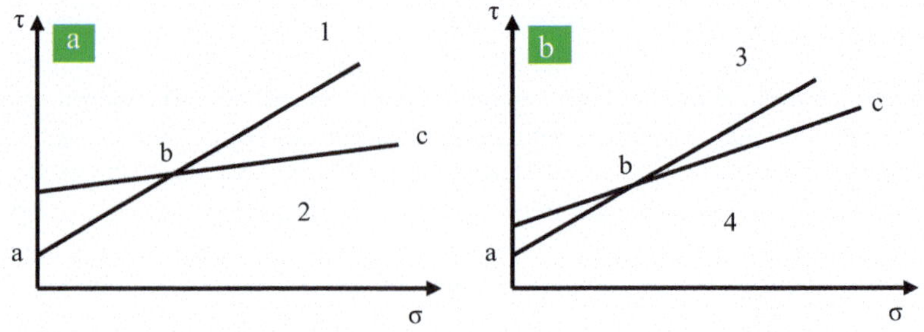

Fig. 8.3 Combination of strength envelopes 1 and 3. Effective strength envelope CD; 2. Undrained shear total strength envelope UU; 4. Consolidated undrained shear total strength envelope CU

and drainage boundary conditions of the materials involved. Conversely, the drained shear strength is used in stability analyses for loading conditions during which shear-induced pore pressures are zero, due to the slow application of the shear stresses by the loading with respect to the permeability and drainage boundary conditions of the materials involved. The determination and application method of shear strength index can be selected according to Table 8.1.

The shear strength indexes should be arranged and adopted according to the following principles:

- When the shear strength test is not less than 11 groups, the shear strength index should be determined by the small value average value.
- The adopted shear strength index can be determined according to test results and similar projects.
- When the total stress method is used to determine the strength envelope of fill, the straight line 2 in Fig. 8.3(a) should be used during the construction period, and the lower envelope a and b and c of CD and CU in Fig. 8.3(b) should be used during the reservoir water level falling period.

8.2.6 Check Calculation of Dam Slope Stability

A comprehensive evaluation involves checking the stability of the dam against sliding along its slope or along potential failure surfaces within the dam and its foundation [18]. This assessment is crucial to prevent catastrophic failure due to inadequate resistance to sliding forces. Dam slope stability review generally employs two main analysis methods:

- Sliding Surface Method: This method evaluates the potential for sliding along well-defined failure planes. It considers factors like shear strength, pore pressure, and external forces acting on the dam.

Table 8.1 Determination and application of shear strength index

Control stability	Strength calculation method	Soil group		Instrumentation	Test method	Index of strength	Initial state of sample
Stage of construction	Effective stress	Cohesionless soil		Direct shear apparatus	Slow shear (S)	c', φ'	Filling soil with filling moisture content and filling bulk density, dam foundation with undisturbed soil
				Triaxial apparatus	Consolidated Drainage Shear (CD)		
		Cohesive soil	Saturation less than 80%	Direct shear apparatus	Slow shear (S)		
				Triaxial apparatus	Undrained shear pore pressure (UU)		
			Saturation greater than 80%	Direct shear apparatus	Slow shear (S)		
				Triaxial apparatus	Consolidated undrained shear pore pressure (CU)		
	Method of comprehensive stress	Cohesive soil	Permeability coefficient less than 10^{-7} cm/s	Direct shear apparatus	Quick cut (Q)	c_u, φ_u	
			Any permeability coefficient	Triaxial apparatus	Undrained shear (UU)		

(continued)

Table 8.1 (continued)

Control stability	Strength calculation method	Soil group		Instrumentation	Test method	Index of strength	Initial state of sample
Stable seepage period and reservoir water level falling period	Effective stress	Cohesionless soil		Direct shear apparatus	Slow shear (S)	c', φ'	Ibid., but saturated in advance, and soil above the saturation line does not need to be saturated
				Triaxial apparatus	Consolidated Drainage Shear (CD)		
		Cohesive soil		Direct shear apparatus	Slow shear (S)		
				Triaxial apparatus	Consolidated undrained shear pore pressure (CU)		
Reservoir water level falling period	Method of comprehensive stress	Cohesive soil	Permeability coefficient less than 10^{-7} cm/s	Direct shear apparatus	Fast consolidation shear (R)	c_{cu}, φ_{cu}	
			Any permeability coefficient	Triaxial apparatus	Consolidated undrained shear (CU)		

Note The shear strength of the total stress method during the construction period in the table is the unsaturated soil of the dam filling soil. For the saturated soil of the dam foundation, the shear strength index should be changed to c_{cu}, φ_{cu}

- Stress–Strain Analysis Method: This approach involves analyzing the stress and strain distribution within the dam and foundation to assess stability. It considers the deformations and strength of the materials.

In regions prone to earthquakes, specific attention is given to sandy foundations or areas with sandy soil within the dam body. Dynamic analysis is conducted to assess the dam's resistance to liquefaction, which can occur when saturated soils temporarily lose their strength due to earthquake-induced shaking. Ensuring that the dam and its foundation do not experience plastic flow is vital. This is achieved through appropriate design of dam materials and effective foundation treatment techniques to control deformations and avoid excessive yielding.

8.2.6.1 The Sliding Surface Method

Sliding surface method is based on dam type, soil and foundation geological conditions, combined with experience to select unfavorable conditions, and assume several possible sliding surfaces. Calculate the sliding resistance and sliding force of each working condition, each sliding surface. The dam body is stable when the anti-sliding force is greater than the sliding force. Compare the stability safety factor to find the minimum value. If the minimum safety factor is lower than the allowable minimum safety factor specified in the code, it can be considered that the safety degree of anti-sliding stability of dam slope does not meet the requirements. The rigid body limit equilibrium method should be used in the calculation of stability against sliding of dam slope [19]. Methods such as the Simplified Bishop method [20] and the Morgenstern-Price method [21] considering the forces between bars should be used in the calculation. When the dam crest is also used as highway, the check calculation for stability of dam slope shall take into account the loads of vehicles.

(1) **Simplified Bishop method**

In 1955, Bishop proposed a simplified method that only considers the horizontal force between soil strips and ignores the vertical force (Fig. 8.4). The safety factor is as follows [22]:

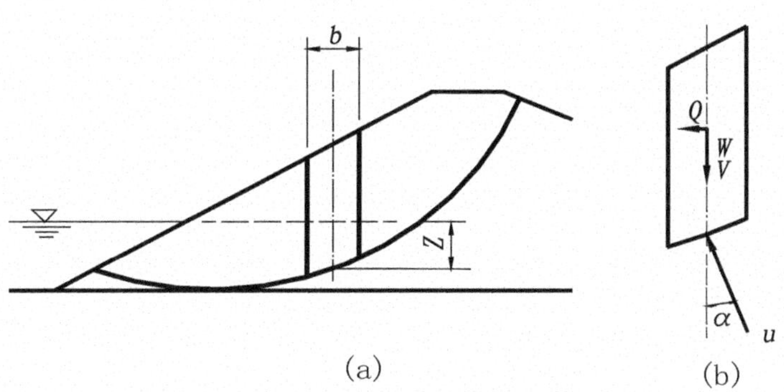

Fig. 8.4 Schematic diagram of circular arc sliding slice method

$$K = \frac{\sum\{[(W \pm V)\sec\alpha - ub\sec\alpha]\tan\varphi' + c'b\sec\alpha\}\frac{1}{1+\frac{\tan\alpha\tan\varphi'}{K}}}{\sum\left[(W \pm V)\sin\alpha + \frac{M_C}{\pi}\right]} \tag{8.2.7}$$

where

- W Weight of soil strip;
- Q, V Horizontal and vertical seismic inertia force (negative upward and positive downward);
- u Pore pressure acting on the bottom of soil strip;
- α The included angle between the block gravity line and the radius passing through the midpoint of the block bottom;
- b Width of soil strip;
- c', φ' Effective stress shear strength index of soil strip bottom;
- M_C Moment of horizontal seismic inertia force to the center of the circle;
- R Arc radius.

Due to its high accuracy, the simplified Bishop method is commonly used for calculating dam slope stability, especially in cases of homogeneous dams, thick core inclined dams, and thick inclined wall dams. This method considers the interaction forces between soil strips and is employed to assess the potential for slope failure.

In the simplified Bishop method, the lateral forces between soil strips are primarily horizontal and adhere to principles of moment and vertical force balance. However, they do not perfectly satisfy horizontal force balance, which can sometimes lead to numerical analysis challenges. Despite this limitation, the safety factor obtained from this method tends to be slightly higher compared to the Swedish method, another common approach for slope stability analysis. If the safety factor calculated using the simplified Bishop method is lower than that from the Swedish arc method, it may indicate numerical analysis issues in the former. In such cases, the Swedish arc method could yield more reliable results. Conversely, when considering circular arc slip crack surfaces, accurate results are usually achieved under all conditions if there are no numerical analysis problems.

(2) **Morgenstern-Price method**

The Morgenstern-Price (MP) method is a widely employed technique for determining the factor of safety of a slope against sliding (Fig. 8.5). This method has been utilized effectively in slope stability analysis for many years. In the MP method, the slope is divided into vertical slices above the potential slip surface, and intricate force and moment integrals are computed within these slices. The soil mass above the potential slip surface is divided into a series of vertical slices [23] This slicing enables the calculation of forces and moments within each slice, contributing to the overall analysis. For each slice, the method involves the computation of force and moment integrals. These integrals account for the gravitational forces, shear forces, and moments acting on the soil within the slice. By analyzing the equilibrium of forces and moments within each slice, the overall factor of safety against sliding can

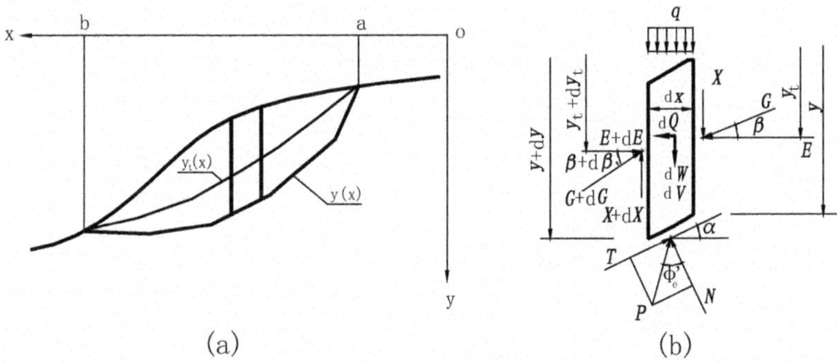

Fig. 8.5 Calculation diagram of Morgenstern-Price method

be determined. The factor of safety is a critical measure that indicates the stability of the slope under the applied loads and conditions.

$$\int_a^b p(x)s(x)dx = 0 \tag{8.2.8}$$

$$\int_a^b p(x)s(x)t(x)dx - Me = 0 \tag{8.2.9}$$

$$\begin{aligned} p(x) = {} & \left(\frac{dW}{dx} \pm \frac{dV}{dx} + q\right)\sin(\phi'_{\mathrm{e}} - \alpha) - u\sec\alpha\sin\phi'_{\mathrm{e}} \\ & + c'_{\mathrm{e}}\sec\alpha\cos\phi'_{\mathrm{e}} - \frac{dQ}{dx}\cos(\phi'_{\mathrm{e}} - \alpha) \end{aligned} \tag{8.2.10}$$

$$s(x) = \sec(\phi'_e - \alpha + \beta)\exp\left[-\int_a^x \tan(\phi'_e - \alpha + \beta)\frac{d\beta}{d\zeta}d\zeta\right] \tag{8.2.11}$$

$$t(x) = \int_a^x (\sin\beta - \cos\beta\tan\alpha)\exp\left[\int_a^\xi \tan(\phi'_e - \alpha + \beta)\frac{d\beta}{d\zeta}d\zeta\right]d\xi \tag{8.2.12}$$

$$M_e = \int_a^b \frac{dQ}{dx}h_e dx \tag{8.2.13}$$

$$c'_e = \frac{c'}{K} \tag{8.2.14}$$

$$\tan \varphi_e' = \frac{\tan \varphi'}{K} \tag{8.2.15}$$

where

dx	Width of soil strip;
dW	Weight of soil strip;
q	Vertical load outside the slope top;
M_e	Moment of horizontal seismic inertia force on the midpoint of the bottom of the soil strip;
dQ, dV	Horizontal and vertical seismic inertia force (negative upward and positive downward);
α	The angle between the bottom of the strip and the horizontal plane;
β	The angle between the resultant force on the side of the soil strip and the horizontal direction;
h_e	the vertical distance from the horizontal seismic inertia force to the midpoint of the bottom surface of the soil strip.

The MP method is applicable to arbitrarily shaped sliding surfaces, and can be used to calculate the stability analysis of dam slopes with weak interlayer, thin sloping walls and thin core wall dams. Meet the force and moment equilibrium and the inter-strip force is arbitrary direction of the strip partition method, is a strict limit equilibrium analysis method, but sometimes encounter numerical analysis between the problem. Once there is no numerical analysis problem, the calculated results are accurate in any case, and the error of the calculated results from the correct answer does not exceed 6% [23].

(3) **The Slip Wedge method**

The slip wedge method is particularly suitable for analyzing slope stability in cases where the potential failure surface takes on a folded or complex shape, resembling a slip crack surface [24]. The inclination angle of the inter-bar lateral force varies according to the specific method, which satisfies the equilibrium of the total force of each soil strip and the landslide body, but not the moment equilibrium, and also has numerical analysis problems. This method's accuracy is significantly impacted by the assumed directions of the inter-slice forces (Fig. 8.6). Inaccurate assumptions regarding these forces can lead to substantial discrepancies between the calculated safety factor and the true value.

When dealing with smooth curve slip crack surfaces, this method can produce safety factors that are in line with more rigorous techniques. However, when the slip crack surface becomes more complex, such as with bending lines or multiple turning points, the calculated results may significantly differ from those obtained using stricter calculation methods. The potential error in the calculated safety factor increases as the number of reversal points in a non-circular slip surface increases.

$$P_i = \sec(\phi_{ei}' - \alpha_i + \beta_i)[P_{i-1}\cos(\phi_{ei}' - \alpha_i + \beta_{i-1}) - (W_i \pm V_i)\sin(\phi_{ei}' - \alpha_i)$$

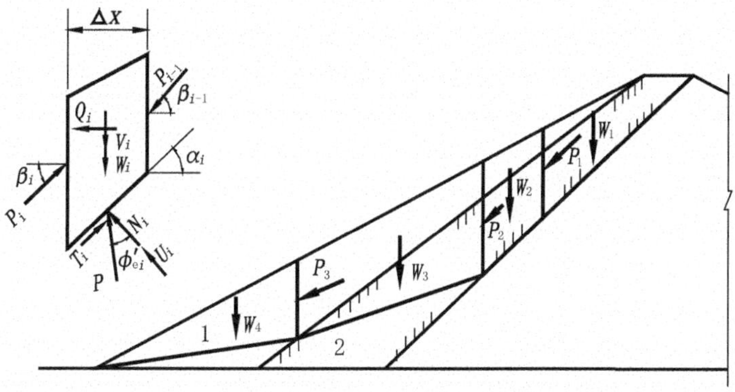

Fig. 8.6 Calculation diagram of Wedge method (1. sand crust; 2. clay)

$$+ u_i \sec\alpha_i \sin\phi'_{ei}\Delta X - c'_{ei}\sec\alpha_i\cos\phi'_{ei}\Delta X + Q_i\cos(\phi'_{ei} - \alpha_i)] \quad (8.2.16)$$

$$c'_{ei} = \frac{c'_i}{K} \quad (8.2.17)$$

$$\tan\varphi'_{ei} = \frac{\tan\varphi'_i}{K} \quad (8.2.18)$$

where

- P_i Anti sliding force at one side of soil strip;
- P_{i-1} The sliding force on the other side of the soil strip;
- W_i Weight of soil strip;
- u_i Pore pressure acting on the bottom of soil strip;
- Q_i, V_i Horizontal and vertical seismic inertia force (negative upward and positive downward);
- α_i The angle between the bottom of the soil strip and the horizontal plane;
- β_i The included angle between Pi on one side of the soil strip and the horizontal plane;
- β_{i-1} The angle between Pi − 1 on the other side of the soil strip and the horizontal plane.

(4) **The General Principles and Regulations**

The shear strength index and pore water pressure required for stability analysis shall be determined by following principles as described below:

- When shear strength parameters representing current state are not available, shear strength index shall be determined by sample boring and through triaxial test for large and important medium sized reservoirs. For general small and medium sized reservoirs, the soil effective strength index shall be determined through direct slow shear test. For soil mass with seepage coefficient smaller than 10^{-7} cm/s or

Table 8.2 Minimum safety factor of anti sliding stability of dam slope

Application conditions	Engineering grade			
	1	2	3	4, 5
Normal operating conditions	1.50	1.35	1.30	1.25
Exceptional application condition I	1.30	1.25	1.20	1.15
Exceptional operating conditions II	1.20	1.15	1.15	1.10

compression coefficient smaller than 0.2 MPa^{-1}, direct quick shear test or consoli dated quick shear test can also be used to determine total stress strength index.

- The pore water pressure in the dam body and dam foundation in the stable seepage period shall be determined in accordance with flow net. The flow net shall be drawn based on pore water pressure monitoring data for those dam of Scale 1 and 2 and important medium sized dams and high dams. Seepage flow finite element method can be also used to calculate the seepage field under the relative high water level and to draw flow net.
- The pore water pressure in the upstream dam shell in the water level drawdown period can be calculated by seepage flow calculation for cohesive less soil so as to determine the location of seepage line inside the dam body in reservoir water level drawdown period, draw instantaneous flow net and determine pore water pressure. For dams of Scale 1 and 2 and high dams, review should be done by using monitoring data. For those ultra-high dams and particularly important projects, finite element method should be adopted for calculation of unstable seepage flow in reservoir water level drawdown period so as to determine the corresponding seepage field and pressure of pore water.

The safety factor for stability against sliding of dam slope calculated by taking into account the inter-bar force method shall not be less than the value specified in Table 8.2. The safety factor for calculating slope stability of concrete face rockfill dam with non-linear shear strength index can be taken as specified in Table 8.2. When stability calculation is carried out by using the sliding wedge method, the safety factor shall comply with Table 8.2 when the force between the sliding wedges is assumed to be parallel to the average gradient of the slope surface and the sliding bottom slope. When the force between the sliding wedges is assumed to be in the horizontal direction, the minimum safety factor for normal operation conditions of Stage 1 dam shall not be less than 1.30, which can be reduced by 8% compared with the value specified in Table 8.2.

8.2.6.2 The Stress–Strain Analysis Method

The stress–strain analysis method involves considering the stress distribution throughout the entire dam body and its foundation. This approach establishes a mathematical model based on a specific non-linear relationship between stress and strain, typically a constitutive relation, and employs numerical techniques like the non-linear finite element method to attain numerical solutions for stress states at various points within the stress field.

The method takes into account the stress distribution across the entire dam body and its foundation. This involves understanding how the internal forces and stresses are distributed within the materials. A mathematical model is formulated based on a non-linear constitutive relation between stress and strain. This relationship captures the response of the materials to applied loads, considering non-linear behavior. Using numerical techniques like the non-linear finite element method, stress states at each point within the stress field are computed. This involves solving a complex system of equations to obtain accurate stress values. The calculated stress values at each point are compared to the corresponding strength of the soil or material at that point. By assessing whether the stress exceeds the material's strength, engineers can identify potential damage areas. By evaluating the stress–strain relationship and computing stress states numerically, this method provides insights into how stresses propagate through the dam and its foundation. It's particularly useful for understanding complex behaviors of materials under different loading conditions and identifying areas where the materials might be prone to failure or damage.

8.3 Evaluation on Structure Safety for Concrete Dam

In evaluation on structure safety of concrete dam, review shall be mainly made on whether the dam strength and stability, dam crest elevation, dam crest width, etc. meet the requirements. See Sect. 7.5 for recheck of dam crest elevation.

8.3.1 Disease Characteristics of Gravity Dam Structure

Gravity dams are generally concrete or masonry structures. After a certain period of operation, the following structural diseases are prone to occur [25]:

- the foundation is not treated seriously, or the foundation deterioration, resulting in the dam along the foundation surface anti-sliding stability can not meet the requirements.
- dam foundation anti-seepage curtain does not meet the requirements, serious leakage, poor drainage, dam foundation uplift pressure exceeds the design value, reduce the safety of the dam stability.
- Hazardous cracks appear on the surface of the dam, corridors, drainage pipes and piers.
- the dam mixed soil serious carbonization phenomenon.
- Dam concrete strength is seriously reduced.

- masonry dam masonry mortar or fine stone concrete is not dense, stone weathering serious, the overall performance of the dam;
- the dam concrete penetrating cracks, resulting in concrete calcification precipitation, dam leakage.

8.3.2 Force Condition of Concrete Dam

In the design of gravity dam structures, there are numerous forces that need to be taken into account. However, some of these forces are inherently complex or uncertain, making their precise determination challenging. Factors that contribute to the complexity of force determination include the dynamic and varied nature of loads such as water pressure, seismic forces, temperature effects, and soil interactions (Fig. 8.7). Additionally, environmental conditions, construction techniques, and long-term behavior of materials can further complicate the accurate prediction of forces.

The consideration of uncertain forces underscores the importance of safety margins, robust design, and ongoing monitoring. Engineers aim to strike a balance between optimizing dam performance and accounting for potential variations and uncertainties in force magnitudes and directions. This iterative process of estimation, analysis, and adjustment contributes to the creation of structurally sound and resilient gravity dam designs [26, 27].

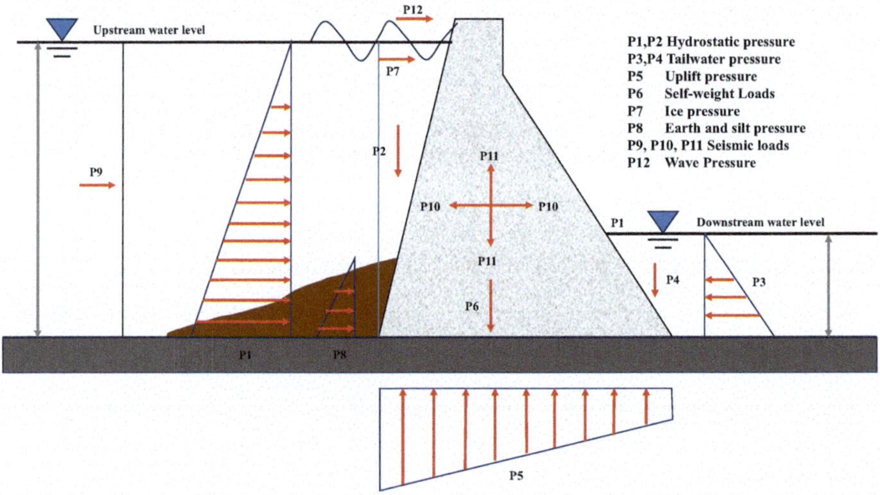

Fig. 8.7 An example of the loads acting on a gravity dam

8.3.2.1 Self-weight Loads

The weight of the structure should be calculated according to the structural design size and material weight. The unit weight of concrete can be determined according to the type and particle size of aggregate, which can be taken as 23.5–24.0 kN/m^3. The unit weight of reinforced concrete can be 24.5–25.0 kN/m^3. The dead weight of permanent equipment located on a spillway building may be taken by the weight of its nameplate.

In the determination of the dead load, relatively small voids, such as galleries, normally are not deducted unless the engineer judges that the voids constitute a significant portion of the dam's volume. The dead loads considered should include weights of concrete and superimposed backfill, and appurtenances such as gates and bridges.

8.3.2.2 External Water Imposed Loads

(1) **Hydrostatic Loads**

Although the weight of water varies slightly with temperature, the weight of fresh water should be taken at 9.81 kN/m^3. A linear distribution of the static water pressure acting normal to the surface of the dam should be applied.

The hydrostatic pressure can be decomposed into vertical direction (i.e. water weight) and horizontal direction for load combination calculation. The hydrostatic pressure acting vertically on a point on the surface of a building (structure) should be calculated as follows:

$$P = \gamma_{\omega} H \tag{8.3.1}$$

where

- P Hydrostatic pressure at calculation point, kN/m^2;
- H Head of action at the calculation point, m, determined by the height difference between the calculated water level and the calculation point;
- γ_{ω} The unit weight of water, kN/m^3, can be used as 9.81 kN/m^3. For sediment-laden rivers, it should be determined according to the measured data. When there is no measured data, it can be 10.5–11.0 kN/m^3.

(2) **Hydrodynamic pressure (nappe forces)**

When water flows through a curved surface (such as an overflow dam surface or an inverted arc of a drain hole), hydrodynamic pressure is generated at the point due to a change in flow direction (Fig. 8.8).

The forces acting on an overflow dam or spillway section are complicated by steady state hydrodynamic effects. Hydrodynamic forces result from water changing speed and direction as it flows over a spillway. At small discharges, nappe forces

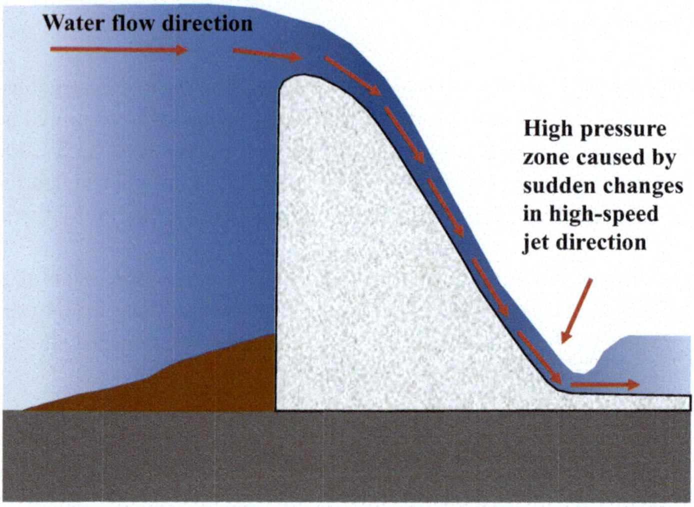

Fig. 8.8 Schematic diagram of hydrodynamic pressure (nappe forces)

may be neglected in stability analysis; however, when the discharge over an overflow spillway approaches the design discharge, nappe forces can become significant and should be taken into account in the analysis of dam stability.

If the tailwater is greater than the conjugate depth, tailwater will fall back against the dam, submerging the jet and lessening hydrodynamic effects. However, unless there is clear evidence that tailwater will be in excess of the conjugate depth, it shall be assumed that tailwater is blown downstream of the dam by the discharge, and that tailwater has no effect on the nappe pressures on the dam. Downstream channel conveyance characteristics are typically not well known for unprecedented discharges. For this reason, it should not be assumed that tailwater will drown out the hydraulic jump without sufficient justification.

8.3.2.3 Uplift Pressure

Calculation of uplift pressure see Sect. 7.4.4.

8.3.2.4 Earth and Silt Pressures

(1) Earth Pressures

Earth pressures against the dam may occur where backfill is deposited in the foundation excavation and where embankment fills abut and wrap around concrete monoliths. The fill material may or may not be submerged. In most cases, at rest earth pressures should be assumed. The rigidity of the foundation and the character of the backfill, along with the construction sequence, may affect this assumption. The unit weight of the backfill and material strength parameters used in the analysis should be supported by site investigations. If the backfill is submerged, the unit weight of the soil should be reduced by the unit weight of water to determine the buoyant weight. Earth backfill on the downstream side of a gravity dam has a beneficial effect on stability, however, if flood conditions can overtop the dam and lead to erosion of the backfill, it can not be relied upon for its stabilizing effects.

(2) **Silt Pressures**

The silt elevation should be determined by hydro graphic surveys. Vertical pressure exerted by saturated silt is determined as if silt were a saturated soil, the magnitude of pressure varying directly with depth. Horizontal pressure exerted by the silt load is calculated in the same manner as submerged earth backfill. For post-earthquake analysis, it's assumed that silt will liquefy under seismic loading. The internal shear strength of silt is taken as zero, unless site investigations confirm that liquefaction is not possible. Silt pressures are considered in the design if suspended sediment measurements indicate that such pressures are expected. The horizontal silt pressure acting on the unit length of the dam surface can be calculated according to the following formula, Vertical silt pressure should be taken into account when the structural retaining surface is inclined.

$$P_s = \frac{1}{2}\gamma_{sb}h_s^2 tg^2\left(45^\circ - \frac{\phi_s}{2}\right) \tag{8.3.2}$$

$$\gamma_{sb} = \gamma_{sd} - (1 - n)\gamma_w \tag{8.3.3}$$

where

P_s Silt pressures, kN/m;
γ_{sb} buoyant unit weight of silt, kN/m^3;
γ_{sd} dry unit weight, kN/m^3;
γ_w Unit weight of water, kN/m^3;
n Porosity of silt;
h_s Silt deposition thickness in front of dam, m;
φ_s Internal friction angle of silt, (°).

8.3.2.5 Wave Pressure

Water surface under the action of wind, the formation of waves and pressure, called wave pressure. For medium and high dams, wave pressure accounts for a small proportion of all loads; for low dams, gate piers, breast walls, surface gates, etc., wave pressure often accounts for a considerable proportion. For the first-stage water retaining structure, when the wave pressure is one of the main loads, it should be demonstrated by model test.

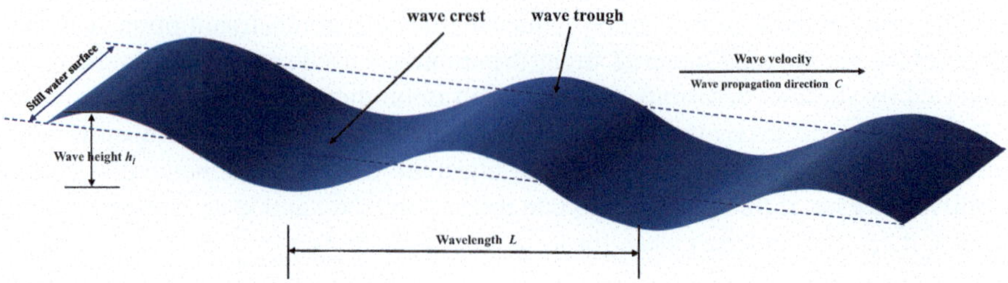

Fig. 8.9 Geometric parameters of waves

(1) **Wave parameters**

The wave action makes the gravity dam bear the wave pressure, and the wave pressure is related to the wave factor and the water depth in front of the dam.

The wave height is h_l, and the wavelength is L. The backwater height caused by the wave center line higher than the static water surface is h_z. Wave height, wavelength and wave height are called the three elements of waves. When the wave advances to the front of the dam, the standing wave is generated due to the reflection of the straight dam surface. The wave height is $2h_s$, and the wavelength remains L unchanged (Fig. 8.9).

The value of each wave factor is related to the size, shape, wind force and topography of the reservoir area, among which the length of the wind area is related to the shape of the water area after the reservoir impoundment. When the waters on both sides of the wind direction are wide, the linear distance from the calculation point to the opposite bank can be used. When there is a local narrowing along the wind direction and the width b of the narrowing is less than 12 times the calculated wavelength, 5b can be used as the length of the wind zone and is not less than the straight-line distance from the calculation point to the narrowing. The average depth of the water area in the wind zone can be obtained by making a topographic profile along the wind direction, and the calculated water level should be consistent with the static water level under the corresponding design conditions (Fig. 8.10).

When the water area on both sides of the wind direction is narrow, or the shape of the water area is irregular, or there are obstacles such as islands, the main ray can be intersected with the water boundary from the calculation point in the windward direction, and then a ray is made on both sides of the main ray every 7.5° to intersect with the water boundary respectively.

As shown in Fig. 8.11, D0 is the distance from the calculation point along the main ray direction to the opposite shore, D_i is the distance from the calculation point along the i ray to the opposite shore, a_i is the angle between the i ray and the main ray, $a_i = 7.5i$ (generally take $i = +1, +2, +3, +4, +5, +6$), and let $a_o = 0$, the equivalent wind zone length D is calculated as follows:

$$D = \frac{\sum_i D_i \cos^2 a_i}{\sum_i \cos a_i}, \quad (i = \pm 1, \pm 2, \pm 3, \pm 4, \pm 5, \pm 6) \tag{8.3.4}$$

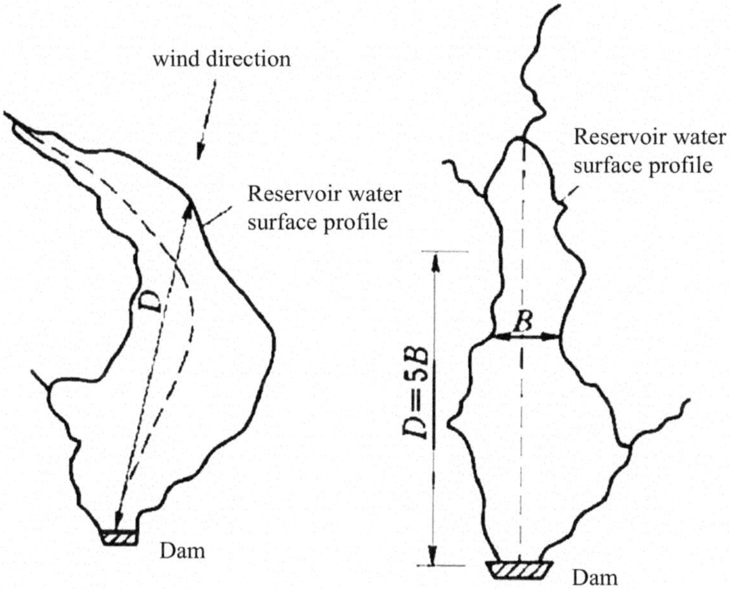

Fig. 8.10 The fetch length

Fig. 8.11 Equivalent wind zone length calculation diagram

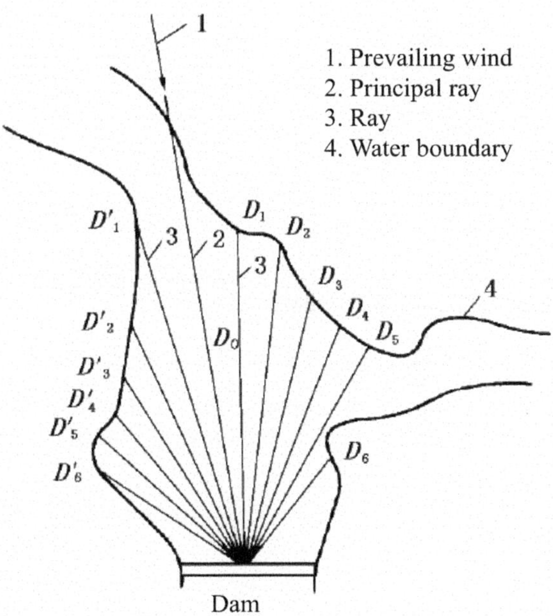

There are many factors affecting the formation of waves. At present, semi empirical formula is generally used to determine wave elements. Common calculation formulas include Putian formula, Hedi formula and Guanting formula.

- For plain and reservoir in coastal area, h_m and T_m should be calculated according to Putian test station formula.

$$\frac{gh_m}{v_0^2} = 0.13th\left[0.7\left(\frac{gH_m}{v_0^2}\right)^{0.7}\right]th\left\{\frac{0.0018\left(gD/v_0^2\right)^{0.45}}{0.13th\left[0.7\left(gH_m/v_0^2\right)^{0.7}\right]}\right\} \tag{8.3.5}$$

where

h_{m} Average wave height, m;
$\mathrm{T_m}$ average wave period, s;
v_0 Calculate wind speed, m/s;
D Wind zone length, m;
H_{m} Average water depth, m;
g Gravity acceleration, 9.81 $\mathrm{m/s^2}$.

- Hilly, plain reservoir, should be calculated according to Hedi reservoir formula $h_{2\%}$ and L_m (suitable for deep reservoir, $v_0 < 26.5$ m/s and $D < 7.5$ km):

$$\frac{gh_{2\%}}{v_0^2} = 0.00625v_0^{1/8}\left(\frac{gD}{v_0^2}\right)^{1/3} \tag{8.3.6}$$

$$\frac{gL_m}{v_0^2} = 0.0386\left(\frac{gD}{v_0^2}\right)^{1/2} \tag{8.3.7}$$

where

$h_{2\%}$ Wave height at 2% cumulative frequency, m;
L_{m} average wavelength, m.

The remaining symbols are the same as before

- Inland canyon reservoir, h and L_m should be calculated according to Guanting reservoir formula (suitable for $v_0 < 20$ m/s and $D < 20$ km):

$$\frac{gh}{v_0^2} = 0.0076v_0^{-1/12}\left(\frac{gD}{v_0^2}\right)^{1/3} \tag{8.3.8}$$

$$\frac{gL_m}{v_0^2} = 0.331v_0^{-1/2.15}\left(\frac{gD}{v_0^2}\right)^{1/3.75} \tag{8.3.9}$$

where

h When $\frac{gD}{v_0^2} = 20-250$, the wave height $h_{5\%}$ of the 5% cumulative frequency; When $\frac{gD}{v_0^2} = 250-1000$, the wave height $h_{10\%}$ of the 10% cumulative frequency.

(2) **Wave pressure calculation formula**

When the water depth in front of the dam is greater than the semi-average wavelength, that is, $H > L_m/2$, the wave motion is not constrained by the bottom of the reservoir, and the wave under such conditions is called deep water wave. The water depth is less than the semi-average wavelength and greater than the critical water depth H_0, that is, $L_m/2 > H > H_0$. Wave motion is affected by the reservoir bottom, called shallow water wave. Water depth is less than the critical depth, namely $H < H_0$. When, the wave breaks, called breaking wave.

The wave pressure acting on a vertical or nearly vertical building should be calculated according to the following three wave states according to the water depth in front of the building.

- When $H \geq H_{cr}$, $H \geq L_m/2$, the wave pressure distribution is shown in Fig. 8.12a. The wave pressure per unit length is calculated as follows:

$$P_\omega = \frac{1}{4}\gamma_\omega L_m(h_{1\%} + h_z) \tag{8.3.10}$$

$$h_z = \frac{\pi h_{1\%}^2}{L_m} cth \frac{2\pi H}{L_m} \tag{8.3.11}$$

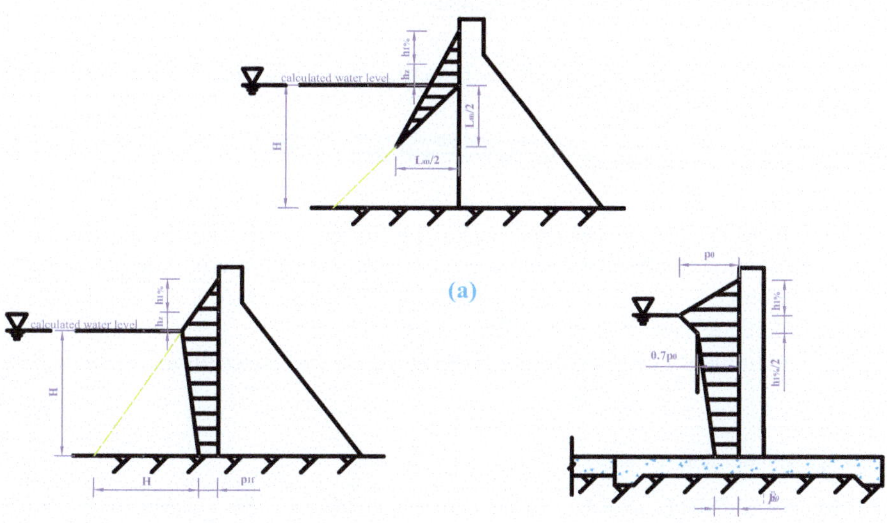

Fig. 8.12 Distribution of wave pressure in straight wall water retaining structures

$$H_{CR} = \frac{L_m}{4\pi} \ln \frac{L_m + 2\pi h_{1\%}}{L_m - 2\pi h_{1\%}} \tag{8.3.12}$$

where

P_k Wave pressure per unit length on the upstream surface, kN/m;
γ_ω Water gravity, kN/m^3;
L_m average wavelength, m;
$h_{1\%}$ Cumulative frequency 1% wave height, m;
H Depth of water in front of water retaining structure, m;
h_z Elevation from wave centerline to calculated water level, m:
H_{cr} Critical water depth for breaking waves, m.

- When $H_{cr} \le H \le L_m/2$, the distribution of wave pressure is shown in Fig. 8.12b, and the wave pressure per unit length is calculated as follows:

$$P_\omega = \frac{1}{2}\left[(h_{1\%} + h_z)\left(\gamma_\omega H + P_{lf}\right) + Hp_{lf}\right] \tag{8.3.13}$$

$$p_{lf} = \gamma_\omega h_{1\%} \text{sech} \frac{2\pi H}{L_m} \tag{8.3.14}$$

where

P_{lf} Residual wave pressure at the bottom of the building, kN/m^2.

- When $H < H_{cr}$, the distribution of wave pressure is shown in Fig. 8.12c. The wave pressure per unit length is calculated as follows:

$$P_\omega = \frac{1}{2} p_0[(1.5 - 0.5\lambda)h_{1\%} + (0.7 + \lambda)H] \tag{8.3.15}$$

$$p_0 = K \bullet \gamma_\omega h_{1\%} \tag{8.3.16}$$

where

λ Reduction coefficient of wave pressure strength on the bottom of building (when $H < 1.7h_{1\%}$, $\lambda = 0.6$; when $H > 1.7h_{1\%}$, $\lambda = 0.5$).
p_0 Calculate Wave Pressure at Water Level, kN/m^2.
K_i The influence coefficient of bottom slope is adopted according to Table 8.3.

Table 8.3 Bottom slope influence coefficient K_i value table

Bottom slope i	1/10	1/20	1/30	1/40	1/50	1/60	1/80	1/100
K_i	1.89	1.61	1.48	1.41	1.36	1.33	1.29	1.25

Note The bottom slope *i* adopts the average value within a certain distance in front of the building

8.3.2.6 Ice Loading

(1) Ice Pressures

Ice pressure is created by thermal expansion of the ice and by wind drag. Pressures caused by thermal expansion are dependent on the temperature rise of the ice, the thickness of the ice sheet, the coefficient of expansion, the elastic modulus and the strength of the ice. Wind drag is dependent on the size and shape of the exposed area, the roughness of the surface, and the direction and velocity of the wind. Ice loads are usually transitory. Not all dams will be subject to ice pressure and the engineer should decide whether an ice load is appropriate after consideration of the above factors. An example of the conditions conducive to the development of potentially high ice pressure would be a reservoir with hard rock reservoir walls which totally restrain the ice sheet. In addition, the site meteorological conditions would have to be such that an extremely rigid ice sheet develops. The existence of a formal system for the prevention of ice formation, such as an air bubble system, may reduce or eliminate ice loadings. Ice pressure should be applied at the normal pool elevation.

(2) Ice/Debris Impact

Certain rivers are susceptible to both ice and debris flow. These events involve the movement of heavy ice sheets and debris downstream, which can pose a threat to dams. The impact of these ice sheets and debris can result in localized damage to dam components like piers, gates, or machinery. In some cases, debris or floating ice can block spillway bays, leading to substantial reservoir surcharges during moderate flood events. It's crucial to take into account the potential effects of ice and debris when assessing the flood-passing capacity of spillways. When the ability of a spillway to pass floods is evaluated, the effect of ice and debris should be considered. The static ice pressure F_d acting on the unit length of the dam surface can be adopted according to Table 8.4 when the ice layer expands with temperature.

- Ice layer thickness takes the average annual maximum.
- For small reservoirs, the static ice pressure value in the table should be multiplied by 0.87; for large plain reservoirs with open reservoir surface, it should be multiplied by 1.25.
- The static ice pressure value in the table is applicable to the case that the reservoir water level is basically unchanged during the freezing period; the static ice

Table 8.4 Static ice pressure values

Thickness of ice layer (m)	0.4	0.6	0.8	1.0	1.2
Static ice pressure (kN/m)	85	180	215	245	280

pressure under the condition of reservoir water level fluctuation during freezing period should be specially studied.

- The static ice pressure value can be interpolated according to the ice thickness listed.
- The static ice pressure acts vertically on the front edge of the structure, and its action point is 1/3 ice thickness below the ice surface. The ice pressure and water pressure in the ice thickness during the freezing period act on the building at the same time.

The dynamic ice pressure generated by the impact of ice on the dam surface during ice movement should be determined by calculation according to the size of the ice, the front shape of the building and the mode of action. The dynamic ice pressure acting on a vertical dam surface or other wide and long buildings can be calculated as follows:

$$F_b = 0.07vd_i\sqrt{Af_{ic}} \tag{8.3.17}$$

where

F_b Dynamic ice pressure generated when ice hits buildings, MN;
v Ice flow rate, m/s, determined by the measured data (when there is no measured data, for the reservoir can be taken over the ice movement period of 3% of the maximum wind speed, and generally should not be greater than 0.6 m/s);
A Ice area, m^2, can be determined by the measured or survey data of local or adjacent sites;
d_i Flow ice thickness, can be used 0.7–0.8 times the local maximum ice thickness, flow ice initial value.
f_{ic} compressive strength of ice MPa (0.3 MPa for reservoir; for rivers, 0.45 MPa can be used in the initial stage of ice flow and 0.3 MPa in the later stage of high water level).

8.3.2.7 Temperature and Aggregate Reactivity

The temperature fluctuations experienced by dam concrete can lead to either expansion or contraction. When these deformations are restricted, they give rise to temperature-induced stresses. The resulting stress, deformation, and displacement of the structure due to temperature changes are collectively referred to as temperature effects. The primary contributors to temperature effects include various heat sources: Ambient Temperature, Solar Radiation, Water Temperature, Hydration Heat.

Volumetric changes caused by thermal expansion and contraction, or by alkali/aggregate reactivity affect the cross valley stresses in the dam. These stresses are important when three dimensional behavior is being considered. Expansion will cause a dam to wedge itself into the valley walls more tightly, increasing its stability. Contraction has the opposite effect. While these effects are acknowledged, the beneficial effect of expansion is difficult to quantify even with very elaborate finite element models because it is contingent on the modulus of deformation of the abutments which is highly variable [28]. For this reason, the beneficial effects of expansion should not be relied upon in three dimensional stability analysis [29]. If it appears that contraction will cause monolith joints to open, and thus compromise force transfer from monolith to monolith, this effect should be considered.

8.3.3 Selection of Calculation Parameters for Structural Safety Analysis

For the analysis and calculation of structural safety for concrete dams, it's important to obtain accurate parameters that reflect the current conditions of the dam and its foundation. This involves conducting drilling tests on the dam body and foundation, especially for high dams. For medium–high and low dams, retesting is also necessary when monitoring data or analysis results indicate elevated stress, significant deformation, or lower safety coefficients.

If monitoring data is available, an inverse analysis approach can be employed [30]. This means using the monitoring data to determine calculation parameters in a comprehensive manner. This method allows for a more accurate representation of the actual behavior of the dam.

Over time, as the dam operates, the concrete structure gradually undergoes aging due to various factors like seepage corrosion, freezing, carbonization, and creepage. These factors can lead to changes in the physical and mechanical properties of the dam body and its foundation. Additionally, statistical methods for calculating parameters may have evolved since the dam's construction.

Considering these factors, it's crucial to reevaluate and redetermine calculation parameters as accurately as possible. This involves a combination of tests, monitoring data analysis, and inverse analysis techniques. By updating these parameters, engineers can ensure that their analysis reflects the current state of the dam and its components, enabling a more reliable assessment of structural safety and performance.

8.3.4 Recheck Calculation of Dam Strength and Stability

In the evaluation of concrete dam structural safety, a combination of different analysis methods is recommended to ensure accuracy and reliability. The results of material mechanics and rigid limit equilibrium methods should be prioritized and used as the primary basis for determining the dam body section. The finite element method can be used as a supplementary tool to assist in the analysis.

Strength review shall mainly include stress review and local reinforcement check calculation. For stability review, the following shall be check calculated: the stability against sliding of the gravity dam and buttress dam along the dam foundation plane, weak interlayer and structural plane with gently inclined angle, anti-slide stability of the arch seats at the two banks of arch dam, lateral stability of buttresses of the buttress dam. For roller compacted concrete gravity dam, stability against sliding on the compacted layer (seam) shall be checked. For arch dam with smaller plane curvature, stability against sliding along the dam foundation surface shall be calculated. The holistic stability of the sloped section of the dam shall be analyzed whenever necessary.

The gravity analysis method assumes the dam to be a 2-dimensional rigid block and assumes linear distribution of foundation pressure. Conducting a gravity analysis is typically a preliminary step before more rigorous analyses are undertaken. In many cases, if the gravity analysis indicates that the dam is stable, there might not be a need for further extensive analyses.

8.3.4.1 Calculation of Anti Sliding Stability of Gravity Dam

The calculation of anti-sliding stability primarily involves assessing the potential for sliding along the dam foundation surface [31]. The safety factor for the anti-sliding stability of the dam foundation surface can be determined using either shear strength formulas or shear strength formulas within the framework of the rigid body limit equilibrium method. The shear strength formulas take into account the properties of the materials involved and provide a basis for evaluating the stability of the dam against sliding. These formulas consider factors such as the inclination of the dam foundation surface, the weight of the dam, the shear strength of the materials, and other relevant parameters. The rigid body limit equilibrium method involves analyzing the equilibrium conditions of forces and moments acting on a rigid body (the dam) to determine its stability against sliding. This method considers the distribution of forces and moments along the foundation surface and assesses whether the dam has sufficient resistance to prevent sliding.

(1) **Formula for calculating shear strength**

$$K' = \frac{f'\Sigma G + C'A}{\Sigma H} \tag{8.3.18}$$

where

- K' Safety factor for shear stability
- f' Shear friction coefficient of contact surface between dam body concrete and dam foundation
- ΣG Normal value of all loads acting on the dam body (including uplift pressure, the same below) on the sliding plane, kN
- C Shear-breaking cohesion of contact surface between dam concrete and dam foundation, kPa
- A Section area of dam foundation contact surface, m^2

ΣH The tangential component of the total load acting on the dam against the sliding plane, kN.

(2) **Formula for calculating shear strength**

$$K_c = \frac{f \sum G}{\sum H} \tag{8.3.19}$$

where

K_c Safety factor of anti-sliding stability along the base surface of control section

f Shear friction coefficient of contact surface between dam body concrete and dam foundation.

(3) **Calculation of stability against sliding of arch dam**

The stability against sliding of dam abutment is analyzed by rigid limit equilibrium method with the formula as follows:

$$K_1 = \frac{\sum(Nf_1 + c_1 A)}{\sum T} \tag{8.3.20}$$

$$K_2 = \frac{\sum Nf_1}{\sum T} \tag{8.3.21}$$

where

K_1 Safety factor of stability against sliding considering cohesion
K_2 Safety factor of stability against sliding without considering cohesion
N Force perpendicular to slip surface
T Force along the slip surface
f_1 Friction coefficient of shear strength
c_1 Shear cohesion
A Calculating the area of the slip surface.

In actual calculation, the arch ring with arbitrary elevation per unit height is calculated as shown in Fig. 8.13.

The axial force at the arch end H_a, the shear force V_a. Axial force at arch end H_a and shear V_a force projected onto the sliding surface:

$$\begin{cases} S = H_a \cos\varphi + V_a \sin\varphi \\ N = H_a \sin\varphi - V_a \cos\varphi \end{cases} \tag{8.3.22}$$

where

φ the angle between the axial force H_a and the sliding surface.

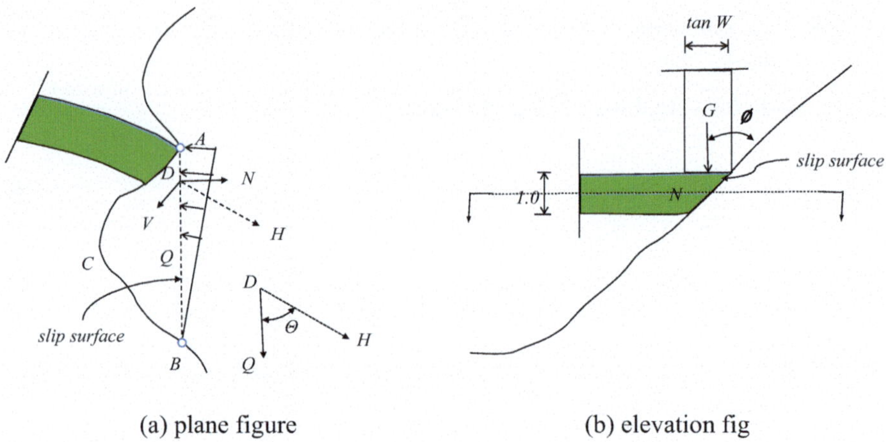

Fig. 8.13 Calculation of local stability of arch dams

In the elevation view, the included angle between the sliding surface and the vertical line is ψ, and then the force N is decomposed into the components of the parallel and vertical arch seating surfaces Q and P:

$$\begin{cases} Q = N\sin\psi - G\cos\psi \\ P = N\cos\psi + G\sin\psi - U \end{cases} \tag{8.3.23}$$

where

ψ the Angle between arch seating surface and vertical line.
G Self weight of cantilever beam on arch end face.
U Uplift pressure of arch end face.

The force acting on the sliding surface S is the sliding force, P is the anti sliding force. Considering the effect of the weight W of the rock mass at the downstream of the arch end, the anti sliding stability safety factor can be obtained as follows:

$$K_1 = \frac{f_1 P + f_2 W \sin\psi + (c_1 L_1 + c_2 L_2)\sec\psi}{S} \tag{8.3.24}$$

$$K_2 = \frac{f_1 P + f_2 W \sin\psi}{S} \tag{8.3.25}$$

where

K_1 Safety factor of stability against sliding considering cohesion
K_2 Safety factor of stability against sliding without considering cohesion
f_1, f_2 Friction coefficient between sliding surfaces and rock
W Weight of rock mass downstream of arch end
l_1, l_2 Length of corresponding sliding surface.

8.3.4.2 Finite Element Methods of Gravity Dam

In most cases, the gravity analysis method discussed above will be sufficient for the determination of stability. However, dams with irregular geometries or spillway sections with long aprons may require more rigorous analysis. The Finite Element Method (FEM) permits the engineer to closely model the actual geometry of the structure and account for its interaction with the foundation. For example, consider the dams in Figs. 8.14 and 8.15.

FEM Method offers the capability to model not only the dam structure but also the underlying foundation rock. A crucial parameter influencing the interaction between

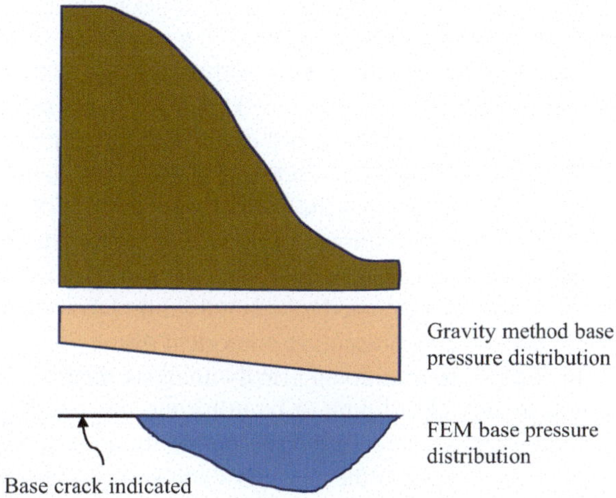

Fig. 8.14 Pressure distribution diagram of rigid body limit balance method and FEM method

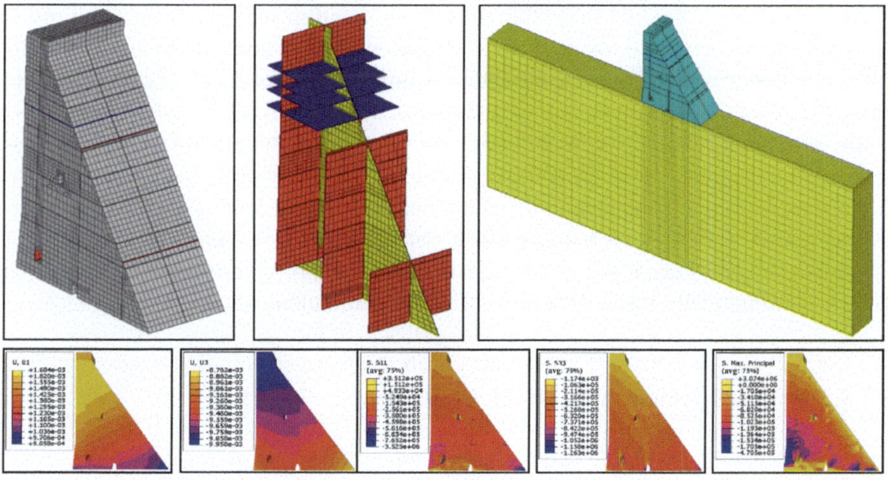

Fig. 8.15 Consider the principle stress plot of the cracked dam

the dam and its foundation is the ratio of the deformation modulus of the rock to the elastic modulus of the dam concrete. This ratio significantly impacts the behavior of the dam-foundation system, particularly in terms of crack propagation. In gravity analysis, the distribution of shear stress on the foundation is not explicitly addressed. Instead, it is tacitly presumed that the shear stress is evenly distributed across the entire base. This presumption lacks precision and is somewhat arbitrary. By utilizing finite element modeling, the distribution of contact stress at the base can gain a deeper understanding.

Three-dimensional (3D) finite element modeling (FEM) should be employed in scenarios where the structure or loading conditions preclude the assumption of plane strain conditions. This includes cases where the geometry of the dam necessitates the consideration of stress distribution parallel to its axis, such as in gravity-arch dams with curved plans, or dams situated in narrow valleys. y utilizing 3D analysis, we can accurately determine the magnitude and location of forces acting on the foundation. Furthermore, when considering 3D behavior, it is imperative to incorporate features such as shear keys or curvature in plane that facilitate horizontal force transfer.

When reviewing finite element analyses for dam structures, it is paramount to maintain a clear focus on the ultimate question: Will the dam withstand the specified loading conditions without failing? Despite the vast array of stress outputs provided by these analyses, it is crucial not to lose sight of this fundamental inquiry. The analysis must be evaluated based on its ability to address various failure mechanisms, rather than merely presenting numerical data.

In contrast to traditional gravity analysis methods, which often presume specific failure modes like sliding or overturning, the standard linear elastic finite element method does not inherently incorporate these mechanisms. Therefore, it falls upon the reviewer to assess the analysis's adequacy in terms of its consideration of potential failure modes. When interpreting stress distributions obtained from finite element analyses, it is imperative to verify that global force and moment equilibrium conditions are satisfied. Additionally, the stress states within individual elements must remain within acceptable material strength limits. For instance, if the analysis indicates tensile stresses at the dam/foundation interface, it is necessary to rerun the analysis, excluding tensile elements from the stiffness matrix. Excessive shear stress at the interface can also pose a challenge. In such cases, there is no frictional resistance present, and all shear stress must be transferred solely through cohesive bonds. If the reviewer doubts the availability of cohesive strength at the interface, the analysis should be rerun with a reduced shear stiffness for those elements, allowing for the redistribution of shear stress. Modern finite element programs often facilitate this process automatically through the use of gap-friction elements.

8.3.4.3 Finite Element Analysis of Arch Dam

The finite element procedure is the numerical method most often used for the structural analysis of arch dams [32]. The finite element analysis of arch dams is based on the same assumptions that underlie all finite element analyses. This being the case, the basic principles that govern element formulation, mesh construction, and load application are as valid in the analysis of arch dams as they are anywhere in structural mechanics (Fig. 8.17). There are, however, certain special with respect to arch dam analysis:

(1) **Dam/Foundation Interface Tension Resistance**: As with gravity dam analysis, the interface between the dam and its foundation should be assumed to be incapable of resisting tension normal to it. This means that any tensile forces normal to this interface should be carefully evaluated and addressed in the analysis.
(2) **Uplift Forces**: Uplift forces acting on potential failure surfaces within the dam structure cannot be overlooked. These forces can significantly influence the stability of the dam, and their effects must be included in the analysis.
(3) **Vertical Contraction Joints and Cracks**: The impact of vertical contraction joints and preexisting cracks on the dam's structural integrity must be taken into account. While it is not necessary to include every crack in the model, those that are significant or oriented in such a way that tensile stress is applied normal to them should be included. Horizontal cracks, for instance, are particularly important and cannot be ignored in the analysis.
(4) **Mesh Density**: The density of the mesh used in the finite element model depends on the specific objectives of the analysis. If the focus is on capturing stress concentrations at openings or corners, a finer mesh is required. However, if the analysis is primarily concerned with global sliding and rocking mechanisms, a coarser mesh may be sufficient (Fig. 8.16).
(5) **Hydrodynamic Coupling**: The interaction between the dam and the reservoir's water has a significant impact on the dam's dynamic behavior [33]. Arch dams, due to their thinner profile, experience less concrete mass acceleration compared to gravity dams. This means that the accelerated water can contribute a larger portion to the total system mass. Therefore, it is crucial to account for hydrodynamic effects in the seismic analysis of arch dams more so than other dam types. Modeling the reservoir directly using finite elements with zero shear strength can lead to numerical instability. To avoid this, alternative methods or approximations may need to be employed to capture the hydrodynamic effects without compromising the stability of the analysis.

8.3.5 Standard for Safety Assessment of Concrete Dam Structures

1. The dam structure can be regarded as unsafe or as hidden troubles exist if one of the following circumstances is found in on-site inspection or observation and further monitoring and analysis shall be carried out:

Fig. 8.16 Three-dimensional (3D) FEM

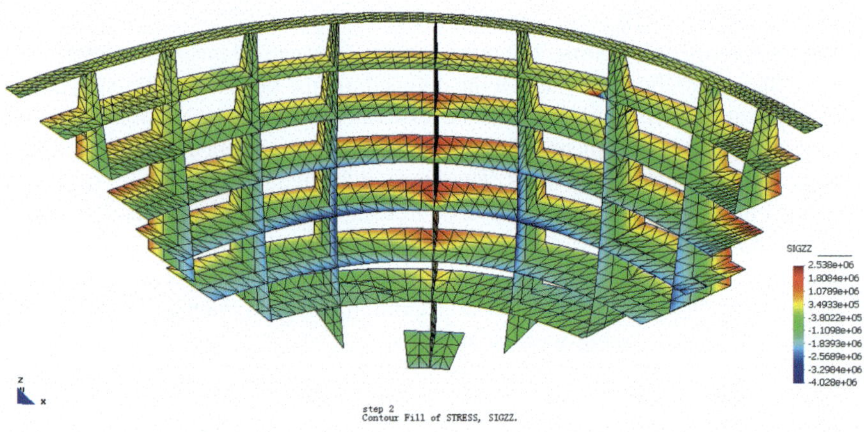

Fig. 8.17 Calculation of displacement and stress in arch dams using multi-arch beam method

- Cracks appeared on dam surface or in holes or discharging pipes and other weakened parts and on some locations like gate piers, which would endanger the structure safety.
- Serious corrosion appeared in dam concrete;

- Concrete on the dam surface or inside the dam body are broken due to compression;
- Obvious displacement of dam body along the foundation surface or obvious inclination of dam body took place;
- Uplifts appeared at the downstream of the dam foundation or obvious displacement occurred on the supporting massif of the two banks;
- Obvious deformation or displacement took place on dam foundation or at arch seat of arch dam and at buttress of the buttress dam;
- Obvious relative displacement appeared at the two sides of the faultage in dam foundation or in arch seat of arch dam;
- Large quantity of water seepage or water burst suddenly appeared in dam foundation or on the supporting massif at the two banks;
- Resonance took place at dam body when overflow dam discharged;
- Obvious water leakage or water jet appeared in the gallery.

2. When the structure safety of the dam is assessed through monitoring data analysis, the rationality of the calculation value from statistical models shall be analyzed and evaluated if great difference exists between the maximum water level of the reservoir that was reached in actual operation and the designed flood level and check flood level. In evaluation of dam structure safety through analysis with monitoring data, the dam structure can be considered as unsafe or hidden troubles exist in the dam structure when one of the following situations occur:

   - The measured values of displacement, deformation, stress, crack openness, etc. exceed the allowed values specified in relevant specifications or in design and tests.
   - The calculated values by using mathematical model under the designed condition or the verified conditions of the displacement, deformation, stress, crack openness, etc. exceed the allowed values specified in relevant specifications or in design and tests.
   - The relation between measured values of displacement, deformation, stress, crack openness, etc. and such factors as load, time and space suddenly changed and these values increased to a larger extent compared with previous ones under the same condition.

3. In evaluation of dam structure safety through calculation and analysis, the control standards for review of the strength and stability of the gravity dam and arch dam shall meet the requirements. The control standard for review of strength and stability of the buttress dam is the same as that for gravity dam.

   - The anti sliding stability safety factor K' value of gravity dam foundation surface calculated according to the shear strength formula (8.3.18) shall not be less than the value specified in Table 8.5.
   - The anti sliding stability safety factor K of gravity dam foundation surface calculated according to the shear strength formula (8.3.19) shall not be less than the value specified in Table 8.6.

Table 8.5 Safety coefficient K' of anti sliding stability of gravity dam foundation surface (shear strength calculation formula 8.3.18)

Load combination		K'
Basic combination		3.0
Special combination	I	2.5
	II	2.3

Table 8.6 Safety coefficient K of anti sliding stability of gravity dam foundation surface (shear strength calculation formula 8.3.19)

Load combination		Dam grade		
		1	2	3
Basic combination		1.10	1.05	1.05
Special combination	I	1.05	1.00	1.00
	II	1.00	1.00	1.00

Table 8.7 Anti sliding stability safety factor K of arch dam under non-seismic conditions

Load combination		Dam grade		
		1	2	3
Calculate according to formula 8.3.20	Basic combination	3.50	3.25	3.00
	Special (non-seismic)	3.00	2.75	2.50
Calculate according to formula 8.3.21	Basic combination	–	–	1.30
	Special (non-seismic)	–	–	1.10

- When the non-seismic working condition is calculated according to Formula (8.3.20) or Formula (8.3.21), the anti sliding stability safety factor of arch seat shall not be less than the provisions in Table 8.7.

8.4 Evaluation on Structure Safety for Masonry Dam

The process, methods, and requirements for evaluating the structural safety of masonry dams are essentially the same as those for concrete dams. The evaluation encompasses various aspects such as stability, strength, and other relevant considerations. Evaluate the potential for sliding along the contact surfaces between blinding concrete and bedrock, between the masonry body and blinding concrete, and between different masonry blocks. If there are weak interlayers or structural planes with gentle inclinations in the dam foundation, consider the anti-slide stability of deeper layers. The stability against sliding of the masonry dam body can be assessed using the same principles as the calculation formula for gravity dams. In terms of calculating the stability against sliding of the masonry dam foundation, utilize the shear

Table 8.8 Safety coefficient K of anti sliding stability of masonry dam foundation surface

Safety factor	Load combination		2 and 3 grade dam
K' (*Calculate according to formula* 8.3.18)	Basic combination		3.00
	Special combination	I	2.50
		II	2.30
K (*Calculate according to formula* 8.3.19)	Basic combination		1.05
	Special combination	I	1.00
		II	1.00

strength formulas or shear strength formula approach [34]. The calculated safety factor should not be less than the value specified in Table 8.8, which outlines the acceptable safety margins. The safety factor of stability against sliding of foundation surface of masonry dam calculated by shear strength formula and shear strength formula should not be less than the value specified in Table 8.8.

8.5 Evaluation on Structure Safety for Discharging and Conveyance Structures

In evaluation on structure safety for discharging and conveyance structures, review shall be mainly made on whether the elevation of the top of the structure (or elevation of platform), discharging safety, structure strength and stability meet the requirements of relevant specifications [35].

8.5.1 Elevation Review of Water Release and Conveyance Structures

The top elevation of the pier and bank wall in the control section shall meet the following requirements:

(1) When releasing check flood, it should not be lower than check flood level plus safety increase value.
(2) The water retaining shall not be lower than the design flood level or the calculated height of normal water level plus wave and the safe elevation value.
(3) When spillway is close to dam abutment, the top elevation of control section should be coordinated with that of dam crest.
(4) Select the lower limit value of safety increase according to Table 8.9.
(5) The safetyfreeboard of intake work platform is the sum of wave calculation height and safety heightening value, in which the safety heightening value

Table 8.9 Lower limit value of safetyfreeboard for spillway control section (unit: m)

Operating conditions	Building level of control section		
	Level 1	Level 2	Level 3
Water retaining	0.7	0.5	0.4
release flood waters	0.5	0.4	0.3

Table 8.10 Safetyfreeboard standard for water inlet working platform (unit: m)

Operating conditions		Level 1	Level 2	Level 3	Level 4, 5
Characteristic retaining level	Design water level	0.70	0.50	0.40	0.30
	Check water level	0.50	0.40	0.30	0.20

should be adopted according to Table 8.10, and the intake arranged as a whole should be coordinated with the water retaining structure.

8.5.2 Force Condition of Discharging and Conveyance Structures

The main loads of discharging and conveyance structures are: self-weight, water weight, horizontal water pressure, uplift pressure, wave pressure, sediment pressure, soil pressure and seismic force. The mainly force condition for evaluation on structure safety for discharging and conveyance structures are the same with that for concrete dam. Seismic force will be described in detail in Chap. 10. But there are some differences in the calculation of hydrodynamic pressure, uplift pressure and earth pressure. This part is briefly described below.

8.5.2.1 Nappe Forces (Hydrodynamic Pressure)

The hydrodynamic pressure acting on a certain area of the flow surface of the spillway building should be calculated according to the resultant force of the hydrodynamic pressure at each point on the area. The hydrodynamic pressure generally only calculates the time-averaged pressure, but for the stilling basin or when the water flow pulsation affects the safety of the structure and causes structural vibration, the pulsating pressure should be calculated. For important or complex hydraulic structures, the hydrodynamic pressure should be determined by model test and comprehensive analysis.

When calculating hydrodynamic pressure, two flow states of steady flow and unsteady flow should be distinguished. For constant flow, different flow patterns such

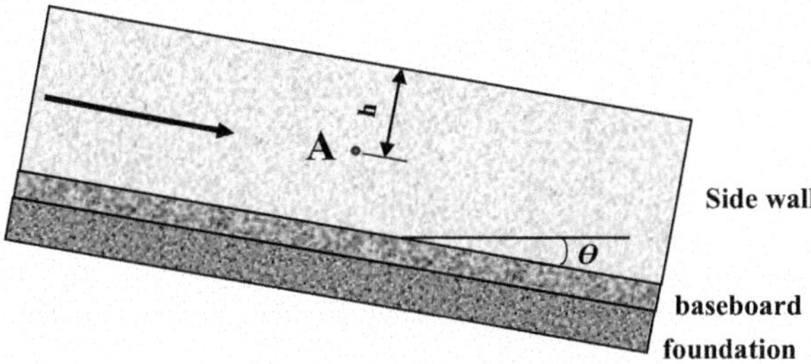

Fig. 8.18 Time-averaged pressure calculation

as gradual flow or abrupt flow should be distinguished and calculated with corresponding methods. Time-averaged pressure gradient flow, according to the corresponding design conditions of the flow conditions, through the calculation or test to obtain the water surface line calculated by the following formula, the schematic diagram is shown in Fig. 8.18.

$$p_{tr} = \rho_{\omega} g h \cos\theta \tag{8.5.1}$$

where

p_{tr} Time-averaged pressure at calculation point on flow surface, N/m^2
ρ_{ω} Density of water, kg/m^3
g Gravity acceleration, m/s^2
h Water depth at point A, m
θ Angle between bottom surface of structure and horizontal plane, °.

The pressure of the hydrodynamic centrifugal force on the bottom surface of the inverted arc section of the spillway is approximately uniformly distributed and can be calculated according to Eq. (8.5.2). As shown in Fig. 8.19, the pressure of centrifugal force acting on the side wall of the anti-arc section is zero at the water surface along the radial profile, and it is p_{cr} at the bottom of the wall, which is approximately linearly distributed. The p_{cr} can be calculated according to the Eq. (8.5.2) and acts vertically on the wall.

$$p_{cr} = q\rho_{\omega} v/R \tag{8.5.2}$$

where

p_{cr} Water flow centrifugal force pressure, N/m^2
q Flow per unit width on the reverse arc under corresponding design conditions, m^3 ρω/(s.m)
g Gravity acceleration, m/s^2
R Reverse arc radius, m
ρ_{ω} Density of water, kg/m^3.

The horizontal and vertical components of the resultant centrifugal force on the anti-arc section can be calculated as follows:

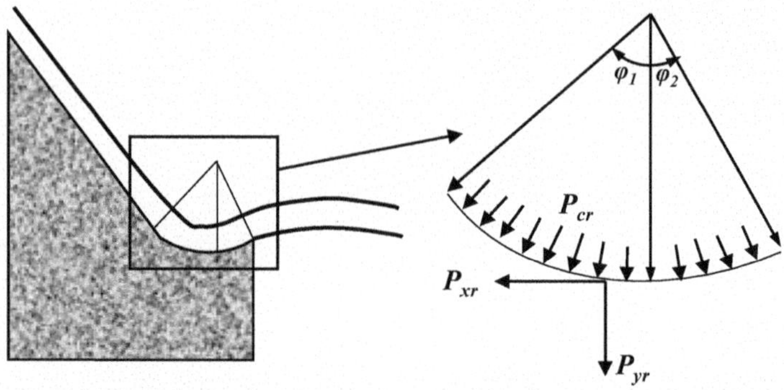

Fig. 8.19 Schematic diagram of centrifugal force of water flow on reverse arc section

$$P_{xr} = q\rho_{\omega}v(\cos\phi_2 - \cos\phi_1) \tag{8.5.3}$$

$$P_{yr} = q\rho_{\omega}v(\sin\phi_2 + \sin\phi_1) \tag{8.5.4}$$

where

P_{xr} Horizontal component of centrifugal force in unit width, N/m;
P_{yr} Vertical component of centrifugal force in unit width, N/m;
φ_1,φ_2 The angle shown in Fig. C.5.4, °, taking its absolute value.

Other symbols are the same as above.

8.5.2.2 Internal Hydrostatic Loads (Uplift)

The uplift pressure acting on the spillway buildings shall be calculated according to the distributed force acting vertically on the total foundation bottom area of the buildings. Different uplift pressure distribution graphs can be selected according to the conditions of anti-seepage and drainage facilities. The uplift pressure distribution graphs acting on the bottom surface of the spillway control weir (gate) can be selected according to the four conditions in Fig. 8.20 (H = H_1–H_2; H_1—upstream water level; H_2—downstream water level; T_1, T_2—osmotic pressure reduction coefficient).

- When the foundation of weir (sluice) is provided with anti-seepage curtain and drainage curtain and the distance between them is not far, the distribution of uplift pressure is shown in Fig. 8.20a, where the uplift pressure coefficient is a = 0.25;
- when the distance between anti-seepage curtain and drainage curtain in weir (sluice) foundation is far (the distance between their center lines is more than 5m), the distribution of uplift pressure is shown in Fig. 8.20b, where a1 = 0.5, a2 = 0.25;

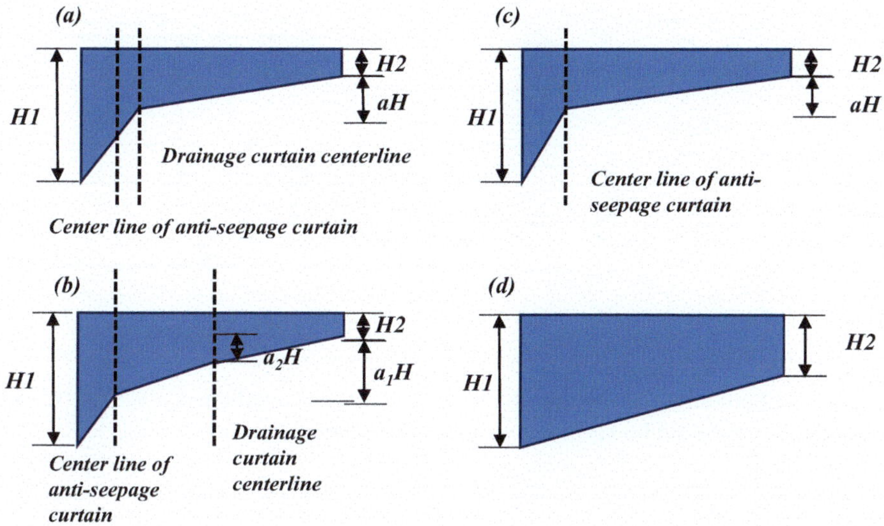

Fig. 8.20 Distribution of uplift pressure

- when the foundation of the weir (sluice) is provided with an anti-seepage curtain without a drainage curtain, or the drainage curtain fails, the uplift pressure distribution graph adopts Fig. 8.20c, at this time a = 0.5;
- when anti-seepage curtain and drainage curtain are not set in weir (sluice) foundation, the distribution of uplift pressure is shown in Fig. 8.20d.

The uplift pressure acting on the bottom surface of the fender can be calculated using the uplift pressure distribution graph shown in Fig. 8.21, where the uplift pressure coefficient a is defined as the ratio of the measured uplift pressure head to the downstream water level head based on the elevation of the drainage hole orifice in the drainage corridor.

- When the apron is not equipped with drainage facilities or the drainage facilities are unreliable, the uplift pressure coefficient $a = 1.0$.
- There is a vertical and horizontal drainage ditch system (with drainage holes in the ditch) under the apron, and its outlet is located below the water surface at the shrinkage depth of the side walls on both sides, or in the negative pressure area of the toe pier at the tail of the chute. When the gate is opened evenly, the uplift pressure coefficient $a = 0.8$–0.9 can be taken.
- There are vertical and horizontal drainage corridors or drainage ditches and drainage holes around and inside the apron. Drainage is provided on the surface of the foundation. Below the elevation of the drainage hole, there is a collecting well and an automatic pumping equipment to ensure drainage. The vertical and horizontal joints of the apron are equipped with reliable water-stopping facilities. When there are anti-seepage curtains and drainage curtains on the downstream

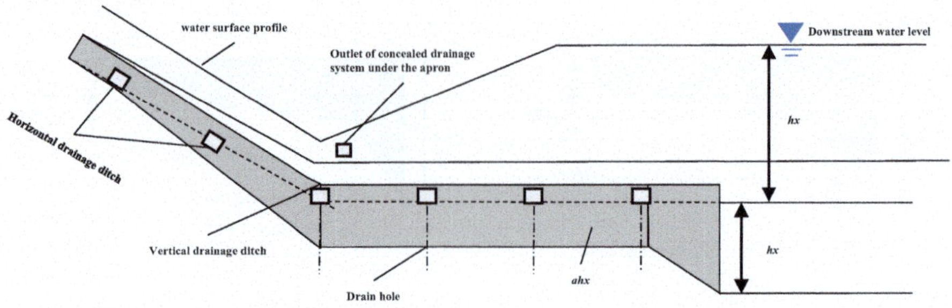

Fig. 8.21 The pressure distribution graph of the apron rising down

side and both sides of the apron, the uplift pressure coefficient $a = 0.2$–0.3. If necessary, it should be verified by electric simulation test.

8.5.2.3 Earth Pressures

When calculating the earth pressure of retaining structure (retaining wall), the active earth pressure can be calculated for the retaining structure moving or rotating outward.

For the retaining structure that remains stationary, it can be calculated according to the static earth pressure. The active earth pressure acting on the back of unit length retaining wall can be calculated as follows:

$$F_a = \frac{1}{2}\gamma H^2 K_a \tag{8.5.5}$$

where

- F_a Active earth pressure, kN/m, applied at the H/3 wall back from the base of the wall at an angle $(\delta + \varepsilon)$ to the horizontal plane (see Fig. C.6.2.1);
- γ Filling weight behind retaining wall, kN/m^3;
- H Retaining wall height, m;
- K_a The active pressure coefficient can be calculated by Eq. (C.6.2.3).

- When there is a uniform load on the filling surface, the load can be converted into an equivalent soil thickness to calculate the active earth pressure acting on the back of the wall. In this case, the active earth pressure acting on the back of the wall should be ladder-shaped distribution.

$$K_a = \frac{\cos(\varepsilon - \beta)}{\cos^2 \varepsilon \cos^2(\varepsilon - \beta + \phi + \delta)}$$
$$\times \{\cos(\varepsilon - \beta)\cos(\varepsilon + \beta) + \sin(\phi + \delta)\sin(\phi - \beta)$$

$$+ 2\eta \cos\varepsilon \cos\phi \sin(\varepsilon - \beta + \phi + \delta) - 2[(\cos(\varepsilon - \beta) \times \sin(\phi - \beta) + \eta \cos\varepsilon \cos\phi)(\cos(\varepsilon + \delta) \sin(\phi + \delta) \times \eta \cos\varepsilon \cos\phi)]^{1/2}\} \quad (8.5.6)$$

$$\eta = \frac{2c}{\gamma H} \quad (8.5.7)$$

where

- K_a coefficient of active earth pressure
- γ- Filling weight behind retaining wall, kN/m^3
- H Retaining wall height, m
- ε Angle between back of retaining wall and vertical surface, °
- β Slope angle of backfill surface behind retaining wall, °
- δ The external friction angle of the backfill behind the retaining wall to the back of the wall can be Table 8.11
- φ internal friction angle, °
- c Filling soil cohesion, kN/m^2.

The φ, c value of the backfill behind the retaining wall should be determined according to the test data.

When the test data is insufficient, the general soil can refer to Tables 8.12 and 8.13.

Table 8.11 External friction angle δ of retaining wall back

Retaining wall situation	δ
Wall back smooth, poor drainage	(0.00–0.33) φ
Rough wall back, good drainage	(0.33–0.50) φ
Wall back very rough, good drainage	(0.50–0.67) φ
Impossible to slide between wall back and fill	(0.67–1.00) φ

Table 8.12 The φ values for gravelly soil G and sandy soil S

Category	Loose condition	Medium dense state	Dense state
Gravel soil G	30°–34°	34°–37°	37°–40°
Sandy soil S	25°–30°	30°–35°	35°–40°

Table 8.13 The φ, c value of fine grained soil

Plasticity index I_p		Void ratio					
		< 0.5	0.5–0.6	0.6–0.7	0.7–0.8	0.8–0.9	> 0.9
< 10	φ (°)	27	25	23	21	19	17
	c (kN/m^2)	10	8	6	4	3	2
10–17	φ (°)	21	19	17	15	14	13
	c (kN/m^2)	18	14	11	9	8	6
> 17	h (°)	17	15	13	12	11	10
	c (kN/m^2)	35	28	22	17	13	10

For a retaining wall with a vertical back and a horizontal backfill surface, the static earth pressure acting on the unit length of the back can be calculated as follows (see Fig. 8.22):

$$F_0 = \frac{1}{2}\gamma H^2 K_0 \tag{8.5.8}$$

$$K_0 = \frac{v}{1 - v} \tag{8.5.9}$$

where

F_0 Static earth pressure, kN/m, applied at H/3 from the base of the wall horizontally to the back of the wall

K_0 Static earth pressure coefficient;

v Poisson 's ratio of backfill behind wall

K_a The active pressure coefficient can be calculated by (8.5.6)

- If the backfill behind the wall is normally consolidated clay, K_0 can also be calculated by:

$$K_0 = 1 - \sin j' \tag{8.5.10}$$

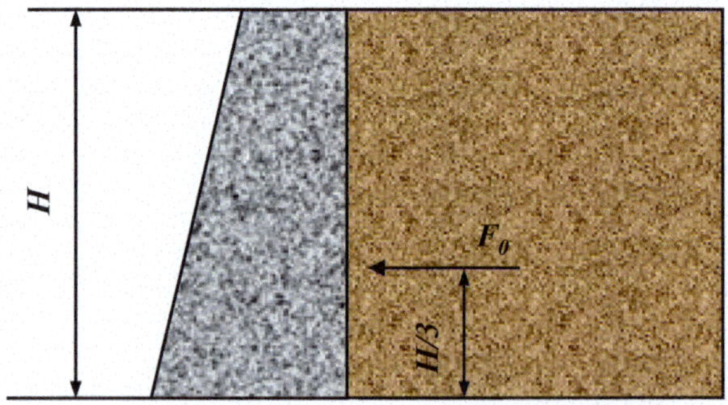

Fig. 8.22 Diagram of static earth pressure

Table 8.14 Static earth pressure coefficient

Category	Soil state	K_0
Gravel soil (G)		0.22–0.40
Sandy soil (S)		0.30–0.60
Low liquid limit silt (ML) Low liquid limit clay (CL)	Hard and rigid plastic	0.40
	Plasticity	0.52
	Soft plastic and flow plastic	0.64
High liquid limit clay (CH)	Hard and rigid plastic	0.40
	Plasticity	0.64
	Soft plastic and flow plastic	0.87

where

φ' Effective internal friction angle of backfill behind wall average wave period, s.

- When the experimental data of v, φ' are not enough, K_0 can be used with reference to Table 8.14.

8.5.3 Structural Strength, Stability Analysis of Water Discharge and Conveyance Structures

The items to be reviewed for discharging safety mainly include discharging ability, height of discharging channel side wall, cross section of passage of pressure less discharging tunnel and energy dissipation and erosion prevention. Suitable calculation methods and calculation models can be selected in accordance and on the basis of architectural style of the structures, characteristics of materials and feature of passage. Anti-cavitations capacity and anti-floating safety of the bottom plate shall also be reviewed in the high speed water flow area.

The items to be reviewed for structure strength and stability of spillway mainly include anti-slide stability, stability against floating, stress and strength at the control section, discharging channel, flip bucket, apron of stilling basin, and along the base faces of the related side walls, retention walls and guiding walls.

For structure safety of the hydraulic tunnels, the stability of tunnel surrounding rock and safety of supporting structure shall be reviewed. Concrete and reinforced concrete lining shall be calculated in the ultimate state of bearing capacity and checked in terms of its function, impermeability, durability and impermeability of surrounding rocks.

For evaluation of stability of the tunnel surrounding rock, the original design and geological survey data disclosed after excavation shall be collected. To analyze and assess the stability of the tunnel surrounding rock, geological survey shall be carried out if necessary. Check calculation of lining structure can be carried out based on the features of the lining structure, load action mode and conditions of surrounding rock and by using suitable calculation methods and calculation models.

The structure strength and stability of the structures at the water inlet shall be reviewed in accordance. Reference to structural calculation of gravity dam.

8.6 Evaluation of Stability of Near Dam Bank Slope

For near dam bank slope that influences the dam safety, the stability of the side slope shall be analyzed and calculated using geological survey and monitoring data. The methods of analysis and control standard shall be in accordance.

The results of the groundwater-surface contour permeability ratio drop and seepage flow rate should be obtained in seepage calculation. The seepage calculation of grade 1 and grade 2 slopes should be carried out by numerical analysis method, and that of grade 3 and below can be calculated by formula.

Seepage calculation parameters should be determined by field test, indoor test, and engineering analogy. For slopes with complex geological conditions, the inversion analysis method should also be used for review and correction. When drainage is set on the slope, the influence of effective and ineffective drainage measures on the seepage field should be considered in seepage calculation.

The limit equilibrium method shall be the basic calculation method for stability against sliding [36]. For the grade 1 slope, the finite element method with a reduced strength index can be used to check its stability against sliding at the same time [37]. For soil slopes and rocky slopes with cataplastic and bulk structures, when the sliding surface is circular, the Simplified Bishop method and Morgenstern-Price method should be used to calculate the stability against sliding. When the sliding surface is non-circular, the Morgenstern-Price method and unbalanced thrust transfer method should be used to calculate the stability against sliding.

For rock slopes with block and layered structures, the Sama method and unbalanced thrust transfer method should be used to calculate the stability against sliding [38]. For slopes with wedge potential sliding mass cut by two or more structural planes such as joints and cracks, the wedge method should be used to calculate the stability against sliding. For Class 1 and Class 2 slopes, the above suitable methods should be used to calculate the stability against sliding and take comprehensive judgment values.

For the new and old slip mass or potential slip mass dear the dam of large sized reservoir, monitoring of deformation and underground water shall be carried out. The monitoring data shall be sorted out and analyzed at regular interval to identify its stability. If conditions permit, mathematic models shall be established for monitoring of the stability of the side slope.

For stability of the rock side slope of category 1 and 2 near the dam, special research and demonstration shall be conducted.

8.7 Conclusions on Structure Safety Evaluation

8.7.1 Conclusion of Structural Safety Review

For review of structure safety, we shall come to the clear conclusions on

- Whether the anti-slide stability of the earth-rock fill dams and the upstream slope protection meet the requirements of specifications and whether the strength and stability of concrete dam and the dams of other materials meet the requirements of specifications;
- Whether the deformation pattern of the dam is normal and whether abnormal deformation exists that would threaten the safety of the dam;
- Whether the discharging safety, structure strength and stability of discharging and conveyance and ship passage structures meet the requirements of specifications;
- Whether the near dam bank slope is stable.

8.7.2 Structural Safety Evaluation Grade

1. The dam structure can be considered safe and can be rated as Class A provided that the strength, stability, discharging safety of the dam and discharging and conveyance and ship passage structures satisfy the requirements of specifications and there is no abnormal deformation and the near dam bank slope is stable.
2. The dam structure can be considered fairly safe and can be rated as Class B provided that the holistic stability and discharging safety of the dam and discharging and conveyance and ship passage structures satisfy the requirements of specifications, there exist local insufficient strength or abnormal deformation but they do not seriously influence the safety of the project and the near dam bank slope is holistically stable.
3. The dam structure can be considered unsafe and shall be rated as Class C when the strength, stability, discharging safety of the dam and discharging and conveyance and ship passage structures do not satisfy the requirements of specifications and abnormal deformation that threatens the safety of the project or the near dam bank slope is not stable.

References

1. Lin J (2018) Hydraulic buildings, 6th edn. China Water and Power Press, 978-7-5170-7791-6
2. Zhou J, Yao S, Wang M, Zhu Y, Chen D (2020) Review on overtopping failure and flood evolution of earth-rock dams. Adv Water Sci 31(2):287–301. https://doi.org/10.14042/j.cnki.32.1309.2020.02.015

3. Foster M, Fell R, Spannagle M (2000) The statistics of embankment dam failures and accidents. Can Geotech J 37:1000–1024. https://doi.org/10.1139/t00-030
4. Al-Riffai M, Nistor I (2010) Impact and analysis of geotechnical processes on earthfill dam breaching. Nat Hazards 55:15–27. https://doi.org/10.1007/s11069-010-9586-6
5. Fell R, Wan CF, Cyganiewicz J, Foster M (2003) Time for development of internal erosion and piping in Embankment dams. Journal of Geotechnical and Geoenvironmental Engineering 129:307–314. https://doi.org/10.1061/(asce)1090-0241(2003)129:4(307)
6. Clague J (2000) A review of catastrophic drainage of moraine-dammed lakes in British Columbia. Quatern Sci Rev 19:1763–1783. https://doi.org/10.1016/s0277-3791(00)00090-1
7. Zhao G, Jiang Y, Qiao J et al (2018) Numerical and experimental study on the formation mode of a landslide dam and its influence on dam breaching. Bull Eng Geol Env 78:2519–2533. https://doi.org/10.1007/s10064-018-1255-0
8. Scaioni M, Marsella M, Crosetto M et al (2018) Geodetic and remote-sensing sensors for dam deformation monitoring. Sensors 18:3682. https://doi.org/10.3390/s18113682
9. Taccedil L (2010) Analysis of dam deformation measurements with the robust and non-robust methods. Sci Res Essays 5:1770–1779. https://doi.org/10.5897/sre.9000180
10. Li S, Li Y, Si Z, Zhang X (2010) A seepage computational model of face slab cracks based on equi-width joint constant flow. Adv Eng Softw 41:1000–1004. https://doi.org/10.1016/j.advengsoft.2010.04.004
11. Sjödahl P, Dahlin T, Johansson S (2009) Embankment dam seepage evaluation from resistivity monitoring data. Near Surf Geophys 7:463–474. https://doi.org/10.3997/1873-0604.2009023
12. Yang X, Chi S (2014) Seismic stability of earth-rock dams using finite element limit analysis. Soil Dyn Earthq Eng 64:1–10. https://doi.org/10.1016/j.soildyn.2014.04.007
13. Cheng L (2005) Study on high pore pressure in clay core wall of earth dam in Xiaolangdi project. J Hydraul Eng 36:219–224. https://api.semanticscholar.org/CorpusID:131008161
14. Wang N, Zhang W, Gu X, Zhang D (2013) Centrifugal model test on seepage characteristics of high core Rockfill dam. Rock Soil Mech 34:2769
15. Xu W-J, Xu Q, Hu R-L (2011) Study on the shear strength of soil–rock mixture by large scale direct shear test. Int J Rock Mech Min Sci 48:1235–1247. https://doi.org/10.1016/j.ijrmms.2011.09.018
16. Yin ZZ (2017) Geotechnical principles. China Water and Power Press, 978-7-5084-4209-9
17. Steven E et al (2015) Physical geology. Victoria, Bccampus, Bc Open Textbook Project
18. Kumar M, Rana S, Pant PD, Patel RC (2017) Slope stability analysis of Balia Nala landslide, Kumaun Lesser Himalaya, Nainital, Uttarakhand, India. J Rock Mech Geotech Eng 9:150–158. https://doi.org/10.1016/j.jrmge.2016.05.009
19. Chen J, You W (2011) Improved rigid body limit equilibrium method for analysis of dam foundation stability against deep sliding. J Hydraul Eng 42:963–969
20. Zheng Y-R (2004) An extension of simplified Bishop method and its application to non-circular slip surface for slope stability analysis. Rock Soil Mech 25:927
21. Sun G, Cheng S, Jiang W, Zheng H (2016) A global procedure for stability analysis of slopes based on the Morgenstern-Price assumption and its applications. Comput Geotech 80:97–106. https://doi.org/10.1016/j.compgeo.2016.06.014
22. Bishop AW (1955) The use of the slip circle in the stability analysis of slopes. Géotechnique 5:7–17. https://doi.org/10.1680/geot.1955.5.1.7
23. Ouyang W-H, Liu S-W, Yang Y (2022) An improved Morgenstern-Price method using Gaussian quadrature. Comput Geotech 148:104754. https://doi.org/10.1016/j.compgeo.2022.104754
24. Ma G-J, Wu C-W, Zhou P (2007) Hydrodynamics of slip wedge and optimization of surface slip property. Sci China Ser G 50:321–330. https://doi.org/10.1007/s11433-007-0034-x
25. Su H, Li J (2018) Review on performance evaluation of structural reinforcement of gravity dams. J Hydroelectr Eng 37:12–25. https://doi.org/10.11660/slfdxb.20180402
26. Lombardi M, Haftka RT (1998) Anti-optimization technique for structural design under load uncertainties. Comput Methods Appl Mech Eng 157:19–31. https://doi.org/10.1016/s0045-7825(97)00148-5

27. Gu C, Su H, Liu H (2018) Review on service risk analysis of dam engineering. J Hydraul Eng 49:26–35. https://doi.org/10.13243/j.cnki.slxb.20170721
28. Bérubé M-A, Duchesne J, Dorion JF, Rivest M (2002) Laboratory assessment of alkali contribution by aggregates to concrete and application to concrete structures affected by alkali–silica reactivity. Cem Concr Res 32:1215–1227. https://doi.org/10.1016/s0008-8846(02)00766-4
29. Leger P, Cote P, René T (1996) Finite element analysis of concrete swelling due to alkali-aggregate reactions in dams. Comput Struct 60:601–611. https://doi.org/10.1016/0045-7949(95)00440-8
30. Omikrine Metalssi O, Seignol J-F, Rigobert S, Toutlemonde F (2014) Modeling the cracks opening–closing and possible remedial sawing operation of AAR-affected dams. Eng Fail Anal 36:199–214. https://doi.org/10.1016/j.engfailanal.2013.10.009
31. Xiang Y, Su H-Z, Wu Z-R (2004) Inverse analysis of mechanical parameters based on dam safety monitoring data. J Hydraul Eng 35:98–102. https://doi.org/10.3321/j.issn:0559-9350.2004.08.018
32. Zhang LH, Zhang LX, Cheng SZ (2014) Stability analysis of gravity dam on soil foundation. Appl Mech Mater 501–504:79–82. https://doi.org/10.4028/www.scientific.net/amm.501-504.79
33. Ren Z (2014) Finite element simulation analysis of concrete hyperbolic arch dam. Appl Mech Mater 488–489:581–584. https://doi.org/10.4028/www.scientific.net/amm.488-489.581
34. Bayraktar A, Sevim B, Can Altunişik A (2011) Finite element model updating effects on nonlinear seismic response of arch dam–reservoir–foundation systems. Finite Elem Anal Des 47:85–97. https://doi.org/10.1016/j.finel.2010.09.005
35. Bretas EM, Lemos JV, Lourenço PB (2014) A DEM based tool for the safety analysis of masonry gravity dams. Eng Struct 59:248–260. https://doi.org/10.1016/j.engstruct.2013.10.044
36. Si C-D (2007) Fuzzy comprehensive evaluation on safe operation of water conveyance structures. Water Resourc Hydropower Eng 38
37. Duncan JM (1996) State of the art: limit equilibrium and finite-element analysis of slopes. J Geotech Eng ASCE 122:577–596
38. Dawson EM, Roth WH, Drescher A (1999) Slope stability analysis by strength reduction. Géotechnique 49:835–840. https://doi.org/10.1680/geot.1999.49.6.835

Chapter 9
Evaluation on Seismic Safety

9.1 Introduction

China is situated within the seismic belts of the Pacific and Eurasian tectonic plates, positioning it as a country prone to earthquakes. This geographical context underscores China's national situation marked by frequent occurrences of powerful earthquakes and the resultant significant seismic disasters [1]. Similar to seismic patterns globally, the genesis and manifestation of earthquakes in China are intricately tied to active fault lines. The seismic activity across mainland China is distinguishable by several key features. These include a high frequency of earthquakes, their considerable magnitude, widespread distribution, shallow focal depths, and notably severe consequences in terms of the destruction caused. The combination of China's geological position and its intricate tectonic interactions makes it susceptible to recurrent seismic events, the country's geographical location renders it particularly prone to powerful earthquakes, and the associated seismic hazards have a substantial impact on various aspects of life, infrastructure, and environment (Figs. 9.1 and 9.2).

When an earthquake occurs in the vicinity of a reservoir dam or within its vicinity, the abrupt surge in seismic forces triggers a series of responses within the dam structure. These responses stem from a combination of internal factors inherent to the dam itself, as well as external influences like the dynamic water pressure exerted by the reservoir [2]. The consequences of an earthquake on dams are manifold:

1. **Deformation and Displacement**: The seismic inertial forces cause the reservoir dam to deform, resulting in relative displacement at its base. This can arise from a blend of internal and external forces acting simultaneously.
2. **Damaging Effects**: Earthquake-induced damage to dams includes cracking, landslides, heightened seepage, liquefaction of sandy foundations, subsidence of the dam's crest, seismic surges, uplift and sliding of slopes, ruptures in buried underground pipes, impairment of wave walls and concrete structures, malfunction of gates and control mechanisms, and potential collapse of embankments within the reservoir region.

Y. Xiang et al., *On-site Inspection and Dam Safety Evaluation*, Hydroscience and Engineering, https://doi.org/10.1007/978-981-97-3987-5_9

Fig. 9.1 Construction joint cracking of Zipingpu dam in China after 5.12 earthquake (concrete faced dam, dam height 156 m)

Fig. 9.2 Opening of transverse joints of Pacoima dam in United States after the earthquake (arch dam, dam height 113 m)

3. **Critical Impact**: Earthquakes measuring 7 or more on the magnitude scale can also induce severe misalignment and sliding of the dam. In situations where reservoir water levels are high, the limited time available for protective measures could lead to the rapid deterioration of the dam's condition. A relatively short period without timely intervention may culminate in dam failure, resulting in extensive and catastrophic losses.

The immense destructive potential of earthquakes in the vicinity of reservoir dams underscores the necessity for robust engineering measures and proactive disaster preparedness. Earthquake-resistant designs, monitoring systems, emergency response plans, and prompt actions in the event of seismic activity are crucial to mitigating the potential consequences of such events and safeguarding lives, infrastructure, and ecosystems.

The purpose of evaluation on seismic safety is to review whether the current dam performance meets the seismic requirements [3]. The following main contents shall be included in the evaluation on seismic safety:

- Review of basic seismic intensity and seismic fortification category of the project site. On this basis, review whether the seismic fortification intensity or the ground motion parameters meet the requirements;
- Review the dam seismic stability and structural strength;
- Review the earthquake induced permanent deformation of the dam and foundations, and the possibility of seismic liquefaction;
- Review the rationality and completeness of the anti-seismic measures for the project;
- Analysis of seismic prototype monitoring data for dams equipped with strong earthquake monitoring array.

The purpose of seismic design criteria is to provide guidelines and procedures for obtaining earthquake ground motion parameters for use in evaluating the seismic response of a given structure or facility [4]. Presently, there are three ways by which the earthquake ground motion parameters can be ascertained: (1) use of local building codes; (2) conducting a deterministic seismic hazard analysis; (3) conducting a probabilistic seismic hazard analysis. Typically, local building codes are intended to mitigate collapse of buildings and loss of life, and do not apply to structures covered in this document. Both deterministic and probabilistic seismic hazard analyses and evaluations are covered in this document [5] (Table 9.1).

Table 9.1 Determination of design seismic intensity or bedrock dynamic peak acceleration of hydraulic structures

Project size or building level	Regional seismic conditions	Method for determining basic intensity or peak dynamic acceleration of bedrock	Design seismic intensity
Building class 2 and below	Commonly	Determination of seismic ground motion parameter area	Basic intensity
Building class 1	Buildings classified as category A	Basic intensity determined by zoning map of ground motion parameters and site seismic risk considered	Increase by 1° based on basic intensity
Dam height greater than 200 m or reservoir capacity greater than 10 billion m^3	Basic intensity 6° and above	Bedrock peak acceleration exceeding probability level: $P_{100} = 0.02$ for backwater structures; $P_{50} = 0.05$ for non-backwater structures	The peak acceleration of bedrock motion is determined according to the results of special seismic risk analysis
Large-scale reservoir with dam height greater than 150 m	Basic intensity 7° and above		

Seismic safety assessments are mandatory for dams whose seismic fortification intensity surpasses a magnitude of 6°. For seismic fortification intensities of 6° or lower, seismic calculations may not be obligatory. However, for Class I hydraulic structures, it is imperative to review anti-seismic measures. In instances where hydraulic structures possess a seismic fortification intensity that exceeds 9° or involve water-retaining structures exceeding a height of 250 m, a specialized investigation and demonstration pertaining to their seismic safety are necessary. The results of investigations are to be submitted to the relevant regulatory authority and industry experts for approval [5]. When the initially designed seismic fortification intensity or the specified ground motion parameters do not align with current specifications, necessary adjustments must be made to the seismic fortification intensity and ground motion parameters. Subsequently, the approval procedures mandated by relevant authorities need to be carried out. These measures underscore the importance of ensuring the seismic resilience of hydraulic structures, particularly in regions prone to seismic activity. By subjecting these structures to rigorous seismic safety assessments and adjustments as needed, their ability to withstand and mitigate potential earthquake-related impacts is greatly enhanced.

The emergency preparedness plan for earthquake prevention and disaster reduction for reservoir dams shall satisfy the provisions with relevant laws and regulations. The safety for opening and closing of the discharging structures is the fundamental condition to guarantee rapid drawdown of reservoir water level. Therefore, special attention shall be paid to the inspection of the seismic safety of opening and closing equipment, standby emergency power supply, and oil and material reserves, to ensure the opening and closing equipment of the water discharge building can be quickly and urgently started after the earthquake [6, 7]

9.2 Review on Design of Seismic Intensity

Seismic fortification intensity stands as a crucial safeguard, ensuring that a dam remains resilient and avoids substantial damage when subjected to earthquakes within the prescribed seismic standards [8]. In the routine evaluation of dam safety, adjustments are necessary for reservoir dams whose existing seismic fortification intensity does not align with mandated criteria. To facilitate these adjustments, the ground motion parameters at the project site and the corresponding seismic fortification intensity must be ascertained, as stipulated in Table 9.2. This recalibrated seismic fortification intensity forms the foundation for subsequent seismic safety evaluations of the dam.

The design seismic acceleration, represented by "a_h" should adhere to the specifications detailed in Table 9.3. This value serves as a critical benchmark for seismic considerations, guiding the dam's structural design and overall preparedness to withstand seismic events. To determine the probability level of the representative value of designed seismic acceleration for those dams with professional seismic risk analysis [9], the exceeding probability within 100 years reference period P_{100} is 0.02 for water

Table 9.2 Seismic fortification standard of dam

Dam grade	Basic seismic intensity of the site J	Seismic fortification category	Design intensity J_C	Calculation method of seismic action effect
1	≥ VI	Class A	$J_C = J + 1$	Dynamic method
2		Class B	$J_C = J$	Dynamic method or quasi-static method
3	≥ VII	Class C	$J_C = J$	
4, 5		Class D	$J_C = J$	Pseudo static method or anti-seismic measures

retaining structures, and the exceeding probability within 50 years reference period P_{50} is 0.05 for non-water retaining structures.

By systematically recalibrating seismic fortification intensity and aligning it with ground motion parameters, dam operators and engineers bolster the dam's resilience and minimize potential damage during earthquakes. This methodical approach to seismic safety ensures that dams can withstand seismic forces while safeguarding both infrastructure and surrounding communities.

Table 9.3 Representative value of horizontal design seismic acceleration a_h

Design intensity	7	8	9
a_h	0.1 g	0.2 g	0.4 g

Notes

(1) use the basic seismic intensity as the seismic fortification intensity

(2) For the hydraulic structures with the seismic fortification category of Class A, their seismic fortification intensity shall be increased by 1° based on the basic seismic intensity according to the hazard of the strong earthquake

(3) For the dam height exceeds 200 m or the reservoir storage capacity is greater than 10 billion m^3 in the areas where basic seismic intensity is 6° or above, and for the large project (scale 1) with dam height exceeds 150 m in the areas where basic seismic intensity is 7° or above, the ground motion parameters shall be taken in accordance with the results of exceeding probability of bedrock peak acceleration provided by professional seismic risk analysis

9.3 Evaluation on Seismic Safety for Earth-Rock Dam

The seismic safety evaluation of earth-rock dams (including soil foundations of other hydraulic structures) primarily entails assessing whether the dam's seismic stability and implemented anti-seismic measures adhere to the prescribed code. If deemed necessary, calculations related to the dam's permanent deformation and the potential for liquefaction should also be conducted (Fig. 9.3). By conducting a thorough seismic safety evaluation with a focus on stability, anti-seismic measures, permanent deformation, and potential liquefaction, dam operators and engineers can ensure the structural integrity and resilience of earth-rock dams [10].

9.3.1 Seismic Stability Review

The pseudo-static method shall be adopted for seismic stability review [11]. For earth-rock dams whose seismic fortification category is Class A, designed seismic intensity is 8°, 9° with the dam height exceeding 70 m, or liquefiable soil is present in the foundation, the following requirements shall be met when reviewing:

- The finite element method shall be used to analyze the dynamic responses of the dam and its foundation, with which comprehensively judgement of dam seismic stability and seismic liquefaction possibilities shall be made.
- The dam permanent deformation caused by earthquake shall be calculated based on dynamic analysis [12]. And then the top elevations of the dam crest, the wave-proof wall and the anti-seepage body shall be reviewed with concerning the permanent deformation to check whether the elevations do meet the requirements.

It is necessary to comprehensively determine whether there is possibility of seismic liquefaction of the dam body and the dam foundation [13].

The seismic effect is calculated by quasi-static method [14]. The representative values of horizontal seismic inertia force acting on particle i along the height of the building are as follows:

$$F_i = a_h \xi G_{Ei} \alpha_i / g \tag{9.3.1}$$

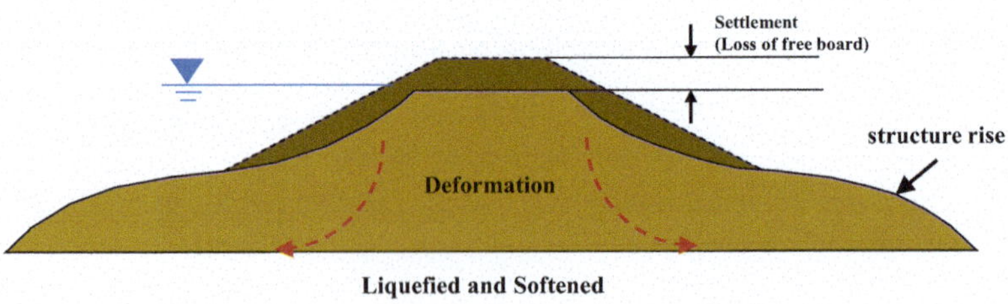

Fig. 9.3 Evaluating the influence of the earthquake safety of the dam

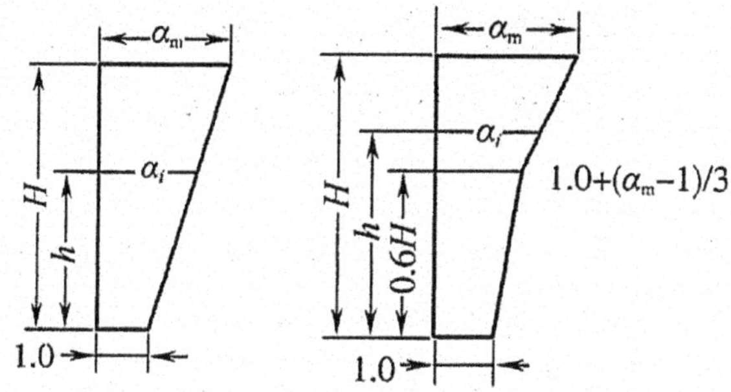

Fig. 9.4 Schematic diagram of dynamic distribution coefficient of seismic inertia force of earth-rock dam body

where

F_i Representative value of horizontal seismic inertia force acting on particle i
a_h Representative values of horizontal design seismic acceleration
ξ The effective reduction factor of earthquake action is 0.25 unless otherwise specified.
G_{Ei} Standard values of gravity action concentrated at particle i
α_i Horizontal design seismic acceleration
g Gravity acceleration.

For earth-rock dams, the values shown in Fig. 9.4 are used for seismic stability calculation by pseudo-static method. In the figure, when the design intensity is 7°, 8° and 9°, α_i are 3.0, 2.5 and 2.0 respectively.

9.3.2 Seismic Fortification Measures

The implemented anti-seismic measures are critically reviewed to ensure they meet the code requirements. These measures encompass engineering designs, structural reinforcements, and any additional provisions aimed at minimizing seismic vulnerabilities.

Selection of Earth-rock Dam Types: When the design intensity is 8° or 9°, specific considerations should be made regarding the design of rockfill dams, and the rigid core wall should not be used for the seepage-proof body. When choosing homogeneous dam, internal drainage system should be set to reduce the saturation line.

Dam axis design: For earth-rock dam construction in seismic area, straight or upstream bending dam axis should be adopted, and downstream bending, folded or *S*-shaped dam axis should not be used.

Dam crest superelevation: According to the design intensity and water depth in front of the dam, the height of seismic surge should be 0.5–1.5 m. The surge caused by large-scale bank collapse and landslide caused by earthquake in reservoir area should be specially studied. When the design intensity is 8° or 9°, the safety superelevation shall be taken into account the additional settlement of dam and foundation under earthquake action [15].

Dam section design: When the design intensity is 8° or 9°, the dam crest should be widened and the section with gentle upper part and steep lower part should be adopted. The dam slope can be compacted by large rocks or reinforced in soil.

Impervious core design: The impervious body of earth-rock dam should be strengthened, especially the top of the dam body, the connection between the dam and rigid buildings such as bank slope or concrete, which are liable to crack during earthquake. The filter layer and transition layer should be set on the upstream and downstream sides of the impervious body, and must be compacted and appropriately thickened [16].

Dam material selection: Earth-rock materials with good seismic performance and permeability stability and good gradation should be selected for dam construction. Uniform medium sand, fine sand, silt and silt are not suitable for dam construction in seismic area.

Embedded pipes: For Class 1 and 2 earth-rock dams, it is not appropriate to lay water conveyance pipe under the dam. When the pipe must be buried under the dam, the reinforced concrete pipe or cast iron pipe should be used, and it should be placed in the bedrock trough, with the top of the pipe flush with the bottom of the dam and the outside of the pipe backfilled with concrete; the seepage prevention and water stopping at the pipe connection should be done well, and the control gate of the pipe should be placed at the upstream end of the intake or the seepage-proof body.

9.4 Evaluation on Seismic Safety for Gravity Dam

The seismic safety evaluation of gravity dams shall mainly review whether the strength of the dam, the overall anti-sliding stability and the anti-seismic measures meet the requirements of the specifications [17]. For roller compacted concrete (RCC) gravity dam, the anti-sliding stability along the compacted layer shall be also reviewed. Under the action of earthquake, the dam turning section, dam joints, discharging openings/holes and the junctions of structures are prone to generate stress concentration effect and high tensile stress zones [18]. Therefore, it is important to review the strength of these parts and whether their anti-seismic measures meet the requirements.

9.4.1 The Basic Information

The selection of earthquake loadings should be a result of careful consideration, taking into account the anticipated accelerations specific to each project site. These accelerations are determined based on various factors such as the geological conditions of the site, the proximity to significant fault lines, and the seismic history of the region as evidenced by existing seismic records [19].

The geological attributes of the site, including soil type, subsurface conditions, and terrain, play a crucial role in influencing the potential seismic impacts. The distance from major fault lines in the vicinity of the project site is a key factor to gauge the potential earthquake effects. Sites closer to active faults are more likely to experience stronger seismic forces. Understanding the historical seismic activity of the region helps in predicting the range of seismic forces that the site might encounter. Analyzing available seismic records provides valuable insights into the magnitude and frequency of past seismic events in the area [20].

For projects lacking comprehensive seismicity studies, seismic risk maps can be employed to establish a probability zone [21, 22]. These maps indicate the likelihood of experiencing seismic events of varying magnitudes. This approach is especially relevant for projects situated in regions without detailed seismic data. Ultimately, the goal is to select earthquake loadings that accurately reflect the potential seismic conditions at each project site.

9.4.2 Review of Seismic Strength of Gravity Dam

In the long-term engineering practice, rich experiences are accumulated in calculation of dam body strength by materials mechanics method. The basic analysis method for review of the strength of gravity dam shall be material mechanics method in which bending and shear deformation under the action of static and dynamic force are considered simultaneously. It is stipulated that the material mechanics method is the basic method for dynamic analysis of gravity dam. For the gravity dam whose seismic fortification category is Class A or whose structures and geological conditions are complex, finite element method is advised to be used at the same time for dynamic analysis [23]. For the important gravity dams in meizoseismal areas, non-linear finite element dynamic analysis can also be used for review of seismic safety.

9.4.3 Review of Anti-seismic and Anti Sliding Stability of Gravity Dam

1. **Review Calculation**

The anti-sliding stability check of the gravity dam shall be calculated by the shear strength formula. When weak interlayer and structural plane with gently inclined angle exist in dam foundation, specialized study shall be made to review the anti-sliding stability of the dam with part of the bedrock [24].

When dynamic analysis of stability against sliding of the gravity dam is carried out, the shear strength formula is difficult to meet the calculation requirements. Therefore, the shear strength formula based on the rigid body limit equilibrium method shall be used for analysis [25]. Considering difficulties and complexity of the seismic stability evaluation for the dam foundation sliding along the weak interlayer or the gently inclined structural plane, while the analysis method, the determination method and standard of material physical and mechanical index, the stability criterion, etc. need further study, therefore, the guideline stipulates that special research should be conducted.

2. **Selection of Calculation Criteria**

When calculating the seismic effect of gravity dam by dynamic method, the mode decomposition response spectrum method should be used [26]. The seismic effects of each order can be combined by the square root method. When the ratio of absolute value of frequency difference between two modes to one of the smaller frequencies is less than 0.1, the seismic effect should be combined by the fully quadratic square root method.

For gravity dams of special importance, supplementary calculation should be carried out by time history analysis method (Fig. 9.5). At the same time, when checking the strength of gravity dam body and the stability against sliding on the dam foundation surface by dynamic method [27], the structural coefficients of compressive and tensile strength should be 2.00 and 0.85 respectively, and that of anti-sliding stability should be 0.60.

When calculating seismic effect of gravity dam by quasi-static method, the representative values of horizontal seismic inertia force [28] of each particle should be calculated according to Eq. (9.4.1). The dynamic distribution coefficient a_i shall be determined as follows:

$$a_i = 1.4\frac{1 - 4(h_i/H)^4}{1 + 4\sum_{j=1}^{n}\frac{G_{Ej}}{G_E}(h_j/H)^4} \tag{9.4.1}$$

where

n Calculate the total number of particles
H Dam height

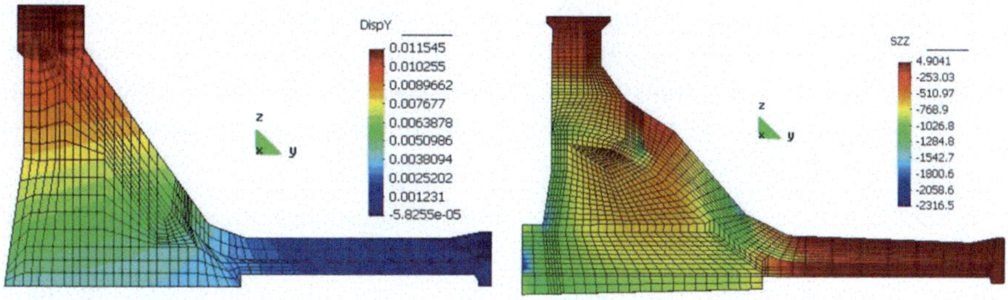

Fig. 9.5 Finite element calculation results under seismic loads

h_i, h_j Height of particle i and j
G_{Ej} Standard value of gravity action concentrated on particle j
G_E Total gravity of buildings generating seismic inertial force.

9.4.4 Seismic Fortification Measures

Seismic measures for gravity dams involve various design considerations to enhance their earthquake resistance and overall stability. These measures include:

1. **Simplified Shape and Slopes**: Gravity dams should have a straightforward shape with gradual slopes to minimize stress concentration. The top slope is best designed as curved to distribute loads more evenly and prevent potential weak points.
2. **Crest Positioning**: To prevent instability and erosion, the dam crest should be positioned away from the upstream face, avoiding placement too close to the water source.
3. **Upper Part Enhancement**: Enhance the upper part of the dam by reducing weight and increasing rigidity. This can involve improving the concrete grade, introducing proper reinforcement, and optimizing the design to better withstand seismic forces.
4. **Foundation Treatment**: Address weak areas like fractures, fractured zones, and weak interlayers in the foundation through engineering treatments. Properly reinforce these sections and raise the concrete grade at the bottom to bolster the foundation's strength.
5. **Light and Simple Crest**: Utilize a lightweight, uncomplicated, and integral auxiliary structure for the dam crest. Avoid heavy bridges or tall tower structures that might introduce instability or stress.
6. **Traffic Bridge Connection**: If incorporating a traffic bridge atop an overflow dam section, strengthen the connection and enhance the lateral stiffness of the piers to effectively manage loads.

7. **Transverse Joints for Abrupt Changes**: In areas where the dam body section experiences sudden changes along its axis or shifts in topography and geology, consider implementing transverse joints. Opt for joint types with ample deformability and water-sealing capabilities to accommodate potential movements and changing water pressures.

By applying these seismic measures, engineers can enhance the seismic resilience of gravity dams, mitigating the potential for damage during earthquakes. By implementing the comprehensive methods outlined above, the safety and long-term stability of these essential structures can be guaranteed even under the challenges posed by seismic stress.

9.5 Evaluation on Seismic Safety for Arch Dam

The seismic safety evaluation of arch dams is primarily centered around three fundamental aspects to ensure the dams' stability and resilience during seismic events, which including the dam strength, the abutment stability, and the anti-seismic measures.

The evaluation of dam strength focuses on reviewing the dam's inherent strength to determine if it aligns with the specified requirements. This assessment ensures that the dam's structural integrity is sufficient to withstand the forces generated by seismic events. The evaluation of the abutment stability includes a comprehensive assessment of the stability of the dam's abutments. These are the natural or engineered supports that flank the sides of the dam. It is crucial to determine if the abutments can effectively withstand the dynamic loads introduced by seismic forces. The assessment examines the anti-seismic measures that have been implemented. These measures encompass a range of engineering strategies, reinforcements, and design considerations aimed at enhancing the dam's ability to resist seismic impacts [29].

By focusing on these core elements, the seismic safety evaluation ensures that arch dams are well-equipped to handle seismic challenges. This approach aligns with established specifications, promotes the structural soundness of the dams, and contributes to the safety of surrounding areas and communities during seismic events.

9.5.1 Review of Arch Dam Seismic Strength

The arch beam load sharing method (dynamic and static) is the foundational analysis approach for assessing the strength of arch dams. This method takes into consideration both dynamic and static forces within the dam structure, offering a comprehensive understanding of its behavior under varying conditions.

For specific scenarios where the seismic fortification category of the arch dam falls under Class A or where the structural and geological conditions are intricate and demanding [30], the finite element method is recommended for conducting dynamic analysis. The finite element method is a powerful computational technique that allows for detailed modeling of complex interactions within the dam, considering factors like material properties, geometry, and dynamic loading. This method offers a more accurate representation of the dam's response to dynamic forces induced by seismic events. The dynamic method or quasi-static method should be used for calculating the seismic effects of dams. For arch dams with seismic fortification categories B and C whose design intensity is less than 8° and dam height is less than or equal to 70 m, the quasi-static method can be used [31].

9.5.2 Review of Anti Sliding Stability of Arch Dam Abutment

Indeed, the stability analysis of abutments in arch dams is a multifaceted and crucial task, influenced by a multitude of factors including the dam site's topography, geological attributes, the behavior of sliding bodies, and the dynamic properties of the rock mass [32]. Given the complexity of this analysis, particularly in relation to seismic events, several aspects require ongoing research and investigation:

1. Seismic Stability Calculation Method: The methods for calculating the seismic stability of abutments are continually evolving as new insights are gained. This includes refining techniques to accurately account for dynamic forces induced by earthquakes.
2. Ground Motion Input: The input of accurate ground motion data is pivotal for realistic seismic stability assessments. Advances in recording and analyzing ground motions contribute to a more precise understanding of potential seismic impacts.
3. Material Parameter Determination: The determination of material parameters plays a vital role in modeling the behavior of abutments during seismic events. Ensuring accurate material properties contributes to reliable analysis results.

Given the ongoing research in these areas, the rigid body limit balance method remains a practical choice for abutment stability analysis. However, for arch dams falling under the Class A seismic fortification category or those with intricate structural and geological conditions, two additional methods are recommended:

1. Finite Element Method: The finite element method is proposed for dynamic stability analysis in such cases. This method's capability to model complex interactions and dynamic responses offers a more advanced and accurate assessment.
2. Physical Model Test: Another valuable approach involves conducting physical model tests. These tests provide a tangible demonstration of the dam's behavior under seismic forces, aiding in the understanding of complex dynamic interactions.

By offering multiple options based on the seismic fortification category, structural complexity, and geological conditions, engineers ensure a well-rounded evaluation of abutment stability. This approach recognizes the evolving nature of seismic analysis methods and aligns with the goal of ensuring the safety and resilience of arch dams under seismic stress.

When calculating the seismic effect of an arch dam by quasi-static method, the horizontal seismic inertia force of each mass point of each layer of the arch ring acts radially, and its representative value should be calculated according to Eq. (9.3.1), in which the dynamic distribution coefficient of dam crest is 3.0, the dam foundation is 1.0 and distributed uniformly along the arch circumference by linear interpolation along the elevation. When using the dynamic method, seismic water pressure under horizontal unit seismic acceleration can be converted into a corresponding radial additional mass of the dam face [33].

9.5.3 Seismic Fortification Measures

The design considerations for arch dams in seismic environments encompass various aspects to ensure their stability and resilience. Here are the key points to consider:

1. Dam Body Shape and Thrust Direction: Opt for a well-selected dam body shape that enhances the thrust direction of the arch abutment. The goal is to minimize tensile stress in the dam body's upper and middle sections and near the foundation during earthquakes.
2. Overhang and Crown Inclination: Check for overhangs toward the upstream side of double curvature arch dams. Consider inclining the crown slightly downstream to distribute forces effectively.
3. Anti-Seismic Stability of Head Bank Slopes: Strengthen the seismic stability of head bank slopes on both sides of the arch dam. Mitigate the impact of differences in lithology and rock mass structure by employing measures such as grouting, concrete plugs, local anchoring, and support to reinforce softer foundation parts.
4. Contact Surface Quality and Anchorage: Ensure strict control over the construction quality of the contact surface between the top arch seat and the bank slope. If needed, consider measures like thickening the arch seat and employing deeply embedded anchorage to enhance stability.
5. Anti-Seepage Measures: Implement anti-seepage curtains and drainage measures for the dam foundation and abutment. Maintain a suitable distance between the pressure tunnel and dam abutment to reduce seepage pressure within the rock mass [34].
6. Dam Body Joint Design: Strengthen the structural design of dam body joints, focusing on water stops, grouting temperature control, and keyway design. Adapt the water stop's shape and material to accommodate multiple openings during earthquakes.
7. Arch and Beam Direction Reinforcement: Arrange proper anti-seismic reinforcement in the tensile area near the arch crown and areas with large local

compressive stress. This can involve enhancing local concrete grades, reducing the top's weight, and increasing stiffness.

8. Light and Integrated Crest Structure: Design a lightweight, well-integrated auxiliary structure for the dam crest, minimizing its protrusion from the dam body.
9. Arch Thrust Transmission: Introduce a structure capable of transmitting arch thrust between the piers of the overflow dam section.
10. Structural Connections: Strengthen the connections of the top traffic bridge and other structures, incorporating measures to prevent detachment during earthquakes.

By addressing these considerations, engineers can create arch dams that are structurally sound and resilient to seismic events, ensuring their stability and safety even in challenging conditions.

9.6 Evaluation on Seismic Safety for Discharging and Conveyance Structures

From the results of the "5.12" Wenchuan catastrophic earthquake investigation, it can be seen that the most severely damaged parts of the discharging and conveyance buildings are high-rise structures such as floodgates and intake towers, and the door-slot deformation causes the opening and closing equipment to fail to operate normally. Therefore, in order to ensure the ability of the post-earthquake reservoir to rapidly reduce the reservoir water level and prevent secondary disasters, special attention shall be paid to the seismic safety review of the floodgates, water conveyance structures, the intake towers, and the deformation calculation of the gate piers [35].

9.6.1 Seismic Safety Review of Spillway

The seismic safety evaluation of spillways encompasses a comprehensive assessment of various critical components to ensure their stability and resilience in the face of seismic events [36]. Here are the key areas of focus in the evaluation process:

1. Seismic Stability: The evaluation begins with a meticulous analysis of the seismic stability of the spillway components. This involves examining the response of floodgates, side walls, retaining walls, guide walls, and other related structures to seismic forces. Engineers assess whether these components can withstand the dynamic loads and ground motions generated by earthquakes without undergoing excessive deformation or failure.
2. Structural Strength: The evaluation also examines the structural strength of key elements within the spillway, such as floodgates, side walls, retaining walls, and guide walls. It ensures that these components are designed to withstand the

dynamic loads induced by seismic forces without compromising their stability. The structural strength of key elements within the spillway is thoroughly assessed. Engineers analyze the materials used, structural configurations, and load-bearing capacities of floodgates, side walls, retaining walls, and guide walls. The goal is to ensure that these structures possess the strength to endure the intense forces exerted during seismic events, safeguarding their integrity.

3. Anti-Seismic Measures: The assessment considers the anti-seismic measures that have been implemented in the spillway design. The engineers scrutinize these measures to ascertain their effectiveness in enhancing the spillway's ability to resist seismic impacts. Any necessary adjustments or improvements are recommended to ensure optimal performance.
4. Compliance with Specifications: Throughout the evaluation, close attention is paid to whether the spillway components meet the requirements outlined in the relevant specifications and guidelines. This ensures that the design and construction align with industry standards for seismic resilience.
5. Dynamic Response Analysis: The dynamic response of floodgates, walls, and other structures to ground motions is simulated and analyzed. Engineers use computational methods to predict how these structures will behave during earthquakes. This analysis provides insights into potential stress concentrations, deformations, and impact on functionality.
6. Interaction with Surrounding Structures: The evaluation considers how the spillway interacts with other dam components and surrounding structures. Engineers assess the potential for cascading failures or adverse interactions that might compromise the overall dam system's integrity. Countermeasures are proposed if needed.
7. Risk Mitigation Strategies: Based on the evaluation's findings, engineers develop risk mitigation strategies. These strategies may involve retrofitting, reinforcement, modification of design parameters, or additional monitoring. The aim is to minimize vulnerabilities and ensure the spillway's continued performance under seismic stress.

By focusing on these aspects, the seismic safety evaluation of spillways aims to guarantee that these crucial components of dam infrastructure are well-equipped to handle the challenges posed by seismic events. Throughout the evaluation, strict adherence to relevant guidelines and specifications is emphasized. Engineers verify whether the spillway's design, materials, and construction practices align with industry standards for seismic resilience. Any deviations or shortcomings are identified and addressed to bring the spillway up to the required safety levels. The ultimate goal is to safeguard public safety, protect surrounding areas, and maintain the functionality of the spillway system during and after earthquakes.

9.6.2 *Seismic Safety Review of Spillway Tunnel and Water Conveyance Tunnel (Culvert)*

The seismic safety review calculation for discharging and conveyance tunnels (culverts) involves a rigorous assessment of various components and aspects to ensure their resilience against seismic events [37]. Here's a detailed breakdown of the review process:

1. Load and Load Combination: The calculation starts with defining the loads and load combinations that the tunnels will experience during seismic events. This includes factors such as static and dynamic loads, as well as their combinations, accounting for the potential impacts of earthquakes.
2. Calculation Methods and Parameters: Engineers employ appropriate calculation methods to analyze the tunnels' behavior under seismic forces. Various parameters, such as material properties, geometry, and seismic inputs, are considered in the calculations to ensure accuracy.
3. Tower Body Strength and Stability: The review encompasses a thorough assessment of the strength of tower bodies, focusing on their ability to withstand seismic forces. Additionally, the overall anti-sliding and anti-overturning stability of intake towers are evaluated, along with the bearing capacity of their foundation at the tower's bottom.
4. Seismic Strength of Tunnel Lining and Surrounding Rock: Engineers examine the seismic strength and stability of the tunnel lining and the rock surrounding it. This involves evaluating the behavior of these components under dynamic loads and ensuring they can endure seismic forces without compromising structural integrity.
5. Seismic Stability of Tunnel Slopes: The review includes an analysis of the seismic stability of slopes at the tunnel's inlet and outlet. The aim is to ensure that these slopes can resist sliding or collapsing under the influence of seismic events.
6. Anti-Seismic Measures: The evaluation scrutinizes the effectiveness of anti-seismic measures that have been incorporated in the design of the tunnels. These measures include reinforcements, anchorages, supports, and other strategies aimed at enhancing the tunnels' ability to withstand seismic impacts.
7. Conformity with Specifications: Throughout the review process, strict adherence to relevant specifications is emphasized. Engineers verify whether the design, materials, and construction practices align with industry standards for seismic resilience. Any deviations or deficiencies are identified and addressed to ensure compliance.

By meticulously addressing these aspects, the seismic safety review calculation for discharging and conveyance tunnels ensures that these critical components of dam infrastructure can withstand seismic forces, safeguard operational functionality, and minimize risks to public safety and the environment during and after earthquakes.

9.7 Conclusion on Seismic Safety Evaluation

When determining the seismic safety level of the dam, it is mainly based on whether the calculation results of the seismic safety review meet the requirements, and also refer to the effectiveness of the engineering and non-engineered anti-seismic measures.

- When the seismic calculation results and the anti-seismic measures meet the requirements, and there is no possibility of earthquake liquefaction, the dam can be considered as seismic safety and rated as **Class A**.
- When the seismic calculation results basically meet the requirements, or the anti-seismic measures are imperfect or there is a possibility of local liquefaction, the dam can be considered to be basically safe and rated as **Class B**.
- When the seismic calculation results and anti-seismic measures do not meet the requirements, or there is a possibility of severe earthquake liquefaction, the dam should be considered to be unsafe for earthquake resistance and rated as **Class C**.

The seismic safety evaluation shall make the clear conclusions as follows:

- Whether the seismic fortification intensity of the project meets the requirements;
- Whether the seismic stability and structural strength of the dam meet the requirements;
- Whether seismic liquefaction probability exist in the earth-rock fill dam and the foundations of structures;
- Whether seismic stability of the near dam bank slope meets the requirements;
- Whether anti-seismic measures and emergency preparedness plan for earthquake and disaster reduction meet the requirements.

References

1. Huang F, Li M, Ma Y et al (2017) Studies on earthquake precursors in China: a review for recent 50 years. Geodesy and Geodynamics 8:1–12. https://doi.org/10.1016/j.geog.2016.12.002
2. Yu R, Hu X, Wen R (2022) Preface to the special issue on ground motion input at dam sites and reservoir earthquakes. Earthq Sci 35:311–313. https://doi.org/10.1016/j.eqs.2022.10.002
3. Ghrib T, Léger P, Tinawi R et al (1997) Seismic safety evaluation of gravity dams. Int J Hydropower Dams 4:126–138. https://doi.org/10.1061/(ASCE)0733-9445(2000)126:4(518)
4. Houqun C, Shengxin W, Faning D (2016) Determination of correlation design seismic motion parameters on dam site. In: Seismic safety of high arch dams, pp 53–81
5. Bozorgnia Y, Hachem MM, Campbell KW (2010) Deterministic and probabilistic predictions of yield strength and inelastic displacement spectra. Earthq Spectra 26:25–40. https://doi.org/10.1193/1.3281659
6. Zhang Y, Weng WG, Huang ZL (2018) A scenario-based model for earthquake emergency management effectiveness evaluation. Technol Forecast Soc Chang 128:197–207. https://doi.org/10.1016/j.techfore.2017.12.001
7. Wieland M (2009) Features of seismic hazard in large dam projects and strong motion monitoring of large dams. Front Archit Civ Eng China 4:56–64. https://doi.org/10.1007/s11709-010-0005-6

8. Li SQ, Yu TL, Chen YS (2019) Comparative analysis of the empirical seismic vulnerability of typical structures in multiple intensity zones. Arch Civ Eng 65:167–183
9. Huang Y, Xiong M (2016) Probability density evolution method for seismic liquefaction performance analysis of earth dam. Earthquake Eng Struct Dynam 46:925–943. https://doi.org/10.1002/eqe.2837
10. Hernández UM, Nava-Tristán OE, Liu XL et al (2008) Seismic guidelines for earth and rock fill dams. In: The 14th world conference on earthquake engineering, Beijing, China
11. Seed HB (1975) Stability of earth and rockfill dams during earthquakes: In: Hirschfeld RC, Poulos SJ (eds) Embankment-dam engineering. Textbook. Wiley, NY, 1973, 30p. Int J Rock Mech Min Sci Geomech Abstr 12:67
12. Ghahraman V, Yegian MK (1996) Risk analysis for earthquake-induced permanent deformation of earth dams. In: Proceedings of the 11th world conference on earthquake engineering, Acapulco, Mexico, p 688
13. Yener Ozkan M (1998) A review of considerations on seismic safety of embankments and earth and rock-fill dams. Soil Dyn Earthq Eng 17:439–458. https://doi.org/10.1016/s0267-7261(98)00035-9
14. Zhang B, Wang X, Zhang JS, Meng F (2017) Three-dimensional limit analysis of seismic stability of tunnel faces with quasi-static method. Geomech Eng 13:301–318. https://doi.org/10.12989/gae.2017.13.2.301
15. Du-bo W, Han-long LIU, Tao Y et al (2013) Seismic fragility analysis for earth-rockfill dams based on deformation. Chin J Geotech Eng 35:814–819
16. Wang F-M, Tulamaiti Y, Fang H et al (2023) Seismic response characteristics of polymer anti-seepage wall in earth dam based on earthquake wave motion input method. Structures 47:358–373. https://doi.org/10.1016/j.istruc.2022.11.060
17. Tinawi R, Léger P, Leclerc M, Cipolla G (2000) Seismic safety of gravity dams: from shake table experiments to numerical analyses. J Struct Eng ASCE 126:518–529. https://doi.org/10.1061/(asce)0733-9445(2000)126:4(518)
18. Yilmazturk SM, Arici Y, Binici B (2015) Seismic assessment of a monolithic RCC gravity dam including three dimensional dam–foundation–reservoir interaction. Eng Struct 100:137–148. https://doi.org/10.1016/j.engstruct.2015.05.041
19. Fenves GL, Chopra AK, University of California, Berkeley (1984) Earthquake analysis and response of concrete gravity dams. University of California, Earthquake Engineering Research Center, Berkeley
20. Arabshahi H, Lotfi V (2008) Earthquake response of concrete gravity dams including dam–foundation interface nonlinearities. Eng Struct 30:3065–3073. https://doi.org/10.1016/j.engstruct.2008.04.018
21. Pacifico A, Chioccarelli E, Iervolino I (2022) Residential code-conforming structural seismic risk maps for Italy. Soil Dyn Earthq Eng 153:107104. https://doi.org/10.1016/j.soildyn.2021.107104
22. Crowley H, Colombi M, Borzi B et al (2008) A comparison of seismic risk maps for Italy. Bull Earthq Eng 7:149–180. https://doi.org/10.1007/s10518-008-9100-7
23. Yazdchi M, Khalili N, Valliappan S (1999) Non-linear seismic behaviour of concrete gravity dams using coupled finite element? Boundary element technique. Int J Numer Methods Eng 44:101–130. https://doi.org/10.1002/(sici)1097-0207(19990110)44:1%3C101::aid-nme495%3E3.0.co;2-4
24. Chen Y, Zhang L, Yang G et al (2012) Anti-sliding stability of a gravity dam on complicated foundation with multiple structural planes. Int J Rock Mech Min Sci 55:151–156. https://doi.org/10.1016/j.ijrmms.2012.07.010
25. Bretas EM, Léger P, Lemos JV (2012) 3D stability analysis of gravity dams on sloped rock foundations using the limit equilibrium method. Comput Geotech 44:147–156. https://doi.org/10.1016/j.compgeo.2012.04.006
26. Zhang S, Wang G, Pang B, Du C (2013) The effects of strong motion duration on the dynamic response and accumulated damage of concrete gravity dams. Soil Dyn Earthq Eng 45:112–124. https://doi.org/10.1016/j.soildyn.2012.11.011

27. Yamaguchi Y, Hall R, Sasaki T et al (2004) Seismic performance evaluation of concrete gravity dams. In: Proceedings of the 13th world conference on earthquake engineering
28. Huang J, Zerva A (2013) Earthquake performance assessment of concrete gravity dams subjected to spatially varying seismic ground motions. Struct Infrastruct Eng 10:1011–1026. https://doi.org/10.1080/15732479.2013.782323
29. Chen H, Wu S, Dang F (2015) Seismic safety of high arch dams. Elsevier
30. Li H, Wang G, Wei B et al (2019) Dynamic inversion method for the material parameters of a high arch dam and its foundation. Appl Math Model 71:60–76. https://doi.org/10.1016/j.apm.2019.02.008
31. Bayraktar A, Sevim B, Can Altunişik A (2011) Finite element model updating effects on nonlinear seismic response of arch dam–reservoir–foundation systems. Finite Elem Anal Des 47:85–97. https://doi.org/10.1016/j.finel.2010.09.005
32. Su H, Li J, Wen Z, Fu Z (2016) Dynamic non-probabilistic reliability evaluation and service life prediction for arch dams considering time-varying effects. Appl Math Model 40:6908–6923. https://doi.org/10.1016/j.apm.2016.02.016
33. Wang H, Li D (2006) Experimental study of seismic overloading of large arch dam. Earthq Eng Struct Dyn 35:199–216. https://doi.org/10.1002/eqe.517
34. Yang Q, Leng K, Liu Y (2011) A finite element approach for seismic design of arch dam-abutment structures. Sci China Technol Sci 54:516–521. https://doi.org/10.1007/s11431-011-4293-4
35. Wieland M (2014) Seismic Hazard and seismic design and safety aspects of large dam projects. Geotech Geolog Earthq Eng 627–650. https://doi.org/10.1007/978-3-319-07118-3_20
36. Ariga Y (2008) Study on earthquake damage mechanism and measures of spillway composed of concrete piers with different shapes and dynamic response properties. In: Semantic scholar. https://api.semanticscholar.org/CorpusID:160005286
37. Chen Z, Yu H, Yuan Y (2012) Full 3D seismic analysis of a long-distance water conveyance tunnel. Struct Infrastruct Eng 10:128–140. https://doi.org/10.1080/15732479.2012.707659
38. Houqun C (2022) Overview of the seismic input at dam sites in China. Earthq Sci 35:410–425. https://doi.org/10.1016/j.eqs.2022.05.006
39. Kojić SB, Trifunac MD (1991) Earthquake stresses in arch dams. I: theory and antiplane excitation. J Eng Mech ASCE 117:532–552. https://doi.org/10.1061/(asce)0733-9399(1991)117:3(532)
40. Ma H, Chi F (2016) Major technologies for safe construction of high earth-rockfill dams. Engineering 2:498–509. https://doi.org/10.1016/j.eng.2016.04.001
41. Yang X, Chi S (2014) Seismic stability of earth-rock dams using finite element limit analysis. Soil Dyn Earthq Eng 64:1–10. https://doi.org/10.1016/j.soildyn.2014.04.007

Chapter 10
Evaluation on Metal Structure Safety

10.1 Introduction

Based on the statistics on the history of dam breaks in China, a significant number of dam break events, such as the Banqiao Reservoir dam break in August 1975, the Gouhou Reservoir dam break in 1993, and the Dahe Reservoir dam break in 2010, have been linked to the failure of timely gate opening, especially when overtopping occurs [1, 2]. Additionally, incidents of penstock instability in hydropower stations have been reported sporadically. The delayed gate opening can be attributed to three main factors:

1. Deformation of the gate structure, including the gate slot.
2. Improper maintenance and repair of the hoist, leading to abnormal operation.
3. The unavailability of a standby power supply, resulting in the inability to open the gate during emergencies.

Hence, it is essential to conduct regular metal structure safety evaluations as part of the dam safety assessment process. The evaluation of metal structure safety shall be carried out on the basis of on-site inspection, the safety detection, the result of calculation and analysis [3]. Key points for evaluation should be focus on identifying the quality defects during manufacture and installation, the weak parts and members revealed during safety detection, as well as abnormality and accidents arising from operation. For simple metal structures of the small and medium sized reservoirs, their safety can be directly assessed with the results of on-site inspection and safety detection.

During the operation of metal structures, various issues such as corrosion, wear, fatigue, deformation may occur. Simultaneously, the designed flood changes for many reservoirs due to the extension of hydrology series. Therefore, parameters and loads for the safety review of metal structures shall be adjusted in accordance with latest information available to ensure a comprehensive and accurate assessment of their safety.

Y. Xiang et al., *On-site Inspection and Dam Safety Evaluation*, Hydroscience and Engineering, https://doi.org/10.1007/978-981-97-3987-5_10

The safety evaluation on metal structure primarily involves reviewing of strength, rigidity and stability of gates, as well as the capacity of hoist and the safety of electromechanical equipment, Additionally, it includes assessing the strength, stability against external pressure of penstock. Through the safety evaluation of metal structures, it can judge whether the metal structures that affect the safe operation of the dam such as trash racks, gates, hoisting facilities and penstocks can operate safely and reliably.

The hoisting equipment is very important for the safe operation of metal structures. It is necessary to ensure that the capacity of hoisting equipment and corresponding electromechanical equipment meet the requirements.

By conducting such evaluations, authorities can identify potential weaknesses, defects, or risks in the metal structures and take appropriate measures to address them, thereby enhancing the overall safety and efficiency of the dam and its associated components.

According to the regulations on the depreciation periods of fixed assets in hydraulic engineering, the specific depreciation periods for metal structures [4] are as follows:

1. Pressure steel pipes: Depreciation period of 50 years.
2. Large gates, valves, and operating equipment: Depreciation period of 30 years.
3. Medium and small gates, valves, and operating equipment: Depreciation period of 20 years.

These depreciation periods define the service life of different types of metal structural equipment in hydraulic engineering. Following these regulations allows for asset management and the formulation of maintenance plans to ensure that the equipment remains in good operational condition within its designated service life.

10.2 Evaluation of Gate Safety

Indeed, a comprehensive review should be conducted on various aspects related to the gate, focusing on the general arrangement of gates, gate type selection, applicable conditions, provision of access doors or emergency doors, hoist room layout, filling mechanisms, ventilation systems, and locking devices. The purpose of this review is to ensure that these components fulfill the specific needs of reservoir operation.

The general arrangement of gates needs to be examined to ensure that it aligns with the dam's design and functions effectively during various discharge scenarios. The selection of gate types should be suitable for the dam's purpose and the prevailing hydraulic conditions. Appropriate conditions for gate application need to be verified to guarantee safe and efficient operation. Providing access doors or emergency doors is crucial to facilitate maintenance and enable swift response in case of emergencies. The layout of the hoist room should be well-planned to support the hoist's functionality and maintenance accessibility. The filling, ventilation, and locking devices must be carefully assessed to ensure smooth and secure operations while considering the specific requirements of the reservoir.

By conducting a thorough review of these factors, dam operators can ensure that the gate meet operational needs and adhere to safety standards, contributing to the overall stability and functionality of the dam system.

10.2.1 On Site Safety Inspection of Steel Gate

During on-site inspections, a thorough examination of various components of the metal structure should be conducted to assess their condition and operational state. Specifically, the following elements need to be inspected:

1. Gate Leaf: Inspect the gate leaf for any signs of deformation, corrosion, or wear that may impact its proper functioning. Ensure that the gate leaf is in good condition and operates smoothly.
2. Support Slipway: Check the support slipway for any damages or wear that could affect the movement of the gate. Make sure it is functioning correctly and providing adequate support to the gate.
3. Water Seal: Examine the water seal for any leaks or deteriorations that might compromise its effectiveness in preventing water leakage.
4. Embedded Parts: Inspect the embedded parts in the metal structure for signs of degradation or detachment from the surrounding material.
5. Filling and Locking Devices: Check the filling and locking devices to ensure they are in good working condition and capable of securing the gate in the desired position.

During the on-site inspection, obvious deformation, corrosion (erosion), or wear is identified in the gate or gate slot, and it impacts the gate's normal operation, or if the depreciable life of the gate as specified has been exceeded, a further safety inspection and analysis become necessary. This additional inspection is essential to evaluate the structural integrity and safety of the gate thoroughly. Based on the findings of this comprehensive assessment, appropriate actions can be taken, such as repairs, replacements, or upgrades, to ensure the continued safe and reliable operation of the metal structure.

To describe and record the appearance of main components of a metal structure, the following steps can be taken using tools like a tape measure, caliper, digital camera, and other relevant instruments:

1. Coating Inspection: Use a tape measure, caliper or ultrasonic thickness measurement to measure the thickness of the coating on the metal components. Record any signs of aging, such as peeling, cracking, or fading, as well as areas with visible corrosion.
2. Corrosion Assessment: Examine the components for signs of corrosion, such as rust, pitting, or discoloration.

3. Deformation, Damage, and Defects: Use the tape measure and caliper to measure any deformations or damages in the metal structure, such as bends, dents, or fractures. Record the location and severity of each defect.
4. Mechanical and Electrical Equipment: Inspect the mechanical and electrical equipment associated with the metal structure, such as hoists, gate mechanisms, and locks. Check for signs of aging, wear, or malfunctions. Use the digital camera to document any issues found.
5. Operation Assessment: Observe the operation of the equipment and mechanisms to ensure they are functioning correctly. Record any abnormal noises, vibrations, or other irregularities during operation.
6. Aging and Wear: Record any signs of aging and wear on the metal components and equipment, such as faded paint, worn-out parts, or degraded surfaces.

During the investigation and recording process, it's essential to take clear photographs using a digital camera to provide visual documentation of the metal structure's condition. This information will help in assessing the overall health of the structure, identifying potential safety risks, and planning necessary maintenance or repairs to ensure its continued safe and reliable operation. If one of the following is found in on-site inspection, it shall be regarded that serious latent defects exist in the gate and further safety tests or analysis shall be carried out.

- Cavitations scouring and caving at the gate slot and the nearby concrete or differential settlement of gate room, which influences hoisting of gate; cracking, denudation and ageing of concrete at gate pier, breast wall or bracket, which affects the gate bearing;
- Abnormal pattern of flow at gate water inlet, near gate slot and at the rear of the gate;
- Stronger gate vibration perception;
- Corrosion of gate or embedded parts at the degree of "relatively serious corrosion";
- Deformation and distortion of gate leaf; damaged, deformed and malpositioned front panel, longitudinal beam, latitudinal beam, support arm; apparent defects on the main stressed weld seams; damage, deformation, loosing and missing of connecting bolts;
- Damage, deformation, parts missing and serious leakage of gate water seal;
- Lifting lug damage and deformation; unsecured connection of lifting tool; damage, deformation, parts missing and rusting of main wheels of the plane gate(slide way), lateral bearing, reverse bearing or hinges of curved gate;
- Damage, deformation, malpositioning and concrete caving of embedded parts like rails, bottom sill, lintel, seal seat plate or hinged support of curved gate;
- Gate equalizing equipment, locking device and de-icing facility not reliable;
- Venthole collapse, blocking or poor ventilation.

10.2.2 Recheck Calculation of Steel Gate

The calculation and analysis should prioritize the thorough examination of the gate's structural strength, rigidity, and stability. When dealing with critical gate structures, it is recommended to employ the finite element method for in-depth analysis [5].

The focus of these calculations and analyses is to ensure that the gate is robust enough to withstand anticipated loads, environmental factors, and potential stressors. The structural strength assessment involves evaluating whether the gate can handle the forces and pressures it will encounter during its operational life. Rigidity analysis aims to determine how resistant the gate is to deformation under different conditions, ensuring its shape integrity. Stability assessment ensures that the gate can maintain its equilibrium and not undergo any undesired movements or displacements [6].

In cases where the gate's structural significance demands a more advanced approach, the finite element method is a powerful tool. This method breaks down complex structures into smaller, manageable elements, allowing for a detailed analysis of stress distribution, deformation, and other critical factors. By employing finite element analysis, engineers can gain precise insights into the gate's behavior under varying loads and conditions [7]. Figure 10.1 shows the constructed three-dimensional model of the gate, and Fig. 10.2 shows the finite element calculation results of the gate.

Ultimately, the combination of thorough calculations and finite element analysis provides a comprehensive understanding of the gate's performance characteristics. This knowledge is invaluable in ensuring the gate's reliability, safety, and longevity throughout its operational lifespan [8].

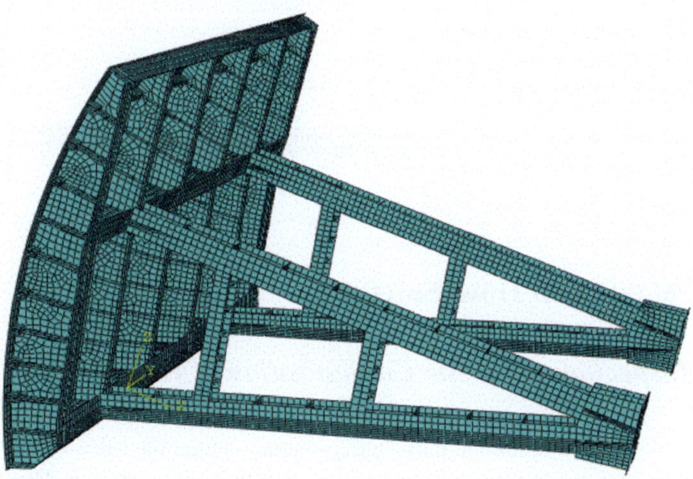

Fig. 10.1 3D modeling of gates

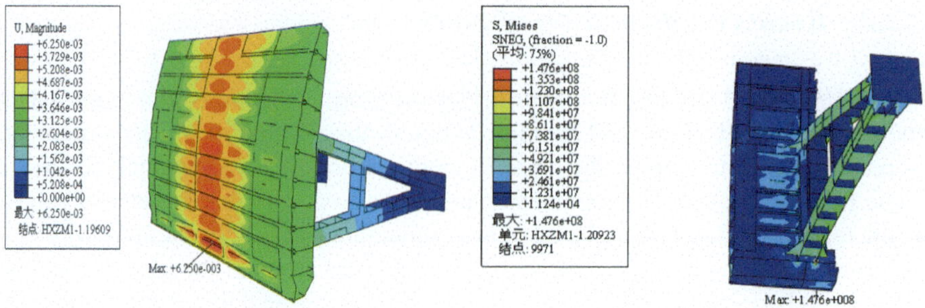

Fig. 10.2 Finite element calculation results of gates

1. **Strength checking calculation**

The calculated stress of each component shall meet the requirements of the following formula from the stress point of view, Where $[\tau]$ is the allowable shear stress and $[\sigma]$ is the allowable tensile, compressive and bending stresses.

$$\sigma < [\sigma]; \tau < [\tau] \tag{10.2.1}$$

2. **Stiffness checking calculation**

Mainly evaluate the member deflection f_{max}, where $[f\ ]$ is the allowable defilection.

$$f_{\max} < [f] \tag{10.2.2}$$

3. **Stability checking calculation**

The ratio of the height h_0 and thickness δ of the main beam web should be checked:

$$h_0/\delta \leq 80\sqrt{235/\sigma_S} \tag{10.2.3}$$

10.3 Evaluation of Hoist Safety

10.3.1 Inspection for Basic Configuration of Hoist

The review process is intended to thoroughly assess whether the selected hoist system aligns with the project's requirements across multiple key criteria. These criteria encompass the hydraulic layout, gate type, number of anchoring points, hoisting style, and hoisting time. Additionally, the review evaluates whether the hoist system's parameters, including hoisting force, lift capacity, span width, and speed, adequately satisfy the operational demands of the gate [9, 10].

A critical aspect of the review is the assessment of the hoist system's reliability. This involves a comprehensive evaluation of the system's design, components, and overall robustness. The goal is to ascertain the system's ability to consistently perform under various conditions over an extended operational lifespan.

10.3.2 On Site Inspection of Hoist

The current detection of mobile hoist shall include the following contents: (1) Detection of door frame (bridge); (2) Brake detection; (3) Reducer test; (4) Drum and open gear pair detection; (5) Drive shaft and coupling inspection; (6) Pulley block detection; (7) Wire rope inspection; (8) Wheel detection; (9) Track detection. The current detection of hydraulic hoist shall include the following contents: (1) Frame detection; (2) Hydraulic cylinder testing; (3) Piston rod inspection; (4)Hydraulic system testing; (5) Hydraulic cylinder leak detection. The current detection of screw hoist shall include the following contents: (1) Engine case and seat detection; (2) Screw and nut, worm and wheel detection; (3) Manual mechanism testing.

The state of appearance and operation of hoist and the state of electric equipment and protective devices shall be inspected on site. In case obvious ageing and wear of hoists are found in on-site inspection, which affects normal opening and closing of the gate, a further safety inspection and analysis is required. If one of the following is found in on-site inspection, it shall be regarded that serious latent defects exist in the hoist and further safety tests or analysis shall be carried out.

- The hoist operates beyond its operation class limit;
- Stronger perception of hoist vibration
- Twin lifting points are not synchronous, affecting operation;
- Corrosion of hoist at the degree of "relatively serious corrosion";
- Damage to and deformation of hoist frame, apparent defects on weld seam and on braking wheel; smaller contact face with braking belt, wheel tooth damage and loose meshing; damaged and cracked reel, cracks and deformation of drive shaft; wear and deformation of pulley block; steel rope wear and broken wire; damage, deformation and apparent weld seam defects on frame or bridge of the movable hoist; wheel wear, cracking, rail deformation, malposition;
- Damage and cracking of hydraulic cylinder of the hydraulic hoist, wear and deformation of piston rod; oil leakage of hydraulic cylinder or oil circuit;
- Wear and cracking of nut of screw hoist; screw wear and deformation;
- Electric control equipment is not complete and cannot be used for normal operation and the insulation protection and grounding protection is not reliable;
- Equipment for load control, stroke control and openness indicator, etc. is not complete and cannot be used for normal operation;
- Dislocation and cracking and rain leakage of hoist room, affecting the normal operation of hoist

10.3.3 Hoist Recheck Calculation

In calculation and analysis, the hoisting capacity shall be reviewed. When necessary, the strength rigidity and stability of the structure members of the hoist shall be reviewed. At the same time, the hoisting force of the hoist shall be rechecked.

(1) The checking of opening and closing force of flat gate shall be carried out according to the following formula:

Closing force:

$$F_w = n_T(T_{zd} + T_{zs}) - n_G G - W_S \tag{10.3.1}$$

Door opening force:

$$F_q = n_T(T_{zd} + T_{zs}) - n_{G'} G - W_S \tag{10.3.2}$$

where:

n_T Safety coefficient of friction resistance, taken as 1.2
n_G Correction coefficient of gate dead weight for calculating closing force, taken as 0.9
$n_G^{'}$ The gate dead weight correction coefficient for calculating the lifting force is 1.1

(2) The hoisting force of radial gate shall be rechecked according to the following formula:

Closing force:

$$\sum M_{\max} = n_T(M_{zd} + M_{zs}) - n_G \sum G_i L_i; \quad F_w = \sum M_{\max}/R_1 \tag{10.3.3}$$

Door opening force:

$$\sum M_{\max} = n_T(M_{zd} + M_{zs}) - n\prime_G \sum G_i L_i; \quad F_Q = \sum M_{\max}/R_2 \tag{10.3.4}$$

where: R_2, R_2—The force arm of the weighting (or downforce) and opening force on the radial gate rotation center.

10.4 Evaluation of Penstock Safety

10.4.1 Basic Configuration of Penstock

The pipeline line shall meet the general layout requirements of the complex, and be determined after technical and economic comparison taking into account the terrain, geology, hydraulics, construction and operation conditions. The number of pipelines shall be determined by technical and economic comparison according to the number of units, length of pipelines, topographic and geological conditions, stages of unit installation, fabrication, installation and transportation conditions, operation mode of power plant and its position in power system, etc. Diameter shall be determined according to technical and economic comparison. Several pipe diameters can be determined by section according to line arrangement and internal pressure variation. However, the number of diameter changes should not be too much.

10.4.2 On Site Inspection of Penstock

On-site inspection of the penstock involves a comprehensive assessment that encompasses various critical aspects. The inspection process covers the following key elements:

1. Visual Inspection and Appearance: A thorough visual examination of the penstock is conducted to identify any visible anomalies, deformations, or irregularities. This includes checking for signs of physical damage, wear and tear, or any abnormalities in the overall appearance of the penstock.
2. Material Evaluation: The material composition of the penstock is assessed to ensure that it meets the required specifications for strength and durability. Any deviations or inconsistencies in the material properties are noted during the inspection.
3. Corrosion Assessment: The penstock's susceptibility to corrosion is evaluated. This involves examining the surface for signs of rust, pitting, or other corrosion-related issues. The inspection aims to determine the extent of corrosion and assess its impact on the penstock's integrity.
4. Non-Destructive Testing (NDT): NDT techniques are employed to assess the penstock's condition without causing damage. Methods such as ultrasonic testing, magnetic particle testing, radiographic testing, or eddy current testing may be used to identify internal flaws, cracks, or defects that aren't visible externally.
5. Stress Monitoring: Stress levels within the penstock are monitored to ensure that they remain within acceptable limits. This is particularly important to prevent overloading and to maintain the structural integrity of the penstock.
6. Vibration Analysis: Vibration monitoring helps identify any excessive vibrations or oscillations that might indicate structural issues. It's crucial for detecting abnormal movement patterns that could lead to operational problems or potential failures.

By conducting a comprehensive on-site inspection that encompasses these various aspects, the condition and operational status of the penstock can be thoroughly assessed. This evaluation ensures the penstock's safety, reliability, and longevity, and aids in identifying any maintenance or repair needs that might arise from issues detected during the inspection.

If one of the following is found in on-site inspection, it shall be regarded that serious latent defects exist in penstock and further safety tests or analysis shall be carried out.

- Leakage of penstock;
- Stronger perception of vibration of penstock;
- Rapid gate valve or emergency gate vale cannot operate normally; Venthole collapse, blocking or poor ventilation
- Penstock surface damage, deformation and apparent defects on weld seam;
- Collapse, displacement and sink of anchor pier and buttress of the open pipe;
- Water pressure outside the embedded pipe exceeding the designed value;
- Corrosion of penstock at the degree of "relatively serious corrosion".

10.4.3 Recheck Calculation of Penstock

By following the rigorous calculation and analysis procedures, the penstock's ability to endure external forces and maintain stability is thoroughly assessed. This approach guarantees the safety, reliability, and longevity of the penstock within the operational context of the hydropower station. A comprehensive assessment of the penstock's ability to withstand external forces is a priority. This evaluation includes examining the penstock's resistance to forces like static pressure, water hammer pressure, seismic loads, construction loads, temperature variations, and other relevant factors. The method used for calculating the penstock's capacity to withstand external forces should adhere to established standards and guidelines. Load combinations, considering multiple force factors acting simultaneously, must be determined to accurately assess the penstock's overall response. For critical penstocks, finite element analysis is recommended. This advanced technique breaks down complex structures into manageable elements, enabling a detailed assessment of stress distribution, deformation, and stability under various loads.

Various loads applied on the penstock need to be re-determined to reflect real-world conditions. This includes re-evaluating static pressure to match flood control standards, determining water hammer pressures based on actual hydropower station operation, and recalculating seismic loads based on updated seismic fortification intensity reviews. Additional loads and stresses such as construction loads, temperature effects, air pressure differences, groundwater pressure, seepage pressure, and differential settlement impacts should be calculated using actual data and safety monitoring results. The dimensions of key load-bearing components should be determined based on safety tests. This ensures that these components can withstand the calculated loads and forces without compromising structural integrity.

10.5 Evaluation of Electrical Equipment

In scenarios where extreme events such as earthquakes, severe storms, and catastrophic floods can lead to interruptions in power supply from the grid, it is imperative to have contingency plans in place. For large and medium-sized reservoirs, the implementation of diesel generators as standby power sources is crucial. This measure ensures the consistent availability of a safe and reliable power supply for the operation of hoists even during such challenging circumstances.

The standby diesel generators serve as a reliable backup to maintain the operation of critical components, such as the hoists. This strategy helps prevent disruptions and ensures that the gates can function as needed, even when external power sources are compromised due to extreme events.

Furthermore, it's essential to conduct a comprehensive review of the power supply mechanisms for hoists associated with discharging gates and other emergency gate systems. This review aims to confirm the robustness of the power supply arrangement for these specific gates. It involves assessing whether the power supply to these gates is guaranteed, both under normal operating conditions and during emergency scenarios.

By incorporating standby diesel generators and thoroughly reviewing power supply arrangements, reservoir operators can enhance the safety and reliability of hoist operations during extreme events. This proactive approach safeguards against power interruptions, reduces potential risks, and ensures that emergency gates can be operated effectively when needed most.

10.5.1 Panelboard Inspection

(1) Check whether the structure of panels and cabinets is complete, and whether there are any phenomena such as deformation, breakage, rust, etc. Check whether the paint surface of cabinet is complete, whether the protection of closure, dust proof, moisture proof and isolation is complete; Check whether the locking device works properly and the action is accurate and reliable; Check whether the cabinet is securely fixed.

(2) Check whether the electrical components in the cabinet are arranged neatly and securely; Check whether the strong and weak current equipment and lines, AC and DC equipment and lines are arranged reasonably, whether the markings are obvious, whether the safety distance and insulation isolation meet the requirements; Check fuses or circuit breakers for grading requirements; Check whether the terminal connections of electrical components are reliable and whether there is heat, rust and hot melt.

(3) Check whether the connection between wires and electrical components is secure and reliable, and whether the identification, number and color of cables or wires are clear and correct; Check cabinet wiring arrangement is neat and

beautiful, conductor insulation is good, insulation is damaged and aging, and wiring harness is protected; Check the signal line for shielding protection.

(4) Check the indicators and recording instruments of the distribution panel to determine whether the parameters such as accuracy of the instruments checked meet the specified requirements when they work under the specified conditions. Check AC/DC voltmeter, ammeter, zero point of power meter, output value, function, indication angle, etc.

(5) Measure the insulation resistance of electrical equipment, circuit breaker, lines, etc. Use insulation resistance tester to select the corresponding test voltage according to different voltage levels of equipment and measure the insulation resistance. The measured insulation structure shall meet the requirements of the design or standard.

(6) Check whether the functions of AC voltage transformer and current transformer used for monitoring in the cabinet work properly. Measure whether the transformer conversion ratio, insulation resistance and excitation characteristics meet the requirements of product standards and specifications; Check whether the switch in cabinet has reliable contact and disconnection functions and flexible operation.

(7) Check whether earthing of cabinets and components is reliable, whether earthing bus is reliable to connect with main earthing network, whether the section size of earthing conductor meets the requirements, whether the conductor is complete and whether the fixing is reliable.

10.5.2 Transformer Detection

(1) Check whether the appearance of the transformer is complete, whether the wiring is reliable, whether the color identification is clear, whether the insulation layer protection is intact, whether the fixation is firm, whether the operation noise is normal, whether the terminal temperature is abnormal, and whether the safety protection measures meet the requirements.

(2) The electrical test of transformer shall be in accordance with regulations. Complete the measurement of DC resistance, insulation resistance, absorption ratio, polarization index, dielectric loss factor and voltage ratio.

- DC resistance: the difference between the windings of each phase shall not be greater than the specification requirements, and the change shall not be greater than the specification requirements when compared with the factory measured value of the product at the same temperature.
- Insulation resistance, absorption ratio and polarization index: the measured value shall not be less than 70% of the factory test value or the corresponding specification requirements.
- Dielectric loss factor: the measured value of winding dielectric loss factor should not be greater than 130% of the product's factory test value.

- Voltage ratio: under rated tapping, the allowable deviation of voltage ratio shall not exceed ±5%.

10.5.3 Circuit Breaker Detection

Circuit breaker electrical test shall be carried out in accordance with regulations. Complete the measurement of circuit resistance, insulation resistance, closing synchronism and bounce time.

- Loop resistance: DC voltage drop method of not less than 100A is used, and the test results meet the specifications of product specifications.
- Insulation resistance: Measure the insulation resistance of the closing coil and the closing contactor coil. The measurement result should not be less than 10 MΩ.
- Opening and closing action characteristics: The breaker opening and closing action characteristics are measured by CT2008D high voltage switch action characteristics tester. The bounce time after contact during closing should not be more than 2 ms for circuit breakers below 40.5 kV and 3 ms for circuit breakers above 40.5 kV. For 10 kV vacuum circuit breakers with current of 3 kA and above, if the jumping time is not less than 2 ms, the technical conditions of the products shall be met. Measurements shall be made at the rated operating voltage of the circuit breaker. The measured values shall conform to the technical conditions of the products.

10.5.4 Pump Detection

(1) Stability test of water pump unit: measure the operating stability parameters such as vibration, noise and swing of water pump. Operating stability of the pump is evaluated in accordance with relevant regulations.
(2) Performance test of water pump unit: measure the current, voltage, power and head when the pump is running, and check whether the operation parameters and performance of the hydraulic system meet the product specifications.
(3) Conduct electrical test on pump motor and circuit breaker switch, and measure DC resistance and insulation resistance of motor winding. The electrical test on circuit breaker is carried out according to the scheme in Sect. 10.5.3.
(4) Insulation resistance of motor winding: the measured value of insulation resistance should not be less than 0.5 MΩ;
(5) DC resistance of motor winding: Each DC resistance value shall not be more than 2% of its minimum value.

10.5.5 Checking Contents of Relay Protection

Relay protection involves transformer protection, 10 kV line protection, circuit breaker protection and motor protection, and verifies parameter setting, signal accuracy and response time of various protection devices. The types of protection checked mainly include short circuit, turn-to-turn short circuit, over current, over voltage, ground fault, over load and low voltage. The main test types are basic performance tests, including contact basic parameters test, coil basic parameters test, accuracy test, time characteristic test, switching input and output test, synchronization performance test. The checking results of relay protection should correspond to the setting values of the design and the error is within the allowable range.

10.6 Conclusions on Metal Structure Safety Evaluation

10.6.1 Conclusion of Metal Structure Safety Assessment

After review of metal structure safety, clear conclusions shall be made on

- Whether the layout of the metal structures is reasonable and their manufacture and installation meet the requirements;
- Whether the strength, rigidity and stability of the metal structures meet the requirements;
- Whether the hoisting capacity of the hoist meets the requirements and whether its operation is reliable;
- Whether the power supply is guaranteed and whether the gate of discharging facility can be started normally under emergency;
- Whether the depreciable life for abandonment is exceeded and whether the operation and maintenance state is good

10.6.2 Safety Evaluation Grade of Metal Structure

1. The metal structure shall be considered as safe and shall be rated as Class A when metal structure is arranged properly and its design, manufacture and installation meet the requirements of specifications; the result of safety inspection is "Safe" and the results of check calculation for strength, rigidity and stability satisfy the requirements of specifications; power supply is safe and reliable; the depreciable life for abandonment is not exceeded and the state of operation and maintenance is good.
2. The metal structure shall be considered as basically safe and shall be rated as Class B when the result of safety inspection is "Basically Safe" and the results of check calculation for strength, rigidity and stability basically satisfy the requirements

of specifications; standby power supply is provided; local deformation, corrosion (erosion) and wear appeared but do not seriously affect normal operation.

3. The metal structure shall be considered as unsafe and shall be rated as Class C when the result of safety inspection is "Unsafe" and the results of check calculation for strength, rigidity and stability fail to satisfy the requirements of specifications; no standby power supply is provided or power supply cannot be guaranteed; maintenance is not well done and deformation, corrosion (erosion) and wear are serious. (The metal structures of the discharging structures are directly related to the dam safety. However, the metal structures of non discharging structures are mainly related to the operation of the reservoir and generally do not influence the safety of the dam. Therefore greater attention shall be paid to the safety of the metal structures of the discharging structures. This will serve as the main basis for rating the metal structure as Class C. Metal structures of non-discharging structures having less impact to project safety are normally not rated as Class C.)

References

1. SHENG J, Li H, Sheng T (2023) Statistical analysis of dam failure and its loss of life in China. In: Hydro-science and engineering, pp 1–15
2. Liu W, Carling PA, Hu K et al (2019) Outburst floods in China: a review. Earth Sci Rev 197:102895. https://doi.org/10.1016/j.earscirev.2019.102895
3. Adamo N, Al-Ansari N, Sissakian V et al (2020) Dams safety: inspections, safety reviews, and legislations. J Earth Sci Geotechn Eng 109–143. https://doi.org/10.47260/jesge/1114
4. Ministry of Water Resources of the PRC (2014) Code for rational service life and durability design of water resources and hydropower projects (SL 654-2014)
5. Zhang X, Lou S, Chen L et al (2022) Study on analysis principle of spatial system method for a hydraulic steel gate. Sustainability 14:14804. https://doi.org/10.3390/su142214804
6. Van Dang T, Rigo P (2022) Risk-based inspection strategies of miter gates based on structural health monitoring. Springer eBooks, pp 135–140. https://doi.org/10.1007/978-3-030-97822-8_15
7. Faridmehr I, Nejad AF, Baghban MH, Ghorbani R (2020) Numerical and physical analysis on the response of a dam's radial gate to extreme loading performance. Water 12:2425–2425. https://doi.org/10.3390/w12092425
8. Erbisti PCF (2014) Design of hydraulic gates. CRC Press. https://doi.org/10.1201/b16954
9. Yi JG, Cai X, Xu JC et al (2011) Hoist safety risk assessment model research. Adv Mater Res 383–390:1225–1230. https://doi.org/10.4028/www.scientific.net/amr.383-390.1225
10. Zhang Y, Zhao S, Dan L (2021) Research on early warning of hoist failure based on big data and parallel simulation. J Phys 2010:012073–012073. https://doi.org/10.1088/1742-6596/2010/1/012073

Appendix
Check List for On-site Safety Inspection of the Dam

Basic Information for On-site Safety Inspection

Name of reservoir and basic information	
Main structures of the water control project	
Objects to be protected in reservoir flood control	
Time of inspection	
Climate	
Reservoir water level at time of inspection (m)	
Reservoir storage capacity at time of inspection (m^3)	
Inspector	
Description of main problems found in on-site inspection	

Note Contents in the list can be added or deleted in accordance with actual circumstances of the project

Y. Xiang et al., *On-site Inspection and Dam Safety Evaluation*, Hydroscience and Engineering, https://doi.org/10.1007/978-981-97-3987-5

On-site Inspection on Water Retaining Structures—Earth-Rock Fill Dam

Inspection area			Records of inspection
Water retaining structures	Dam crest	Dam crest road surface	
		Dam crest water drainage facility	
		Wave wall	
	Dam body	Dam filling	
		Appearance of dam body	
		Upstream slope protection facility	
		Upstream bedding material	
		Upstream filter material	
		Upstream water drainage facility	
		Downstream slope protection facility	
		Downstream bedding material	
		Downstream filter material	
		Downstream drainage facility	
	Dam foundation	Upstream dam foundation	
		Downstream dam foundation	
		Foundation catch drain (wall)	
	Dam shoulder	Left shoulder	
		Right shoulder	
	Downstream ground	Drainage trench	
		Drainage channel	
	Near dam reservoir bank	Left reservoir bank	
		Right reservoir bank	
	Others		

Note Contents in the list can be added or deleted in accordance with actual circumstances of the project

On-site Inspection on Water Retaining Structures—Concrete Dam and Cemented Masonry Dam

Inspection area			Records of inspection
Water retaining structures	Dam crest	Dam crest road surface	
		Dam crest water drainage facility	
	Dam body	Dam body concrete or cemented masonry	
		Dam body appearance	
		Upstream dam surface	
		Downstream dam surface	
		Dam body drainage facility	
		Gallery inside the dam body	
	Dam foundation	Upstream dam foundation	
		Downstream dam foundation	
		Impervious curtain of dam foundation	
		Dam foundation drainage	
	Dam shoulder	Left dam shoulder	
		Right dam shoulder	
	Downstream ground	Drainage trench	
		Drainage channel	
	Near dam reservoir bank	Left reservoir bank	
		Right reservoir bank	
	Others		

Note Contents in the list can be added or deleted in accordance with actual circumstances of the project

On-site Inspection on Discharging Structures—Spillway

Inspection area			Records of inspection
Discharging structure	Water intake section	Side wall at left bank	
		Side wall at right bank	
		Bottom plate	
	Control section	Side wall at left bank	
		Side wall at left bank	
		Gate pier	
		Bracket	
		Bottom plate	
		Overflow weir	
	Gate	Trash rack	
		Access gate	
		Access gate slot	
		Working gate	
		Working gate slot	
		Vent hole	
	Hoist	Hoist room (tower)	
		Hoist	
		Hoist control facility	
		Hoist power supply	
		Standby power supply	
	Discharge chute section	Side wall at left bank	
		Side wall at right bank	
		Bottom plate	
	Energy dissipation facility	Flip bucket	
		Stilling pool	
		Bottom plate	
		Energy dissipation falling sill	
	Tail water	Tail water channel	
		Downstream river channel	

(continued)

(continued)

Inspection area			Records of inspection
	Transport facility	Working bridge	
		Transport bridge	
	Bank slope	Left bank side slope	
		Right bank side slope	
	Others		

Note Contents in the list can be added or deleted in accordance with actual circumstances of the project

On-site Inspection on Discharging Structures—Overflow (Discharging) Tunnel

Inspection area			Records of inspection
Discharging structure	Water intake section	Side wall at left bank	
		Side wall at right bank	
		Bottom plate	
	Tunnel section	Gate shaft	
		Tunnel top	
		Two sides of the tunnel wall	
		Tunnel bottom plate	
	Gate	Trash rack	
		Access gate	
		Access gate slot	
		Working gate	
		Working gate slot	
		Vent hole	
	Hoist facility	Hoist room (tower)	
		Hoist	
		Hoist power supply	
		Standby power supply	
	Outlet section	Left bank side wall	
		Right bank side wall	
		Bottom plate	

(continued)

(continued)

Inspection area			Records of inspection
		Energy dissipation facility	
	Tail water	Tail water channel	
		Downstream river channel	
	Others		

Note Contents in the list can be added or deleted in accordance with actual circumstances of the project

On-site Inspection of Water Conveyance (Diversion) Structure

Inspection area			Records of inspection
Conveyance (diversion) structure	Intake section	Side wall at left bank	
		Side wall at right bank	
		Bottom plate	
	Tunnel (culvert) section	Gate shaft	
		Tunnel top	
		Two sides of the tunnel wall	
		Tunnel bottom plate	
	Gate	Trash rack	
		Access gate	
		Access gate slot	
		Working gate	
		Working gate slot	
		Vent hole	
	Hoist facility	Hoist room (tower)	
		Hoist	
		Hoist power supply	
		Standby power supply	
	Outlet section	Left bank side wall	
		Right bank side wall	
		Bottom plate	

(continued)

(continued)

Inspection area			Records of inspection
		Energy dissipation facility	
	Tail water	Tail water channel	
		Downstream river channel	
	Others		

Note Contents in the list can be added or deleted in accordance with actual circumstances of the project

On-site Inspection of Management Facility

Inspection area			Records of inspection
Management facility	Management organization	Constitution of organization	
		Competent authority of the organization	
	Management team	Administrative staff	
		Technical staff	
	Management system	Type of management system	
		Implementation of management system	
	Office rooms	Area of office rooms	
		Structural safety	
	Office facilities	Computers	
		Printers	
		Monitoring equipment	
		Office furniture	
	Water and rain monitoring and reporting facilities	Water monitoring and reporting facility	
		Rain monitoring and reporting facility	
	Safety monitoring facility	Deformation monitoring facility	
		Seepage and leakage monitoring facility	

(continued)

(continued)

Inspection area			Records of inspection
		Stress and strain monitoring facility	
		Temperature monitoring facility	
		Earthquake monitoring facility	
		Environmental variables monitoring facility	
		Other monitoring facilities	
		Processing and analysis of monitoring data	
	Transport roads	Road leading to dam for flood control	
		Transport roads for connection with the external transport	
	Vehicles and ships	Office vehicles	
		Vehicles for flood control and emergency rescue	
		Ships used for flood control and emergency rescue	
	Reserves for flood control and emergency rescue	Earth and stone material	
		Timber pile	
		Steel wire (rebar)	
		Woven bag	
		Lighting for flood control and emergency rescue	
		Others	
	Communications facility	Fixed line telephone	
		Satellite phone	
		Radio station	
		Mobile phone	
	Alarming system	Upstream alarming facility	

(continued)

(continued)

Inspection area			Records of inspection
		Alarming facility in water control project area	
		Downstream alarming facility	
	Power supply and lighting facilities	Power supply for water control project area	
		Lighting for water control project area	
	Equipment and materials for maintenance and repair	Mechanical equipment for maintenance and repair	
		Materials/goods for maintenance	
	Operation plan	Contents of compilation	
		Training	
	Emergency preparedness plan	Contents of compilation	
		Flood risk map	
		Effectiveness and feasibility	
		Publicity, training and exercise	
	Manuals of operation, maintenance and monitoring (OMS)	Contents of compilation	
		Training	
	Others		

Note Contents in the list can be added or deleted in accordance with actual circumstances of the project

Inspection Upstream and Downstream the Reservoir

Inspection area			Records of inspection
Reservoir area	Water resources and hydroelectric projects built upstream	Reservoir	
		Hydropower station	
		Water gate	
		Pump station	
		Hilly pond	
		Warping dam	
	Seepage and leakage in reservoir area		
	Underground water in reservoir area		
	Transport roads in reservoir area		
	Near dam bank slope		
	Slip mass (sliding deformation) in reservoir area		
	Silt deposit in reservoir area		
	Ice in reservoir area		
	Residential quarters in reservoir area		
	Source of pollution in reservoir area		
	Vegetation in reservoir area		
	Others		
Downstream	Water resources and hydroelectric projects built downstream	Reservoir	
		Hydropower station	
		Warping dam	
		Hilly pond	
		Dike	
		Water gate	
		Pump station	
		Flood detention basin	
	Cross section of river channel		
	River crossing bridge		
	River crossing pipelines		
	Downstream villages distribution		
	Downstream urban towns distribution		
	Downstream industrial and mining enterprises distribution		

(continued)

(continued)

Inspection area			Records of inspection
	Downstream polluting enterprise		
	Downstream schools and hospitals		
	Downstream natural and historic landscape		
	Downstream roads distribution		
	Evacuation shelter		
	Others		

Note Contents in the list can be added or deleted in accordance with actual circumstances of the project

The manufacturer's authorised representative in the EU is Springer Nature Customer Service Centre GmbH, Europaplatz 3, 69115 Heidelberg, Germany. If you have any concerns regarding our products, please contact ProductSafety@springernature.com

Printed and bound by CPI Group (UK) Ltd, Croydon, CR0 4YY
30/09/2026
02225298-0013